# Antarctica

## Map of Antarctica

(Joseph Hill)

# Antarctica

## The Battle for the Seventh Continent

Doaa Abdel-Motaal

PRAEGER™
An Imprint of ABC-CLIO, LLC
Santa Barbara, California • Denver, Colorado

**Library of Congress Cataloging-in-Publication Data**

Names: Abdel-Motaal, Doaa, author.
Title: Antarctica : the battle for the seventh continent / Doaa Abdel-Motaal.
Description: Santa Barbara, California : Praeger, 2016. | Includes bibliographical references and index.
Identifiers: LCCN 2016027484 (print) | LCCN 2016038132 (ebook) | ISBN 9781440848032 (hard copy : alk. paper) | ISBN 9781440848049 (ebook)
Subjects: LCSH: Antarctica—Environmental conditions. | Antarctica—Economic conditions. | Global environmental change.
Classification: LCC GE160.A6 A33 2016 (print) | LCC GE160.A6 (ebook) | DDC 341.4/209989—dc23
LC record available at https://lccn.loc.gov/2016027484

ISBN: 978-1-4408-4803-2
EISBN: 978-1-4408-4804-9

20 19 18 17 16 1 2 3 4 5

This book is also available as an eBook.

Praeger
An Imprint of ABC-CLIO, LLC

ABC-CLIO, LLC
130 Cremona Drive, P.O. Box 1911
Santa Barbara, California 93116-1911
www.abc-clio.com

This book is printed on acid-free paper ∞

Manufactured in the United States of America

*To Karim*

# Contents

Acknowledgments ix

ONE: Introduction 1

TWO: The Case for Antarctica 7

THREE: Who Owns Antarctica? 53

FOUR: The Conquest: Science and Minerals 83

FIVE: An Antarctic Economy: Lessons from the Melting Arctic 119

SIX: Lines in the Ice: An Orderly Division of Antarctica 147

SEVEN: Toward Greater Environmental Protection in Antarctica 201

EIGHT: The World Is Not Waiting for 2048 261

Appendix 1: The Antarctic Treaty 265

Appendix 2: The Svalbard Treaty 273

Abbreviations 279

Bibliography 281

Index 309

# Acknowledgments

A work of this nature cannot be undertaken without the extensive support of a cadre of academics, policymakers, legal practitioners, climatologists, geologists, and other scientists, as well as a network of professional and amateur mariners, travelers, and experts on Antarctica and the Arctic. The research underlying this book took two years, with months of field work and on-site visits. No work, no matter how noble its intentions or thoroughness, can be exhaustive, however. I am certain that there are areas of the literature, some information, and many opinions that I have failed to reflect, either inadvertently or simply in an attempt to keep the work manageable and focused. All errors and omissions are mine alone, as are any opinions, and there are many—some controversial.

I want to start by thanking Dr. Alan Hemmings, of the Gateway Antarctica Centre for Antarctic Studies and Research, whose extensive and seminal work on Antarctica inspired me and a generation of researchers to pursue this subject. His detailed review of this book is as humbling in its generosity as it is appreciated. I also want to thank Dr. Patrizia Vigni, of the University of Siena for reviewing this work and offering her time, support, and guidance throughout this project. Dr. Rasmus Gjedssø Bertelsen, of the University of Tromsø was instrumental in guiding me toward the exploration of the Svalbard model—a critical insight, as will become apparent to the reader.

I am grateful to Michel Rocard, former prime minister of France; Aleqa Hammond, former prime minister of Greenland; Jonas Store, former foreign minister of Norway; and the Governor of Svalbard for their invaluable contributions and input. I also want to thank Dr. Kevin Hughes, of the British Antarctic Survey, for his critical insights on Antarctica's environmental governance; and Dr. John Turner, of the British Antarctic Survey, for patiently teaching me Antarctica's complex climate system. I am indebted to Dr. Adam Grydehoj, of Island Dynamics, for so generously discussing with me the sociology of Svalbard, Dr. Francesco Orrego Vicuna, of the University of Chile, and Dr. Tullio Treves, a professor at the University of Milan and a judge in the International Tribunal of the Law of the Sea, for helping me untangle the complexities the of the Law of the Sea and legal principles such as the Common Heritage of Mankind. I extend my thanks to all of them for agreeing to provide invaluable interviews for this book.

In addition, I am grateful to the Antarctic Treaty Secretariat, the Scientific Committee on Antarctic Research, and the International Seabed Authority for their assistance with all background documentation. I want to thank the Museo Nazionale dell'Antartide in Siena for allowing me to see Antarctic gold and other mineral deposits firsthand and for connecting me with geologists, experts on Antarctic mineral resources, and the Programma Nazionale di Ricerche in Antartide in Italy. I want to thank Dr. Laura Crispini of the University of Genova for walking me through the complex geology of Antarctica, its mineral resources, and her work on gold deposits. I want to thank the U.S. Geological Survey for their collaboration and all the material that they shared so generously to enable this work. The dedication, expertise, and generosity of these institutions speaks volumes about humanity's capacity for collaborative work and noble intentions on matters affecting our common planetary heritage. I remain humbled by the encouragement and support that I received from professionals in these organizations, whose expertise in their field and time constraints dwarf mine.

This work would not have been possible without *National Geographic Expeditions* and the Earthwatch Institute, who enabled me to travel throughout the Antarctic and the Arctic. My work with the Earthwatch Institute on climate change in the Canadian sub-Arctic allowed me to appreciate firsthand the relentless pace of climate change and the challenges and opportunities for human habitation of remote, ice-covered lands. Last but not least, I want to thank cartographer Joseph Hill, whose endless patience enabled the preparation of all the maps for this book; Bill Watkins of Literary Solutions, who lifted this work to a higher editorial standard; and Jessica Gribble from Praeger, my publisher.

# ONE

# Introduction

To many experts, there has not been, and will never be, a battle for Antarctica. This continent is remote, inaccessible, and covered in ice, it may not have natural resources, and it has no indigenous inhabitants. Moreover, it is governed by the Antarctic Treaty System (ATS), which is widely celebrated as one of the most successful international legal regimes. This regime brought peace and security to the continent and preserved its pristine environment through a ban on mining.

Emerging from the International Geophysical Year (IGY) of 1957–1958, a year in which the international community came together to further the aims of science, was the 1959 Antarctic Treaty, which is the cornerstone of the ATS. This treaty dedicated the continent to peace and scientific inquiry. The notion that there could one day be a battle for Antarctica, therefore, will seem outlandish to many—a theory peddled by scaremongerers with a limited understanding of Antarctic exceptionalism. This exceptionalism was reinforced as recently as the 2007–2008 International Polar Year (IPY), which reconfirmed the primacy of science.

The ATS has been the envy of the Arctic. Academics and policymakers alike have looked at Antarctica with bewilderment. If a regime such as the ATS is possible down south, why has it not found a parallel in the north? Why is Antarctica covered by the Protocol on Environmental Protection to the Antarctic Treaty (a 1991 addition to the ATS), and its wilderness left intact, when the Arctic is solely governed by the Arctic Council, a loose forum for dialogue among the main eight Arctic powers?[1]

This book will demonstrate that this conventional wisdom is not only wrong—it is dangerous. To believe that Antarctica is a battle-free scientific playground is to ignore not just current developments in the Antarctic, but also its history. Commercial exploitation lay at the origin of Antarctica's discovery. Among its first conquerors were sealers and whalers, to whom the world owes a debt of gratitude for Antarctica's mapping. After all, American sealer Nathaniel Palmer was one of the first to lay eyes on the Antarctic Peninsula in 1820. The arrival of fishers gradually depleted some of the most precious living resources of the Southern Ocean.

The battle's complexity will be exacerbated by the relentless pace of global warming. On March 24, 2015, a temperature of 17.5°C (63.5°F) was recorded at Esperanza Station on the northern tip of the Antarctic Peninsula, making it the highest temperature ever recorded on the continent. As Antarctica warms and starts to become more habitable, many other parts of the globe will become increasingly uninhabitable. The growth of environmental and climate refugees—a problem that has now been internationally acknowledged—if intelligently managed, could find part of its solution in Antarctica. The incentives for this to happen are certainly already in play.

The ATS has been built on land that lacks a clear title. Faced with the slightest tremor, it will collapse like a house of cards. This is especially the case if that tremor were to come with the discovery of diamond, gold, uranium, or any other precious resources that Antarctica is now known to harbor. Negotiated by a small group of countries, the Antarctic Treaty was bedeviled by the issue of territorial sovereignty. It froze all seven unilateral territorial claims made on the continent in the early part of the 20th century and left it supposedly unowned, the object of a collective management system. However, the 12 original signatory nations to the treaty were quick to find creative ways to lay claim to the continent's resources.[2] This is without considering the implicit claims later brought by the rest of humanity.

The Antarctic Treaty in fact turned science (or, rather, the pretense of science) into the entry ticket to its exclusive club. Consultative party status, which confers decision-making powers on its holder, is obtainable upon demonstration of "substantial scientific research activity." This term has been defined as the establishment of a scientific research station, no matter how small or immaterial its research program, or even the mere dispatch of a scientific expedition. "Science" has gradually become a code word for uncovering the mineral potential of the continent and laying claim to it. The conquest of Antarctica is well under way, with the continent now dotted with research stations established by recent entrants such as China, India, and South Korea, in addition to the traditional players like the United States, Russia, the United Kingdom, Argentina, Chile, and Australia. Antarctica is now effectively under scientific occupation, with increasing calls being made for its decolonization.

The United Nations has long fought the Antarctic Treaty and its associated agreements by placing the so-called Question of Antarctica on its agenda for well over two decades. It warned against a small subset of players monopolizing the continent and sounded the alarm when parties started secretly negotiating a minerals convention. Its efforts succeeded when the 1988 Convention on the Regulation of Antarctic Mineral Resource Activities (CRAMRA) collapsed, giving way to the 1991 Environment Protocol. Accompanied by a gradual widening of the membership of the Antarctic club (discussed later in this book), the protocol banned mining on the continent until 2048 at the earliest.

The international community hailed these developments as a victory. It was considered a victory for the Antarctic Treaty, since the continent would continue to be dedicated to science; a victory for the environment, since no harmful mining would take place; and victory for humanity at large, since the wilderness of the continent would continue to belong to all mankind

Little did the world know at that stage that the collapse of CRAMRA was in large part driven by concerns over territorial sovereignty in Antarctica, with some of the original seven territorial claimants reluctant to let others mine in the areas of their claims. The ATS had inadvertently strengthened the quest for sovereignty in Antarctica. Under the guise of devotion to scientific inquiry, Chile and Argentina also went ahead and built experimental human settlements on the peninsula and the South Shetland Islands of Antarctica. Today, scientific research stations have multiplied across the continent to the point of being the norm, not the exception. Commercial activities such as fishing, tourism, and bioprospecting (the process of discovery and commercialization of new products based on biological resources) have all flourished. A race is well under way, ironically by the countries with the greatest current capabilities to stake claims and establish implicit sovereignty.

In fact, the ATS has subjected Antarctica to one of the worst cases of the "tragedy of the commons," a situation in which individual countries act in their own self-interest, rather than the best interest of either the collective or the continent. The term "illegal, underreported, and unregulated" fishing (abbreviated as IUU, a common acronym in international law), would be coined first in the Antarctic region to describe the plight of the Southern Ocean. Antarctica has become one of the world's least environmentally protected continents, with only 1.5% of its ice-free surface designated as protected. It has also fallen victim to the weak institutional underpinning of the treaty, whose secretariat was only established in 2004, and whose only decision-making forum after nearly 60 years of existence continues to be the Antarctic Treaty Consultative Meeting (ATCM).

This book makes the case that the solution to Antarctica's current dangerous and unsustainable predicament is to embrace the colonization and exploitation of the continent. It proposes a system for the division of its spoils that benefits from lessons learned during the negotiation of the United Nations Law of the Sea (UNCLOS), the Outer Space Treaty, and the Moon Treaty. In doing so, it draws on the invaluable experience of regulating the Arctic. This may seem controversial compared to the understandable and politically correct desire to maintain the pretense of a global or regional commons where none in practice exists. Ironically, however, it is the path most likely to guarantee the goals of environmentalists and humanitarians who oppose it.

The book hopes to demystify Antarctica by demonstrating the multifaceted nature of the silent battle for its spoils that started with its discovery and has continued without interruption ever since. It aims to initiate a structured conversation on the fate of the seventh continent, urging policymakers to take the lead in dismantling the ATS instead of awaiting its disorderly demise. Lessons should be learned from the Falklands War, whose underlying territorial conflict remains unresolved, and the increasing militarization now being witnessed in the Arctic. Instead of proposing that the Arctic look to the Antarctic for solutions, this book suggests that the answer to the Question of Antarctica could very well lie in looking north, to the Arctic.

The battle for Antarctica is about to intensify, and the world must prepare itself. It will be one of the most complex and truly international contests for habitable space (and mineral resources) of modern times. It will be a battle in which an entire continent will be up for grabs and which will take place against the complex historical background of the creation of the ATS. The good news is that a realistic resolution to the continent's current predicament exists. Counterintuitively, it is based on embracing the commercial and other incentives that many find deleterious, but which can have much more desirable economic and environmental outcomes. Most important, it is a solution with practical precedent in human experience—namely, the history of a little-known Norwegian Arctic outpost called Svalbard.

## NOTES

1. These would be Canada, Denmark (including Greenland and the Faroe Islands), Finland, Iceland, Norway, Russia, Sweden, and the United States.

2. The original signatories to the ATS are the 12 countries that were active in Antarctica during the International Geophysical Year of 1957–1958

and then accepted the invitation of the U.S. government to participate in the diplomatic conference at which the treaty was negotiated in Washington, DC, in 1959. They include Argentina, Australia, Belgium, Chile, France, Japan, New Zealand, Norway, South Africa, the Soviet Union, the United Kingdom, and the United States.

# TWO

# The Case for Antarctica

The world's coldest continent is beginning to thaw, with the western side of the Antarctic Peninsula already enjoying a similar climate to currently inhabited Greenland. By the end of this century, vast quantities of sea ice surrounding the continent are also expected to melt, making it far more accessible. As Antarctica's climate becomes milder and its habitation more feasible, significant changes will be taking place around the globe. Climate change is expected to make many other parts of the world less habitable, leading to a dramatic rise in the number of environmental and climate refugees by the end of the century. This, together with increased knowledge about the mineral wealth of the continent, could renew the discussion on the ownership of Antarctica. The various research stations that now exist across the continent have demonstrated that it is possible to inhabit Antarctica, that sustainable housing is feasible, and that renewable energy has a viable future in what is the world's windiest landmass. All of these elements combined will gradually strengthen the case for Antarctica.

## COMMERCIAL ENTERPRISE DRIVES ANTARCTICA'S DISCOVERY

To the general public, Antarctica is a daunting continent. After all, it is the coldest, windiest, most isolated, and most deserted part of the planet. It is the southernmost continent, and to reach it, one must brave the treacherous waters of the Southern Ocean, the world's most turbulent body of water.[1] To the general public, living in Antarctica is completely unimaginable. In fact, those who point to the possibility of life on the seventh continent find themselves having to go back in history 300 million to 180 million years ago—a time when Antarctica continued to form part of a supercontinent called Gondwanaland that is said to have enjoyed a relatively mild climate.[2]

The history of Antarctica's discovery, however—as well as of the discovery of most other continents—demonstrates that humans can brave the harshest of conditions, and take the biggest of risks, to explore new frontiers. Human interest in Antarctica was driven by a desire for geographical and scientific discovery, the commercial exploitation of resources, adventure, and most recently tourism.[3]

In fact, Antarctica existed in the minds of the ancient Greeks long before it was physically discovered. They believed that the Earth had to be in balance, and that there had to be an anti-Earth, or *Antichthon*, that moves unseen around the other side of the sun. As there were large masses of land to the north of the planet, an equal counterweight had to exist to the south. They were the first to name Antarctica; they called it the Unknown Southern Land, which translated into Latin as *Terra Australis Incognita*.[4]

However, proof of the continent's existence would not actually be documented until the 1772–1775 voyage of Captain James Cook.[5] Upon finding Antarctica, he wrote:

> Land doomed by nature to everlasting frigidness and never once to feel the warmth of the Sun's rays, whose horrible and savage aspect I have no words to describe; such are the lands we have discovered, what may we expect those to be which lie more to the South, for we may reasonably suppose that we have seen the best as lying most to the North, whoever has resolution and perseverance to clear up this point by proceeding farther than I have done, I shall not envy him the honor of discovery but I will be bold to say that the world would not be benefitted by it.[6]

Ironically, despite his dismissal of Antarctica as a continent doomed by its climate and destined for isolation and confinement, Captain Cook's report of plentiful seals on the island of South Georgia would open the way for the first phase of the commercial exploitation of the seventh continent.[7] In fact, in *Mapping Antarctica: A Five Hundred Year Record of Discovery,* Clancy et al. (2014) demonstrate how our understanding and charting of Antarctica mostly evolved through successive sealing and whaling expeditions.[8] It evolved so dramatically that by the early 20th century, the assessment of Antarctica's potential changed fundamentally. In his voyages to Antarctica in the 1920s and 1930s, Admiral Richard Byrd, who initiated the involvement of the United States in the continent, had by then already been describing Antarctica as a place that "God had set aside as man's future—an inexhaustible reservoir of natural resources."[9]

The sealing industry was at the origin of commercial activity in and around Antarctica. British sealers traveled to South Georgia in 1778, and their American counterparts followed in the 1790s.[10] Fur seals were the first types of seals to be exploited, especially when important markets for their skin opened up in China, and were followed by elephant seals, which were taken for their oil.[11]

An examination of the evolution of sealing operations around Antarctica demonstrates that unsustainable fishing practices gradually killed off the industry and forced governments to step in with heavy regulations. As sealers exterminated the seal populations of South America and the Falkland Islands, they began to venture farther south, toward Antarctica.[12] In fact, it is said that the sealers were the first to set foot on the Antarctic continent.[13]

Sealing operations were geographically focused on the sub-Antarctic islands and the islands on the periphery of the Antarctic peninsula.[14] Both fur and elephant seals were vulnerable targets since they breed onshore in large numbers, in colonies to which they return each spring. By the 1820s, it is said that 1.2 million fur seals had been taken for their skin on the island of South Georgia. As an order of magnitude of the size of sealing operations, in 1820–1821, 91 fishing vessels were said to have arrived to the island, with a total of 3,000 men employed onshore.[15] These sealing operations contributed significantly to the physical discovery of the continent. On his 1820–1821 voyage to Antarctica, for example, American sealer Nathaniel Palmer was one of the first to catch a glimpse of the Antarctic Peninsula, and happened to cross Russian captain Thaddeus Bellingshausen on the very first circumnavigation of the continent.[16]

As the fur seal populations declined, the industry's emphasis gradually shifted to elephant seals. These were hunted for their blubber oil, which was used for lighting, lubrication, and the preparation of leather. These seals too were pursued mercilessly, and it was in 1972 that the Convention on the Conservation of Antarctic Seals (CCAS) was adopted to establish catch limits and conservation measures for different types of seals and locations.[17]

The evolution of the sealing industry demonstrates how rival companies competed fiercely in Antarctica and how global the industry had become. The best-known trading house was that of Samuel Enderby & Sons, founded by Samuel Enderby, whose interest in sealing and whaling in Antarctica grew after his normal Atlantic trade was interrupted by the American Revolution. His commercial interests were combined with an expectation that his captains would find and claim new lands for the British Crown. In fact, it was Enderby company ships that discovered Auckland Island, south of New Zealand, in 1805.[18]

The grandson of Samuel Enderby, Charles Enderby, later founded the Royal Geographic Society in 1830. Before dying, he hired Captain John Biscoe to lead exploratory sealing trips. Biscoe was the first to sight land on the Australian side of Antarctica, as well as parts of the Antarctic Peninsula.[19] Another sealing captain who contributed to Antarctic discovery before the decline in the number of seals terminated the industry was Peter Kemp, who worked for the rival company Daniel Bennet & Sons. Aware of Biscoe's discovery, in the 1830s he was one of the first to sight land just east of that area, which is now called Kemp Land.[20]

The sealing industry gradually became global, as European and American explorers caught seals and shipped them off to China, where tremendous skill existed in their deskinning, and where seal fur was considered a high-fashion item in the harsh winter. Processed seal products manufactured in China often traveled back toward European and American markets.[21]

The Heroic Era, between around 1895 to 1915, saw remarkable progress in the understanding of Antarctica's geography, the scientific study of the continent, and the quest for its resources. The era was supported by a growing number of industrial and commercial entrepreneurs willing to finance expeditions to the continent.[22] Much like the space race of the 1960s, the world became fascinated with news of a new continent, wanting to field discoverers to its pristine and untouched lands.[23] This era also ushered in the second main phase of the continent's commercial exploitation, particularly commercial whaling. The first whaling station, for example, was established in South Georgia in 1904, allowing the whaling industry to expand exponentially, with whale oil quickly becoming a strategic resource during World War I.[24]

Much like the sealing industry's contribution to the discovery of Antarctica, the gradual evolution and sophistication of the whaling industry also contributed to the world's understanding of the Southern Ocean and its complex ecosystem. Because the fishery was based on the catching of whales in their feeding grounds when they developed their thickest layer of blubber, a process that took place in the Austral summer months, it expanded to various locations in the Southern Ocean where the fodder for whales, krill, was to be found.[25]

The industry's expansion was also determined by the locations where it could more readily establish onshore stations or floating factories, and by the gradual technological development in processing and handling techniques, which allowed it to cover ever-wider geographical areas. The industry eventually spread to the high seas of the Southern Ocean as it became technologically feasible to do so.[26] For most of its lifespan, it was driven by demand for whale fat, used in the production of soap, margarine, lard compounds, industrial waxes, lubricants, cosmetics, and, of course, lighting. By the end of the 19th century, however, whale oil was replaced by mineral oil for lighting and lubrication.[27]

From 1925 onward, the whaling industry started to benefit from the introduction of bigger, more far-ranging factory ships. These ships allowed the industry to peak in the 1937–1938 season, reaching a record total of 33 expeditions that caught 46,000 whales. Norwegian and British whalers were at the origin of most of the catch.[28] Factory ships were in fact created before World War I, with their design improving over time. The largest factory ship to exist at the time was the *Guvernoren*, which was owned by Norwegians and was the most sophisticated processing ship of the day. It could break up whale meat and bone as it cooked them, making the quality of the oil that it collected lighter and less acidic. The ship caught fire in 1915, however, with all 85 of the crewmen saved but 2,600 tons of whale oil lost.[29]

As with the sealing industry, the overexploitation of whales, ushered in stricter governmental regulation. The International Whaling Commission, a regulatory body established in 1946, placed a moratorium on all commercial whaling until the reassessment of all stocks.[30] The moratorium remains in force until today, but it has been the subject of much controversy. While countries had ceased whaling by the mid-1960s, Japanese and Russian whaling expeditions continued under the pretext of "scientific whaling."[31] In 2014, on the basis of an Australian complaint, the International Court of Justice (ICJ) ruled that Japan's whale hunting was not scientific and should cease.[32] Despite this, a number of Japanese fishing vessels were headed for Antarctica for the purpose of whale hunting in the December 2015 season.[33]

Commercial enterprise, therefore, has been closely intertwined with Antarctica's discovery, as successive sealers and whalers have braved its harsh climate in search of new lands and resources. Their thirst for scientific discovery and for the riches of the new continent made their difficult voyages all the more bearable. While unsustainable fishing practices drove a decline in seal and whale populations and led to new international regulations, interest in other types of Southern Ocean fish continued to grow. However, the continent's harsh climate shielded it from any onshore commercial enterprise of note. Interestingly, though, this harsh climate has been slowly changing.

## ANTARCTICA'S CLIMATE IS CHANGING

To understand Antarctica's climate, one must first understand its geology. Antarctica's geology is the outcome of the formation (500 million years ago) and subsequent breakup (180 million years ago) of the supercontinent known as Gondwanaland. Fragments of Gondawanaland spun off and assembled in the form of a high continental landmass around the South Pole, split by an unbroken high mountain chain. As the supercontinent dispersed, oceans opened up on all sides of Antarctica, with circumpolar ocean currents expanding into what became known as the Southern Ocean. These currents regulate the climate of the continent, more evenly distributing the cold from its higher to its lower latitudes.[34] The sharp contrast in temperature between Antarctica's ice sheet and the ocean results in one of the world's strongest wind systems, known as *katabatic winds*. These circumpolar westerly winds produce violent storms that encircle the continent.[35]

Approximately 98 percent of Antarctica is covered in ice, with high mountain peaks and volcanoes protruding from its snowfields and glaciers. It is divided into two main geographical regions, East Antarctica and West Antarctica, which are separated by the Transantarctic Mountains.[36] The line of division corresponds approximately to the division between the Eastern and Western hemispheres, with East Antarctica making up approximately two-thirds of the continent.[37] The East Antarctica bedrock, which largely lies above sea level, is composed of much older and geologically stable rock formations than West Antarctica, which is predominantly below sea level and contains the continent's only active volcanoes. As Scott Harris Friend puts it, "both regions consist largely of flat expanses of snow and ice, providing little sense of scale or direction."[38]

Antarctica's ice cover contains about 70 percent of the world's freshwater. Its ice sheet rises rapidly from the coast to high elevations that create two separate and prominent ice domes: the East Antarctica Ice Sheet (EAIS) and West Antarctica Ice Sheet (WAIS). Friend describes this ice cover as the "frozen library of our planet's past going back over one million years." Ice breaks off the outer edges of Antarctica every summer, in billions of tons, resulting in the creation of massive icebergs. Because of the dynamic nature of the Antarctic ice sheet, most the continent's surface is uneven and has large fractures and crevasses.[39] The Transantarctic Mountains traverse the entire continent and constitute one of the longest mountain ranges on the planet. Most of Antarctica's exposed rock lies within this range, which also includes the continent's only dry valleys. These are the only ice-free portions of the continent due to their low humidity and precipitation levels.[40]

Being covered in ice, Antarctica retains little of the sun's warmth due to the ice's reflectivity and accumulates cold in its dark winters. This makes it the coldest continent on Earth. In fact, the lowest temperature on the planet was recorded at the inland Russian Vostock Station in 1983, and was approximately −89°C (−128°F).[41] Almost all precipitation on the continent falls as snow, except on the coastal fringes and the sub-Antarctic islands,[42] which also means that Antarctica is as much of a desert as the Sahara.[43]

Antarctica's climate is defined by its long polar nights and polar days, as it experiences periods of either total darkness or light. In summer, the South Pole enjoys a splendid climate, with continuous sun for 186 days. In winter, however, the Pole plunges into darkness. Because of its long and dark winters, the continent undergoes what are referred to as "coreless winters" (in other words, winters in which there is no specific month of particularly low temperatures). The length of polar nights and polar days decreases with the distance from the Pole.[44] Figure 2.1 provides an indication of temperature variation across the continent as recorded by its various weather stations. It shows that the peninsula is the continent's warmest area.

**Figure 2.1 Temperature Variation across Antarctica**

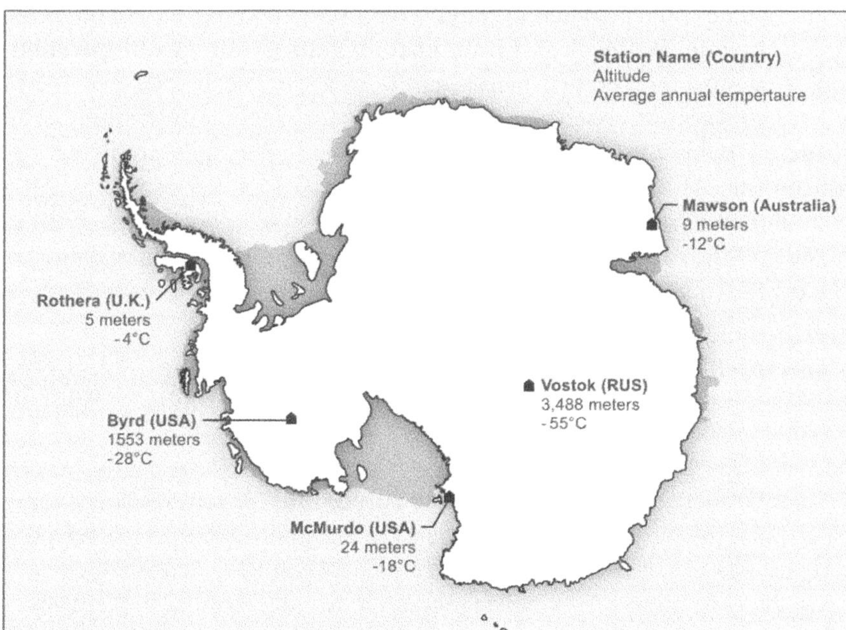


(Joseph Hill)

The climate in Antarctica has been undergoing important change over the course of the last century. According to prominent climatologists, temperatures have been rising in West Antarctica and the Antarctic Peninsula in particular since 1950, with smaller and generally insignificant changes in East Antarctica.[45] In preparation for the IPY 2007–2008, the Scientific Committee on Antarctic Research (SCAR) undertook a major review of the state and current understanding of Antarctic and Southern Ocean climate science. The project, entitled "Antarctic Climate Change and the Environment (ACCE)," led to the production of a 526-page report that became one of the most comprehensive reviews of climate change in the Antarctic continent.[46] The ACCE report established that the largest warming trends in Antarctica have occurred in the Antarctic Peninsula.[47]

In 2013, the scientists involved in ACCE published an update of the report in the *Polar Record*. The update discussed the profound impact of the ozone hole on the Antarctic environment in the past 30 years, explaining that it appears to have shielded the continent from the full effects of global warming. However, it stated, "This effect is not expected to persist. Over the next century, ozone concentrations are expected to recover and, if greenhouse gas atmospheric concentrations continue to increase at the present rate, temperatures across the continent are projected to increase by several degrees and sea ice will be reduced by about one third."[48]

The update confirmed that the largest warming trends have been observed in the Antarctic Peninsula, with weather stations in this part of the continent showing a temperature increase of 0.54°C per decade between 1951–2011. The warming mostly took the form of a reduction in the number of extremely cold winters in the peninsula.[49] The update also predicted significant surface warming in all of Antarctica by 0.34°C per decade by 2100 over land and ground ice sheets. This increase, however, will not result in inland melting, with coastal temperatures falling the fastest and the most.[50] The degree to which the planet will warm will depend on global efforts to curb greenhouse gas emissions.

The authors pointed to the serious climate change that has already occurred in the sub-Antarctic islands—a change now creeping toward the continent itself. They wrote:[51]

> Some formerly snow and ice-covered covered sub-Antarctic islands are now increasingly snow-free during the summer. Brown Glacier on Heard Island lost 29% of its surface area between 1947 and 2004, causing the formation of several lagoons. On South Georgia, 97% of the 103 coastal glaciers surveyed from the 1950s to the present have retreated and the average rate of retreat has increased . . . Further south in the

South Shetland Islands, the ice-covered area of King George Island decreased by 1.6% during 2000–2008 . . . , at a loss similar to that of the period 1956–1995, when the ice-covered area decreased by 7%.

With respect to sea ice, while the update concluded that it will fall by about one-third over the next decade, it nevertheless debated what is currently an unexplained phenomenon: the increase in Antarctic sea ice that has been witnessed in recent years. It wrote, "For the first half of the 20th century, ship observations suggest that the extent of sea ice around the Antarctic was greater than seen in recent decades, although the validity of such observations has been questioned." It later added that this matter continues to be under debate.[52] These observations, however, did not change the overall conclusion reached by the authors, which is that by the end of the century, the continent will become far more accessible.

The ACCE was followed by the Intergovernmental Panel on Climate Change's (IPCC) Fifth Assessment Report in 2014. This new report confirmed the warming trends previously described. It wrote that "Over the period 1992–2011, the Greenland and Antarctic ice sheets have been losing mass (high confidence), likely at a larger rate over 2002 to 2011," with glaciers continuing to shrink worldwide.[53]

Warning members of the Antarctic Treaty of the rapidly warming seventh continent, SCAR wrote in 2015: "Since the 1950s, surface air temperatures have risen significantly at many of the stations on the Antarctic Peninsula, although warming has slowed markedly over the last decade. Nevertheless, on 24 March 2015, a temperature of +17.5 degrees Celsius was recorded at Esperanza station in the northern part of the Peninsula . . . This is the highest temperature ever recorded in the Antarctic continent."[54]

Table 2.1 provides an indication of the average monthly temperature of the Antarctic Peninsula, as recorded by the Rothera Weather Station.

A prominent climatologist, and one of the main contributors to the ACCE report, is interviewed in Box 2.1. This person (who asked to remain anonymous) points to the complexity of climate science and the many uncertainties that surround our understanding of the climate system governing Antarctica. He speaks of a warming of the Antarctic Peninsula since the 1950s and a cooling of the rest of the continent, which he attributes to natural climate variability and the ozone hole, rather than global warming per se. "Under the medium emissions scenario of the IPCC, we expect the world to warm by 3-4 degrees by the end of the century. The same will happen in Antarctica," he states. He adds that the Antarctic Peninsula could be habitable in its coastal areas, particularly on its Western side, as it enjoys a climate comparable to the currently inhabited Greenland.

**Table 2.1 Average Temperature in the Antarctic Peninsula* (Degrees Celsius)**

| | Jan | Feb | March | April | May | June | July | August | Sep | Oct | Nov | Dec |
|---|---|---|---|---|---|---|---|---|---|---|---|---|
| Rothera Weather Station | 1.0 | 0.1 | −1.6 | −3.7 | −6.8 | −8.8 | −12.6 | −11.8 | −9.4 | −7.2 | −3.3 | 0.2 |

*Cool Antarctica, "Antarctica Climate and Weather," http://www.coolantarctica.com/Antarctica%20fact%20file/antarctica%20environment/climate_weather.php.

**Box 2.1 Interview of Prominent Climatologist[a]**

1. *What are the main determinants of climate change in Antarctica?*
First of all, it is important to bear in mind that our climate record for Antarctica is short. Most weather stations were only established in the 1950s, and information on Antarctica's climate prior to then [has] been derived from ice cores and other proxies. Currently, only 1 percent of Antarctica is ice free even in summer, with most of it covered in an ice sheet 2–4 km thick. In summer, only the Antarctic Peninsula is ice free, but in winter, it gets covered by a lot of snow.

There is currently no evidence of the impact of increasing greenhouse gases in Antarctica. Overall, since the 1950s, there has been a warming in the Antarctic Peninsula and a small cooling in the rest of the continent, which are attributable largely to natural climate variability rather than greenhouse gas concentration increases, although the loss of stratospheric ozone has played a part in this pattern of climate change. The warming of the Antarctic Peninsula stopped in the late 1990s and is something that we are beginning to write about today.

Natural climate variability is larger in polar regions than in other parts of the globe. What we had underestimated as scientists was the profound effect which the ozone hole would have on the climate of the entire southern hemisphere. In Antarctica, the ozone hole has warmed the peninsula a little but cooled the rest of the continent. However, the part played by the ozone hole in warming the peninsula will disappear, as we are expecting the hole to recover by 2060–2070. Over the same time, greenhouse gas concentrations are expected to rise, and these will counteract each other. This said, the peninsula started to warm even before the hole in the ozone layer developed in the early 1980s. The rest of the continent has been getting cooler since 1980. The continent's climate is also affected by El Nino and by volcanic activity. Nothing is simple in climate science.

2. *Does Antarctica have an isolated climate system or does its climate impact the rest of the globe?*
Antarctica's climate is very closely coupled with the rest of the climate system. The El Nino phenomenon sends signals into both hemispheres and particularly impacts the Antarctic Peninsula. It leads to more storms to the west of the peninsula and generally colder conditions. There is also a strong coupling between Antarctica and the rest

of the climate system through the oceans. Ocean change, however, occurs on [a] much longer time scale than change in the atmosphere.

3. *Is melting ice from Antarctica leading to global sea level rise?*
There are two main ice sheets in Antarctica, the EAIS and the WAIS, divided by the Transantarctic mountains. There is also a much smaller ice sheet covering the Antarctic Peninsula. The peninsula is very long (about 2,000 kilometers) and over 2 kilometers high for much of its length. Sea levels have been rising by approximately 3 millimeters per year globally over recent decades, with 10 percent of that rise coming from the melting around the Amundsen Sea Embayment on the edge of the WAIS. What we also know is that West Antarctica has around 5 meters of sea level rise locked into it.

4. *How much warmer can we expect the Antarctic Peninsula to be in the future?*
Under the medium emissions scenario of the IPCC, we expect the world to warm by 3–4 degrees by the end of the century. The same will happen in Antarctica. Across most of Antarctica, the temperature will remain below freezing. The Antarctic Peninsula would mostly be habitable along the coast. Inland it is very cold and the mountains very high. The west side of the peninsula is warmer, and that is why most research stations are on that side. Last year, the highest temperature recorded in Antarctica was recorded near the tip of the peninsula.

5. *Is current weather on the Antarctic Peninsula comparable to any other part of the planet that is currently inhabited? What other region has the most comparable climate to the Antarctic Peninsula?*
Current weather on the Antarctic Peninsula is comparable to Greenland. However, the peninsula is only inhabitable in coastal areas. Climate change has affected people in Greenland. During the Little Ice Age (approximately 1550–1850), people have had to leave Greenland after having settled there.

6. *How much melting of sea ice is required to significantly increase the year-round navigational accessibility of the Antarctic Peninsula, and by when is it expected?*
Around one-third of sea ice is expected to disappear by the end of the century in Antarctica. However, climate models have difficulty

representing sea ice, as seen by the fact that 95 percent of IPCC models had sea ice decreasing over the period 1979—2005, while in reality sea ice increased in extent. More snowfall in Antarctica as a whole is likely to counteract sea level rise by a small amount. In the Arctic, sea ice is located close to the pole, whilst in Antarctica it is further Equatorwards. Strangely, sea ice around the Antarctic has increased over the past 35 years, which is thought to be a result mainly of natural climate variability.

7. *Could the Antarctic Peninsula become habitable? Are there other parts of the globe that are likely to become less habitable?*
Globally, there is an unequivocal warming signal, but at the regional level, this signal is more challenging to detect. The Antarctica Peninsula is 100–200 kilometers wide, and the resolution of our models is too small to pick that up. Virtually all in situ measurements of temperature taken in Antarctica are from weather stations that are along the coast; the elevation rises rapidly inland, and there are only three long-term stations there. The sub-Antarctic islands are already habitable now. For example, the weather in Bird Island and South Georgia is very mild, with more rain than snow. The eastern side of the Antarctic Peninsula will continue to be inhospitable in future because of multiyear sea ice, and will continue to need icebreakers for access. On the western side, however, we will lose ice progressively.

At the same time, there will be extreme warming in other parts of the globe. Desert areas will get hotter the most. Areas of the edge of deserts will get even dryer. These extremes will be really dangerous. It has been shown that as temperatures reach the high 40s degrees Celsius, deaths increase rapidly, in particular the elderly and those without air conditioning. A temperature rise of an extra few degrees may not sound like much, but it can have terrible consequences in terms of extreme events.

[a]Interview conducted by the author on October 31, 2015, The climatologist has asked to remain anonymous.

The warming of Antarctica, which will make it increasingly habitable, will take place along with a warming across the rest of the planet that could make certain parts of the world uninhabitable. In fact, the climatologist explains how a 1°C temperature increase in certain parts of the planet could make a difference between life and death, especially for the elderly and anyone living in a non-air-conditioned environment.

As can be seen from this interview, the gradual thawing of Antarctica has started, with the warming of the peninsula in particular in full swing. The peninsula appears to have warmed the fastest as a result of the ozone hole—a somewhat unexpected outcome of the ozone problem. It is also forecast to heat up further in the future, at higher rates than the rest of the continent, with the entire continent becoming more accessible by 2100 with the melting of sea ice. All this will take place in tandem with the warming of the rest of the globe, which may make other portions of the planet uninhabitable.

With the conservative side of the climatologist's prediction for the warming of Antarctica (i.e., a 3° rather than a 4°C rise) by 2100 added to the average monthly temperature chart of the peninsula, its weather begins to look much more bearable (Table 2.2). The western side of the peninsula could already be inhabited today. It has a similar climate to Greenland.

With the gradual thawing of Antarctica, and in particular of the peninsula, the question becomes who would move to Antarctica and why. Part of the interest in Antarctica may well be fueled by the quest for its mineral resources.

## ANTARCTICA IS A MINERAL-RICH CONTINENT

One of the main ways in which the mineral potential of Antarctica has been assessed is similar to how mineral deposits have been found in the surrounding continents, the continents that together once formed Gondwanaland. Its mineral picture, therefore, is largely constructed—or reconstructed—by geological association with the surrounding continents from which it broke off (the study of plate tectonics, essentially).[55] In other words, current knowledge of its mineral potential is only partially based on actual drilling or mining operations.[56] The main exception consists of the U.S. Deep Sea Drilling Project off the continental shelf in the Ross Sea area in 1973, which encountered traces of petroleum and natural gas.[57]

Drilling and mining have been limited for a number of different reasons. The first of these are the climatic conditions of the continent, which were seen as prohibitive both economically and physically for any serious resource development throughout the 20th century. With no native population, Antarctica is far from labor markets and from the markets to which any minerals would need to be transported for final consumption.[58] The second was the world population level, which was much lower in the beginning of the

**Table 2.2 Potential Average Temperature in the Antarctic Peninsula by 2100* (Degrees Celsius)**

| | Jan | Feb | March | April | May | June | July | August | Sep | Oct | Nov | Dec |
|---|---|---|---|---|---|---|---|---|---|---|---|---|
| Rothera Weather Station | 4.0 | 3.1 | −0.1 | −0.7 | −3.8 | −5.8 | −9.6 | −8.8 | −6.4 | −4.2 | −0.3 | 3.2 |

*Original table from Cool Antarctica, "Antarctica Climate and Weather," with the predictions of the climatologist interviewed by the author added.

20th century than it is today, making it unnecessary to develop resources in such an isolated part of the planet, especially with many other corners not having been fully explored. The third, as will be explained later in this book, was the 1991 decision by the international community to ban mineral development in the continent on environmental grounds. The ban itself was made possible by the fact that no immediate urgency was seen for resource exploitation in the most remote part of the planet.[59]

In 1974, the U.S. Geological Survey (USGS) concluded, "Large accumulations of minerals very probably occur in Antarctica, for no other continent is void of mineral deposits. The question is whether these can be found and exploited economically."[60] Prior to the imposition of the 1991 ban, there is ample evidence to suggest that a number of countries had initiated serious geological research on the mineral potential of the continent. In other words, despite its prohibitive climate, these countries were nevertheless curious to discover what Antarctica had to offer.[61]

In 1975, for example, it is known that the Union of Soviet Socialist Republics (USSR) conducted mineral research off its Druzhnaya Base in the Weddell Sea. In 1983, the search for oil, gas, uranium, iron ore, and other mineral resources was part of the West German Antarctic Program. Geological projects conducted by Argentina and Chile around the same period had as their focus the Andean-type mineralization of the continent. It is also known that in 1984–1985, two U.S. research science projects in Antarctic were being planned for uranium resource assessment.[62]

Much of the onsite research in Antarctica has focused on the portions of the continent with exposed rock. As Broughton (2011/2012) states, "Roughly one percent of Antarctica is exposed rock and these are the only areas considered viable for mineral prospecting or mining." West Antarctica has more exposed rock than any part of the continent, particularly along the peninsula. The mineral potential lying beneath thick ice sheets is considered far less economical to obtain.[63]

The information available to date points to a continent that is fairly rich in mineral resources. East Antarctica is known to harbor deposits of iron (large but low-grade), titanium, uranium, niobium, tantalum, and a great number of industrial minerals (such as graphite, fluorite, and rock crystals), as well as coal.[64] In addition, it has been said that East Antarctica may contain rare earths, and that while diamonds have not been found, "ultramafic rocks similar to diamond-bearing rocks of Africa and Australia have been mapped in the Prince Charles Mountains and other locations in East Antarctica."[65] Crispini et al. (2011) have also ascertained the presence of gold in northern Victoria Land (namely, the Dorn Gold Deposit). There is no mention, however, of the precise extent of this gold.[66]

The Transantarctic Mountains are known to have small deposits of tin and rare earth minerals, copper, zinc, molybdenum, and large quantities of coal.[67] De Wit (1985) states that while a "resource and radioactivity survey during the 1976–1977 field season conducted in the Dry Valley region found no significant uranium concentration . . . reconnaissance field work in the Transantarctic mountains has located uranium pegmatites."[68]

Finally, in West Antarctica, the Antarctic Peninsula is known to hold copper, lead, zinc, silver, iron, chromium, nickel, cobalt, and other deposits.[69] In addition, the best hydrocarbon prospects appear to be in the parts of the Ross and Weddell Sea basins that are free of permanent ice.[70] Of all the resources presented here, however, based on current knowledge, the most promising are said to be coal reserves along the Transantarctic mountain chain and the iron ore deposits in East Antarctica.[71]

As Gjelsvik (1983) put it, "It may be safe to say that the Gods of the environment have put the mineral deposits in Antarctica in a deep freezer for the benefit of our grandchildren or even more distant generations."[72] In fact, in a 1998 mineral resource map of the circum-Pacific region of the Antarctic ice sheet (which includes southern portions of both East and West Antarctica), the USGS sets out the full scope of mineral occurrence, demonstrating that ferrous metals (iron and its alloying metals) have been reported in 29 locations across this region, base metals in 19, precious metals in 6, nuclear fuels in 2, and nonmetallic minerals in 9.[73]

In addition, the USGS has pointed to the mineral potential of the Southern Ocean's floor, the potential for geothermal energy generation in Antarctica because of volcanic activity in its Western portion, and the ample freshwater resources that the continent stores in its infinite icebergs.[74] Figure 2.2 provides some indication of where the continent's main minerals are positioned.

One of the most discussed regions of Antarctica in relation to minerals has been a zone that straddles the Transantarctic mountain chain, which is known as the Dufek Massif. This massif is composed of dark, mafic, igneous rock that occupies almost the entire northern part of the Pensacola Mountains. The massif was discovered in the 1957 International Geophysical Year (IGY) and was since studied by Russian and American geologists and geophysicists for over 18 years.[75]

The Dufek Massif has been compared to the highly valuable Bushveld Complex in South Africa due to its geological similarity, and it is thought to hold platinum group elements,[76] copper, nickel, tin, vanadium, and chromium. As Broughton states, "Almost all known layered mafic igneous complexes of this nature contain metal deposits which are economically exploitable and apart from the Bushveld Complex, the Dufek is believed

**Figure 2.2 Map of Principal Resources in Antarctica**

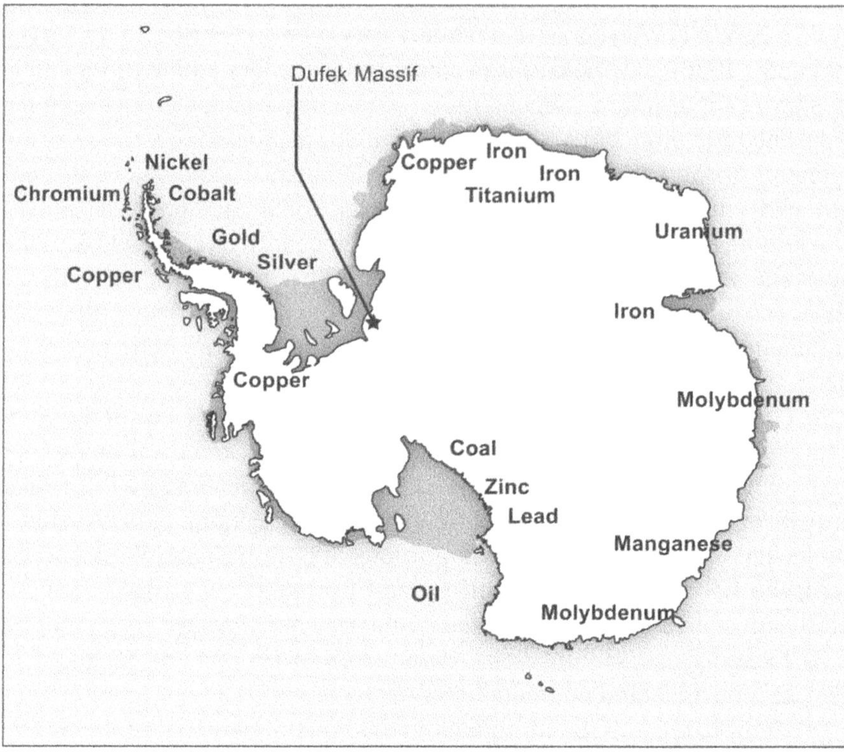


(Joseph Hill)

from interpreted geophysical data to be larger than any other known complex of this type on Earth." It is said to represent one of the main concentrations of platinum group elements on a global scale. These elements represent a highly valuable set of minerals used in a range of industries, from jewelry all the way to automobiles, electrical appliances, pharmaceuticals, and glass. However, only a fraction of the massif currently consists of exposed rock that would allow for the exploitation of these minerals.[77]

A number of feasibility studies have been conducted on Dufek's potential exploitation, with different assumptions made as to the type of power supply, equipment, personnel, infrastructure, and transportation that would be used.[78] The most prominent is a study conducted by De Wit in 1985, which concludes that while Dufek would be a risky project, its economic viability would nevertheless depend on the existing world price of platinum and the specific levels of platinum concentration in the massif.[79]

However, perhaps most interesting is the connection that the author draws between mining technology used in the Arctic back in 1985 and the mining of Dufek. He concludes that with the technology prevailing at the time, Dufek could have already been economical to exploit. In doing so, De Wit compares Dufek to the 35 operational mines in the Arctic at the time, and discusses the technological solutions that were found to overcome each and every type of problem that was encountered. He concludes that continued technological progress would no doubt facilitate the mining of Dufek even further.[80]

As is clear from this discussion, with the gradual thawing of Antarctica and its greater accessibility through melting sea ice, the temptation to exploit its mineral resources could certainly rise. The continent is mineral rich, with its coal, iron, and platinum reserves being particularly promising. As De Wit states, "unfortunately, there is the inescapable fact of heterogeneous distribution of mineral riches in the world. Platinum deposits, for example, are geographically very limited in their occurrence. South Africa has inherited about 85% of the world's platinum reserves outside Antarctica."[81] Therein may lie the strategic value of the seventh continent. The world may not need to run out of platinum for interest in Antarctica to grow. Just a few countries losing access to this valuable resource could prompt even more interest in the seventh continent because of its extremely uneven distribution across the globe.

In general, it also important to bear in mind that what is known about Antarctica's mineral potential today is constrained by the ban that was imposed on minerals development in the continent back in 1991, a ban that will be discussed in detail in coming chapters of this book. Therefore, all the main studies undertaken in this area predate the ban, and much reconnaissance work would need to take place when the ban comes up for review. With the extent of the technological development that has taken place since the 1990s, and the experience which the international community has acquired through mining in the Arctic, the world may very well discover that Antarctica is even richer than it once thought.

## THERE WILL COME A NEED FOR MAJOR LAND REDISTRIBUTION

If Antarctica is mineral rich, if its climate is changing, and if humans already showed themselves capable of braving its harshest climate as early as the 18th century in search of a better livelihood and commercial enterprise, is this land likely to remain uninhabited? The next chapter will describe how the territorial claims to Antarctica made by several countries were suspended by the 1959 Antarctic Treaty. Of course, while nothing would rule out the prospect that these territorial claims be revived and that new ones be tabled to capture the mineral riches of the continent, and that the competition for resources becomes one of the principal drivers of its occupation, the physical migration and resettlement of populations in Antarctica may be determined by yet another force. The world's population is exploding, and the climate crisis is likely to make environmental migration one of the main imperatives for land redistribution worldwide. With Antarctica being the only remaining continent in which no territorial rights have been recognized, it may emerge as an important part of the solution to the climate migration crisis by 2050–2100.

In 2015, the world population amounted to approximately 7.3 billion and is expected to rise to around 11 billion by 2100—a near 40 percent increase.[82] According to the State of World Population Report 2014, the world's "less developed regions"—with about 6 billion inhabitants—dominate today's global population. These regions also are growing the fastest. While the average global fertility rate was 2.5 children per woman in 2014, in East and Southern Africa it was 4.8, and West and Central Africa 5.6. In addition, roughly a third of the population in these regions is between the ages of 10–24.[83] In fact, the report warns against the increasingly large numbers of adolescents and youth, all of whom will be in search of viable economic opportunities.

While the bulk of the world population today (approximately 60 percent, or 4.3 billion) is Asian, it is the African continent that is expected to grow exponentially by 2100, with the Asian population rising relatively little. More precisely, Africa's population will grow about fourfold, from today's 1.1 billion to nearly 4.4 billion. The world population will be largely African and Asian, with these two continents alone coming to represent nearly 80 percent of humanity.[84]

Produced by the International Organization for Migration (IOM), the World Migration Report 2010—the most comprehensive report to date on global migration—states that if the migrant population continues to increase at the same pace as the last 20 years, the number of international migrants could rise to 405 million by 2100. It estimated that in 2008, 36 million people had been displaced by sudden-onset natural disasters, of which 20 million were specifically linked to climate change. The Sichuan earthquake alone displaced 15 million people. While migration is driven by a number of other factors, such as conflict and the "brain drain" from developing to developed countries, the report represented the first time that the IOM had spoken out this extensively on environmental migration.[85]

In its Migration Portal, the IOM states that while there are no perfectly reliable estimates of climate-induced migration, it is clear that environmental change is now leading to substantial population movements. The number of storms and droughts has increased threefold over the last 30 years. Its forecast for future environmental migration varies between 25 million to 1 billion environmental migrants by 2050, with 200 million being the most widely cited estimate.[86]

The IOM warns that "international migration is likely to transform in scale, reach, and complexity due to growing demographic disparities, the effects of environmental change, new global political and economic dynamics, technological revolutions and social networks." These transformations will be associated with increased opportunities if the migrant population is utilized effectively, but they also could exacerbate existing problems and generate new challenges.[87] It has called upon the international community to begin to explore "carefully managed migration."[88]

In its Fifth Assessment Report, the IPCC also tackled the problem of climate migration. While a highly controversial subject in climate negotiations, it increasingly became unavoidable. The IPCC wrote:[89]

> Climate change is projected to increase displacement of people (*medium evidence, high agreement*). Populations that lack the resources for planned migration experience higher exposure to extreme weather events, particularly in developing countries with low income. Climate change can indirectly increase risks of violent conflicts by amplifying well-documented drivers of these conflicts such as poverty and economic shocks (*medium confidence*).

It added that climate change will endanger food and health security, particularly for the most vulnerable populations.[90] At the 16th Conference of the Parties of the United Nations Framework Convention on Climate Change (UNFCCC) held in Cancun in December 2010, governments already started to recognize the problem of climate-induced migration, displacement, and relocation as an adaptation challenge, and they agreed to increase their cooperation on the matter. Shortly after, in 2012, a coalition known as the Nansen Initiative was created to develop a more coherent approach to the issue. Norway and Switzerland led the initiative, which currently includes a number of developed and developing countries that are together fighting to place the matter under international spotlight.[91]

In the runup to the Paris Climate Summit in December 2015, the Pacific Island nations, which are some of the countries most vulnerable to climate change, held a summit on the island of Kiribati, during which they called upon the world "to prepare our people for migration with dignity."[92] These island states face a serious risk of drowning with sea level rise.

At a press conference held in the runup to the 2015 Paris Climate Summit, the IOM, the United Nations High Commissioner for Refugees (UNHCR), and other institutions urged governments to "prepare for potential large-scale human mobility in the face of climate change."[93] The Advisory Group on Climate Change and Human Mobility of the UNFCCC, which is composed of number of different UN agencies, think tanks, and other groups working on refugee issues, called on governments to reduce climate change–related population displacement through careful adaptation and mitigation strategies.[94]

The Nansen Initiative expressed as a major source of concern that while refugees from war and persecution are protected by various international conventions, currently there are no clear international laws that offer protection to people displaced as a result of extreme weather events.[95] It called upon the Paris Climate Summit to provide guidance and assistance in planning for and facilitating voluntary internal and cross-border migration as a climate adaptation strategy.[96]

All these efforts indeed culminated in a small step forward at the Paris Climate Summit, in which a "Loss and Damage" section featured in the final report spoke of the need to examine population displacement due to climate change. The section essentially called for a further examination of the issue and the development of recommendations, but it explicitly stated that it does not provide any guidelines for liability or compensation.[97]

While the Pacific island states have been the most vocal in the international climate negotiations on this issue, according to Germanwatch's Global Climate Risk Index, Honduras, Myanmar, and Haiti were the countries most affected by extreme weather events between 1994 and 2013. In general, the index shows that the countries most affected by such events are low-income developing countries. It finds that more than 530,000 people died as a result of approximately 15,000 extreme weather events in that period, with economic losses amounting to nearly $2.2 trillion. In 2013, however, the Philippines, Cambodia, and India topped the list of the most affected nations,[98] demonstrating that middle-income countries are also suffering. Countries with low-lying lands and deltas are particularly vulnerable to climate change.

The Notre Dame Global Adaptation Index 2014 reached a similar overall conclusion—that is, that some of the world's poorest countries are the least prepared to adapt to the climate challenge. However, it also demonstrated that some very large and populous countries are even less prepared than island states such as Fiji. While the index placed Fiji as the world's 103rd least-prepared country, India was the 120th, Pakistan the 126th, and

Bangladesh the 140th. Outside Asia, countries such as Yemen, Kenya, and Zimbabwe have also been featured among the least prepared.[99]

In terms of national legislation, the issue of climate migration has started to become more prominent since 2007. That year, the Green Party of Australia launched an initiative to establish a "climate refugee visa" in its immigration legislation. The initiative would have created a "Climate Refugee Visa Class," a visa program to move 300 climate refugees from Kiribati, Tuvalu, and other parts of the Pacific each year, and which have given the government a mandate to work with the United Nations and other international forums to create new frameworks for environmental and climate refugees. But the program was defeated. Facing similar pressure, New Zealand introduced a new "Pacific Access Category," under which 75 people from Tuvalu, 75 from Kiribati, and 250 from Tonga could emigrate to New Zealand each year. While the category did not specifically speak about climate change, it was certainly be motivated by the risk of displacement that it creates.[100]

Strong pressure was also applied by the now-ousted president of the Maldives, Mohammed Nasheed, who announced in 2008 that the Maldives would establish a sovereign wealth fund for the purchase of a new homeland for his people. He stated that "this trust fund will act as a national insurance policy to help pay for a new homeland, should future generations have to evacuate a country disappearing under the waves."[101] In 2009, he convened an underwater cabinet meeting to draw international attention to the plight of his people.[102]

The president of Kiribati prepared his country along similar lines. In 2014, he completed the purchase of one of the islands of Fiji, about 2,000 kilometers from his country. François Gemenne, a specialist on migration at the Versailles–Saint Quentin University in Paris, has stated that Kiribati is one of the countries that has done the most to prepare for climate migration, launching a "Migration with Dignity" policy to allow its people to apply for jobs in neighboring New Zealand, along with other measures, to avoid a full-scale humanitarian crisis.[103]

Kenneth Weiss, a Pulitzer Prize–winning author who writes about science, the environment, and public health, published an article in *Foreign Policy* in 2015 entitled "The Making of a Climate Refugee," which grabbed international attention. The article discussed the plight of a citizen of Kiribati attempting to apply for climate refugee status in neighboring New Zealand, with his request denied. Few countries are prepared to accept a new legal category of refugees that could quickly add to millions, he wrote, with the concept of a "disappearing state"—or a modern-day Atlantis—becoming a new and serious challenge for policymakers.[104]

Today, if policymakers were told to consider Antarctica as the potential new homeland for people displaced by climate change, they would treat it as an outlandish proposal. However, in light of the numerous warnings about climate migration, the annual deaths that already occur as a result of extreme weather events, and how ill prepared some of the world's poorest countries are for the onslaught of climate change, the placement of Antarctica's future on the international negotiating table may start to become inevitable. This, coupled with the growing attractiveness of the continent's mineral wealth, may lead to an entirely new debate on the fate of the seventh continent.

## A MESSAGE FROM GREENLAND TO ANTARCTICA

As the climatologist stated in the interview in Box 2.1, "current weather in the Antarctic Peninsula is comparable to Greenland." He spoke of how Greenland was inhabited, and then disinhabited, in the Little Ice Age (around 1550–1850). What makes Greenland the most logical counterpart to the Antarctic Peninsula is the fact that it is the only landmass, outside Antarctica, that is covered in a permanent ice sheet.[105] If life currently exists in Greenland—with Greenland inhabited despite its ice cover—could the Antarctic Peninsula also be occupied someday? What can Greenland tell us about potential life in Antarctica?

Greenland is the world's biggest island, and the third-largest country in North America. A total of 80 percent of its surface area is covered in ice so heavy that it has depressed its central land area into a basin of about 300 meters below sea level.[106] Greenland has been inhabited, on and off, during the last 4,500 years by Arctic people.[107] Currently, a total of 57,700 Greenlanders inhabit the country, mostly living along its ice-free Western coastline,[108] with a few scientific research and expedition stations located farther inland. The northeastern part of the country is home to the world's largest national park—the Northeast Greenland National Park.[109]

Around 88 percent of the country's inhabitants are Inuits, with over 40 percent being of working age. The population of the capital, Nuuk, has been growing steadily and now includes 17,000 of the island's inhabitants.[110] As in other parts of the Arctic, Greenland has seen its population congregating around large settlements, experienced increased urbanization, and seen a gradual depopulation of its more remote areas.[111]

In its June 2010 edition, entitled "Greenland: Ground Zero for Global Warming," *National Geographic* described how one civilization replaced another on the island, with the climate acting as one of the principal historical determinants. It also described how the island's ice sheet—once considered too big to melt—has been slowly disintegrating. This phenomenon is likely to continue, and even accelerate with global warming, changing Greenland's geography and causing the sea level to rise by an amount sufficient to inundate coastlines around the planet. In fact, Greenland's ice sheet is said to contain approximately 7 percent of the world's freshwater, and to have reached a record 40 percent melt by 2007. Its continued disintegration will lead the island's central plateau to eventually rebound as the weight of the ice is lifted.[112]

With Antarctica's warming, parts of its ice sheet are also disappearing and its geography evolving. In fact, a 200-meter-thick iceberg that measured 78 kilometers by 37 kilometers (known as Larsen A) was lost in 1995 to the Larsen Ice Shelf.[113] Another part of the ice shelf (known as Larsen B) disintegrated and collapsed in 2002, covering an area comparable to the U.S. state of Rhode Island.[114]

In 2013, the National Aeronautics and Space Administration (NASA) released an animation of what Antarctica would look like without its ice sheet. "Before we had a regional overview of the topography, but this new map, with its much higher resolution, shows the landscape itself: a complex landscape of mountains, hills, and rolling plains, dissected by valleys, troughs, and deep gorges," Peter Fretwell, from the British Antarctic Survey, said in a press release.[115] Water has also collected below Antarctica's ice sheet, with the continent now known to have large numbers of subglacial lakes, the largest of which is Lake Vostock, near Russia's Vostock Station.[116]

In the 18th century, Denmark and Norway affirmed their sovereignty over Greenland. However, as Norway's power gradually waned, Greenland became a Danish colony in 1814 and formally became part of Denmark in 1953. Greenland joined the European Community (now the European Union) with Denmark in 1973 but withdrew about a decade later over stringent fishing quotas, with fish being its main food supply. It was granted self-government in 1979 by the Danish parliament, which evolved into greater responsibility for its internal affairs when the Act on Greenland Self-Government was signed in 2009. Today, Denmark continues to exercise control over Greenland's foreign affairs, security, and financial policy in consultation with the country's Self-Rule Government.[117]

A major part of Greenland's recent history has been defined by its quest for independence and the desire to control and exercise sovereignty over its mineral wealth. In fact, in its February 1983 edition entitled "People of the Arctic," *National Geographic* described how the struggle for territorial rights, self-government, and control of resources has been shared across all Arctic populations. They all sought to develop and to raise their standards of living. The magazine wrote, "The people of the Arctic . . . have stopped apologizing for themselves. They are not merely unfinished products of the civilization process."[118] Greenland's experience in self-rule would in fact bring its government into contact with any other populations across the globe in search for self-determination, receiving, for example, requests for meetings from the Palestinian Liberation Organization.[119]

Over the years, Greenland has acquired greater rights over its mineral wealth. In 1998, negotiations between its government and the Danish Ministry of the Environment and Energy resulted in an agreement to transfer the administration of mineral resources to Greenland. With the introduction of self-government in 2009 and the passage of the Mineral Resources Act in 2010, Greenland eventually assumed responsibility over all of its mineral resources. However, it received an annual block grant from the Danish government to cover the expenses associated with activities for which it took control.[120]

Greenland's Mineral Resources Act currently requires all mining applicants to complete Environmental Impact Assessments and a Social Sustainability Assessment for their activities. Companies are required, through these processes, to send reports for public consultation to project stakeholders. Citizen participation and consultation is, therefore, a core component of the authorization process, even if there has been criticism of how well these have been carried out.[121] In the meantime, the country has sought to become self-sufficient in energy by developing its renewable energy resources. Almost 70 percent of Greenland's electricity is now generated by greenhouse gas emission–free hydropower, which harvests its melting ice and expanding lakes.[122]

As its struggle for independence proceeded, together with its quest for greater economic development, Greenland—like many other parts of the Arctic—battled difficult social problems linked to violence, alcoholism, and a high suicide rate.[123] Its response has been more education, as well as greater economic development and job creation. It has continuously fought to define the Arctic identity of its people (rather than let others define it), to refine its governance arrangements, and to enhance local participation, control, and ownership of natural resources. In this endeavor, it was joined by the many other indigenous populations of the Arctic.[124]

Herein lie two key messages for Antarctica. First, any new Antarctic population will likely face some of the same challenges that the indigenous population of Greenland has. It will want economic growth, jobs, and quite likely some degree of self-determination, autonomy, and control over Antarctica's natural resources and environment. Second, the process of establishing effective and workable regimes for the governance of natural resources will take time.

However, with the continent having no indigenous population of its own, a future Antarctic population may be able to avoid some of the pitfalls that Arctic populations have suffered. Depending on who moves or is moved to Antarctica and how, the continent may find itself benefiting from a highly educated, highly skilled workforce right from the start, with a very clear sense of what it is doing. This could provide a major boost for economic development in the seventh continent, avoiding some of the social ills that have befallen the Arctic.

In Box 2.2, the former prime minister of Greenland, Aleqa Hammond, is interviewed. Asked to send a message to the future population of Antarctica, she says, "Decide your own destiny and your own future; do not let others decide what should happen to you. Everyone talks about conserving and preserving natural resources. You will have your own life, you will live there, and you should be the ones to decide."

## LIVING IN EXTREME CONDITIONS: WILL ANTARCTICA PROVE TO BE AVANT-GARDE?

In the interview in Box 2.2, the former prime minister of Greenland expressed the sentiment that Greenland is a "freezing" but not "hostile" environment. For her, renewable energy is central to Greenland's development. Because the country is surrounded by ice, the transportation of fuels to Greenland is complicated and costly, so renewable energy is giving her people greater hope of survival and independence. As Antarctica is far more isolated geographically, the transportation of fuel is even costlier.[125]

Greenland has found ways to meet the demands of its environment, which in some aspects is more challenging than Antarctica. Antarctica has neither polar bears nor Arctic foxes, but friendlier penguins and seals—even

### Box 2.2 Interview of Aleqa Hammond, Former Prime Minister of Greenland[a]

1. *Tell me about your homes in Greenland. Where do you build your homes, and do you worry about their safety with changes to the ice sheet? What kind of building materials do you use?*

All our building material is wood. Nobody builds houses on soil. Homes are built on rock; therefore, our homes are not affected by climate change and the thawing permafrost. We also dig deep to build our houses. Today, 70 percent of the energy we use to heat our homes comes from hydropower. In fact, Greenland is working very hard to achieve energy self-sufficiency. Ten years back, we received an international loan to develop our renewable energy sector, and paying it back has proven easier than we thought. With the cost of heating our homes having fallen, we have now changed our housing policy. Our people are now allowed to build bigger houses with bigger windows.

2. *Greenlanders are often described as living in a "hostile environment"? Do you see your environment as hostile? How does the average Greenlander feel about ice?*
Our environment is "freezing," yes, but "hostile," no. I see myself as living in a pristine, magnificent, and beautiful environment. People tell me, "you live far." I say, "I live right in the center of the world." If Greenland's environment were hostile, it would not have been inhabited for thousands of years. I was born in the Arctic, raised in the Arctic, and have always lived in the Arctic. Like every other country, Greenland has seasons. That, I think, is how the average Greenlander feels about his/her environment.

3. *Is there a sense of "community" amongst the population of Greenland? How does your environment, your remoteness, the ice you are surrounded with, impact your way of life?*
Greenland has a small population of roughly 58,000 people, of which non-Greenlanders or nonindigenous people are approximately 10,000. Our rate of growth is very low, and there is a reason for this. We live in the midst of the world's smartest animals—the polar bear and the Arctic fox. To live well, we exercise precaution, we plan our activities very carefully, in particular on stormy, snowy days, when we are immersed in darkness. Existing in Greenland also means carefully planning our foodstock. Other parts of the world have more children. We know that the fewer we are, the higher our chance of survival, hence the existing regulations limiting the number of children. If we grow too much, we know that winter may kill us.

4. *How will climate change impact Greenland, and do you see it as a threat or an opportunity?*
Glaciers are melting faster than ever before; rivers are now bigger than ever before. Hydropower is an enormous opportunity. But also, the retreating ice has led to the creation of new fjords that ships have

never been to before. We are now seeing some of the most beautiful places on Earth, that man was not privy to before. This is attracting large amounts of tourists: "climate tourism." People are coming because they want to see the icebergs, they want to see the melting, the calving.

5. *How would you like to see the Greenland economy evolve in future?* The Arctic is not a museum, and my people are not planning to stay cute. We want to develop. I would like to see all Greenlanders employed; I would like to see jobs of economic value. I would like to see renewable energy grow. I would like to see more hydropower, but also more exploitation of our mineral resources. I would like to see good housing and a higher standard of living. I would like to see my country enjoy even greater economic and political independence.

6. *Do you have a message for the future population of Antarctica?* Decide your own destiny and your own future; don't let others decide what should happen to you. Everyone talks about conserving and preserving natural resources. You will have your own life, you will live there, and you should be the ones to decide.

[a]Interview conducted by the author on December 12, 2015.

if seals can on occasion be hazardous. In particular, in terms of construction material, urban planning, and food storage, Greenlanders seem to have found a suite of solutions to address the special challenges they face.

As she mentioned, food storage in a frozen part of the world requires careful planning. While Greenland is an important exporter of fish, it is nevertheless obliged to import most of its fresh food supply.[126] In fact, today, much research is being devoted to urban sustainability in the extreme climatic conditions of the Arctic—research from which Antarctica could eventually come to benefit.[127]

So how would a built environment look like in Antarctica, and what would it take to live in its extreme conditions? Before any economy can be established in a place, housing must become feasible. In fact, as this book will show, there is already life in Antarctica, in the form of the multiple scientific research stations built by various countries across the continent, especially along its coastline. How have these stations been constructed, how have they evolved, and can they provide insight into the potential occupation of the seventh continent?

The U.S. McMurdo Station, built during the course of the IGY 1957–1958, was one of the first research stations on the continent, and today it has become its largest, housing over 1,000 staff.[128] It is located on the southern tip of Ross Island, the New Zealand–claimed portion of the continent. It serves as the Antarctic science facility for the United States. All personnel and cargo traveling to the U.S. station that is much farther inland—the Amundsen-Scott South Pole Station—first pass through McMurdo. The United States began the occupation of the site in December 1955, establishing a naval air facility called McMurdo as a logistics base from which construction would start. Operation Deep Freeze mounted the first prefabricated structures that formed the beginnings of the station.[129]

In 1963, Rear Admiral David Tyree, Commander of the U.S. Naval Support Force, Antarctica, in charge of Operation Deep Freeze from 1960–1963, published an article in *National Geographic* describing how the technology of the early 1960s had already revolutionized McMurdo. Entitled "New Era in the Loneliest Continent," the article described how nuclear power, project cargo planes on skis, and new construction material that allowed McMurdo to survive buried under the snow marked major advances in the conquest of Antarctica.[130] Huskies "face technological unemployment," he wrote, as machines displaced dogs for transportation on the continent. For example, the two-passenger-seat Polaris motor toboggan was introduced, which could pull a 1-ton load of cargo across a glacier.[131] A ski-equipped plane, the *Lockheed Hercules,* could "leap into the air in 20 seconds" to become the main supply vehicle for stations inland.[132]

He also described Antarctica's first atomic reactor, which was transforming life at the station. The new power source could provide four times more output for "Antarctica's metropolis" than did the traditional diesel generators that were previously used, and which required fuel to be shipped from other neighboring continents. "The reactor promises to reduce drastically the costly shipment of fuel oil, and it will also run a plant that desalts sea water, ending dependence on melted snow for a drinking supply." To Admiral Tyree, nuclear power meant "permanency" in Antarctica, confirming his realization that the United States was on the continent to stay.[133]

Forecasting a new era for Antarctica, he wrote, "I believe the time is near when women may be accommodated at McMurdo (so far as I know, only six women have ever set foot on the continent). I foresee also the coming of tourists to view the majestic mountains, glaciers and dry valleys, the comical penguins. . . ."[134]

Little did he know how true his predictions would prove to be, with women now employed in numerous research stations across the continent, and the tourism industry having taken off. In fact, women were not allowed to work with the U.S. Antarctic Program until 1969. Soon after, however, they broke the barrier and were followed by women carpenters, cooks, computer programmers, equipment operators, engineers, meteorologists, plumbers, janitors, water plant technicians, and many more professionals.[135]

He also could not have predicted how technological progress would not only spur the development and expansion of McMurdo, but that of many other stations across the continent. McMurdo's nuclear power plant in fact became a failed experiment, with renewable energy now gaining ground in Antarctica. Back in the 1950s and 1960s, the arguments for nuclear power had been strong. The station had significant fuel demands for its operations. Its aircraft, diesel-electric generators, heating units, and land vehicles were consuming over half of the total fuel freight delivered to the continent in the 1950s. Furthermore, supply ships could access the continent only during the austral summer months. However, after 438 malfunctions during its operational lifetime between 1964 and 1972, the reactor proved unreliable, having been available only 72 percent of the time. It was decommissioned in 1972, achieving only half of its expected design life of 20 years.[136]

McMurdo continued to grow and evolve. While heavily criticized during much of its existence for its localized environmental impact, it began to demonstrate greater environmental stewardship gradually. Because of its central location, large surface area, trained personnel, and specialized facilities, today it receives and processes waste from multiple U.S. facilities in Antarctica. At McMurdo, the waste is sorted, identified, packed for processing, and shipped to the United States, with much of it recycled.[137] As Klein et al. (2008) write, the station evolved on a "project by project basis" in terms of physical structures, with gradual improvements made to its functioning.[138] Its fuel storage system (its internal and surrounding road network, for example) was slowly redesigned to contain the growth of its environmental footprint.[139] The station also broke new ground with the production of fresh vegetables in a greenhouse that allowed its residents to eat salad several times a week and have a much more comfortable existence, particularly in the dead of winter, when outside supplies were not forthcoming.[140]

However, as will be shown next, McMurdo was surpassed by newer and more sustainable architectural design across Antarctica, and it required an overhaul. In 2015, The U.S. National Science Foundation (NSF) presented a $300 million plan to upgrade the flagship station on which it solicited input from other agencies.[141] Writing for the *Journal of the American Institute of Architects*, Hill (2013) wrote:[142]

> Like an old shopping mall or a neglected urban neighborhood, McMurdo needs a serious makeover, and not just because of the aesthetics of the place. Last year, a blue-ribbon panel commissioned by the National Science Foundation and the White House Office of Science and Technology Policy concluded that most of McMurdo's 105 buildings are "in imminent need of repair and replacement." Moreover, the station's "somewhat haphazard arrangement," as the report gingerly puts it, "leads to wasted resources and also raises serious safety questions." The panel warned that McMurdo's "atrophying logistics infrastructure" needs to be upgraded or replaced. "Failure to do so will simply increase logistics costs until they altogether squeeze out funding for science." All of which explains the report's rather wonkish title: "More and Better Science in Antarctica Through Increased Logistical Effectiveness."

With research stations having existed in Antarctica for over 60 years now, major progress has been made in their choice of location and building design. As Friend (2009) explains, when choosing a site for a research station, a number of parameters must be taken into account. The station should ideally be based on exposed rock to provide stable and permanent anchoring, although some stations cannot meet this requirement when their goal is to conduct research in some of the more remote, inland parts of the continent. They should have neighboring space for small windfarms and photovoltaic arrays to generate renewable energy. Locations near or within mountains can benefit from a natural shelter from katabatic winds, and those in proximity to blue ice fields may find themselves with natural aircraft landing strips. In addition, stations with access to soft snow that can more easily be melted can find their freshwater supply enhanced.[143]

Much research has also gone into the ideal interior design features for human survival in cold, isolated environments. For example, studies on confinement, which began with the *Apollo* space missions, found that the longer the confinement, the larger the living space that is required and the bigger the need for privacy. Using light-emitting diode (LED) lighting systems, which change the color of light to simulate daily and seasonal cycles, can improve mental health during long periods of confinement. Special thermal indoor protective heating systems can also make life much more comfortable.[144]

As Antarctica does not belong to any set of nations, it has no specific building codes, and research into how best to build in the seventh continent is still evolving. Currently, research stations or facilities only respond to the environmental regulations of the Antarctic Treaty, especially its Madrid Protocol, with individual countries applying modified or sometimes simplified versions of their own national building regulations.[145]

The Madrid Protocol requires that countries consider the entire lifecycle of a station upon construction, including its removal once decommissioned, and it imposes an Environmental Impact Assessment requirement.[146] The Council of Managers of National Antarctic Programs (CONMAP) has also issued guidelines to assess and reduce the impact of projects being developed in Antarctica, encouraging reduced fossil fuel consumption for energy generation.[147]

Since McMurdo was built, research stations have become far more sophisticated, with many of them now subject to architectural design competitions. While some of the older stations have been rebuilt, others have been renovated or repurposed. However, it is the entirely new stations that demonstrate the greatest progress in their design. Stations that were renovated or repurposed have seen changes over much longer time scales.[148]

In 2013, the Ice Lab Exhibition in Glasgow showcased the evolution of architectural design in Antarctica, featuring five recent Antarctic stations, celebrating "the exciting new ideas that can emerge at the intersection of science, art, architecture, and engineering."[149] The exhibition demonstrated the scope of the challenge with which architects are faced in Antarctica. With snow levels rising by over a meter annually, the continent's climate has caused several stations to be buried over the space of 10 years. For instance, phase VI of Halley Station, owned by the United Kingdom, envisaged a structure supported by skis that would be fully relocatable.[150]

The exhibition also celebrated Bharati Station, operated by India, whose eco-credentials include that it was constructed from the very shipping containers that carried the building equipment to the site. The *Guardian* wrote, "Compared to the primitive cabins of the first Antarctic explorers, these buildings are as high-performance machines, the architectural equivalent of zip-up Gore-Tex to Shackleton's tweed."[151] At the Ice Lab, architect David Garcia exhibited a plan for a subterranean station inspired by the way that igloos use compacted snow for isolation. If his project were to materialize, it would entirely do away with the need to transport foreign building materials to Antarctica.[152]

Station design in Antarctica has evolved from stations being built above the surface to being designed for burial, to being elevated on stilts, all the way to combining both overground and underground aspects to maximize a station's resistance, prolong its lifespan, make it safer and more comfortable, and minimize its environmental footprint.[153] Elevated stations—stations built on legs that can be raised to counter the accumulation of snow—are now quite common in Antarctica. The Amundsen-Scott South Pole Station, for example, was abandoned in 1975 to be rebuilt on stilts after it was buried by 8 meters of snow.[154]

Station design has also evolved in terms of fire protection, with fires spreading quickly in Antarctica's dry climate—the "white Sahara." In fact, several stations are known to have entirely burned down, such as Brazil's Comandante Ferraz, or to have had parts of them melt, such as a storage facility at McMurdo.[155]

However, Belgium's Princess Elisabeth Station, a zero-emissions station that opened in 2009, represents the state-of-the-art architecture in Antarctica. Located in Queen Maud Land at the north of the continent, it sits on a small, exposed granite ridge. The station harnesses the power of wind and sun to sustainably fuel the entirety of its operations. Its near-full energy autonomy has made the transportation of fuel to the station unnecessary. The station is unique in that it was launched as a private initiative, built by the International Polar Foundation, and only subsequently donated to the Belgian state in 2010.[156]

The station's location was carefully chosen to allow it to harness renewable energy year round, and to benefit from soft and flat surrounding snowfields that are only minimally affected by ice sheet movements. These fields generate a constant stream of freshwater for the station and allow small aircrafts to land. The station is equipped with an automated system that can be remotely activated to prepare for its occupation preseason, allowing its managers to switch on its heating, for example. It utilizes black-gray water treatment systems to allow a portion of wastewater to be recycled for use in toilets and other nonpotable utilizations and has very sophisticated insulation.[157]

That said, even developing countries in Antarctica are now taking their stations to new levels. South Africa's SANAE IV was elevated and positioned in such a way that it would avoid any snow accumulation, which is expected to extend its lifetime significantly. Its rounded corners have created improved aerodynamics, making it more wind resistant and optimizing its performance. The station uses waste heat produced by its diesel generators to heat its buildings and water supply. It also enjoys high-speed Internet from a continuous satellite connection to the South African National Antarctic Program in Cape Town.[158]

Greater energy-efficiency measures, as well as the use of renewable energy has also been explored across a range of other stations, such as the Ecuadorian Pedro Vicente Maldonado Station[159] or the new Brazilian

Antarctic station.[160] With the abundance of solar energy in the austral summer months, an increasing number of stations are now equipping themselves with solar panels or wind systems that can harness the full power of the continent's katabatic winds.[161] Today, wind is the most-utilized renewable energy source in Antarctica.[162] In fact, the architectural progress made in Antarctica may gradually come to provide lessons in sustainability for the rest of the planet. The ban on mining in the continent clearly has obligated research stations to look for creative alternatives, and it is working.

The revolution seen in Antarctica's architecture is now also visible in transportation to and from the continent. The first commercial Boeing 757 passenger plane landed in Antarctica in December 2015. The Loftleidir Icelandic Airliner had set out to prove "the feasibility of landing commercial passenger airlines at Union Glacier," said a spokesman for Antarctic Logistics & Expeditions, a company that provides flights and logistics services for scientific and other expeditions to the continent. Union Glacier Camp, located in the Ellsworth Mountains, is the only private seasonally occupied camp in Antarctica.[163]

The Australian Antarctic Division and Royal Australian Air Force showcased their enhanced logistical capabilities in November 2015, flying a C-17A Globemaster to Antarctica. The aircraft covered a distance of 3,450 kilometers from Australia to the Wilkins Aerodrome on the western side of the continent in just over five hours, carrying 12,340 kilograms of cargo and equipment, at the start of the austral summer season. The Wilkins Aerodrome is a single-runway aerodrome operated by Australia, and the C-17A Globemaster is now the largest aircraft to have flown to it to date.[164]

In December 2015, China sent a fixed-wing aircraft—the *Xueying*—to the South Pole as a test. Fixed-wing aircraft are designed for fast transport and emergency rescue and are particularly suitable for research purposes. So far, only the United States, Germany, and the United Kingdom own multifunctional fixed-wing aircraft in Antarctica.[165] As can be seen from this discussion, therefore, Antarctica is becoming far more accessible, with a revolution in transportation to the continent also well underway.

With the housing and logistics revolution currently underway in Antarctica, the continent is slowly becoming more habitable. As global warming continues to sweep through the planet, rendering various parts of it uninhabitable, will the international community be able to resist the addition of yet another continent to the world's living space? Will it be able to resist the exploitation of the known mineral riches of the seventh continent? The case for Antarctica will only get stronger.

## NOTES

1. Landsat Image Mosaic of Antarctica, NASA; last accessed March 15, 2016, http://lima.nasa.gov/antarctica/.

2. For a description of the breakup of Gondwanaland, see Peter Cattermole, *Building Planet Earth: Five Billion Years of Earth History* (London: Cambridge University Press, 2000).

3. James D. Hanson and John E. Gordon, *Antarctic Environments and Resources: A Geographical Perspective* (Essex, UK: Addison Wesley Longman Limited, 1998), 175.

4. John Harrison, *Forgotten Footprints: Lost Stories in the Discovery of Antarctica* (Cardigan, UK: Parthian Books, 2014), 28.

5. Scott Harris Friend, "Sustainability and Habitation in Antarctica," master's thesis, Department of Architecture, Montana State University, 2009, 10.

6. Hanson and Gordon, *Antarctic Environments and Resources,* 177.

7. Ibid.

8. Robert Clancy, John Manning, and Henk Brolsma, *Mapping Antarctica: A Five Hundred Year Record of Discovery* (Chichester, UK: Praxis Publishing, 2014). See also Helen S. Wright, *The Seventh Continent: A History of the Discovery and Explorations of Antarctica* (Boston: Gordon Press, 1918).

9. Friend, "Sustainability and Habitation in Antarctica," 16.

10. For American sealing operations in Antarctica, see Kenneth J. Bertrand, *Americans in Antarctica, 1775–1948* (New York: American Geographical Society, 1971).

11. Hanson and Gordon, *Antarctic Environments and Resources,* 177.

12. Ibid., 198.

13. Ibid., 178.

14. Ibid., 198.

15. Ibid., 198–99.

16. See Friend, "Sustainability and Habitation in Antarctica," 10; Hugh Robert Mill, *Siege of the South Pole* (New York: Stokes Company, 1905); Alan Gurney, *Below the Convergence: Voyages Towards Antarctica, 1699–1839* (New York: Norton, 1997); and David Day, *Antarctica: A Biography* (Oxford: Oxford University Press, 2013).

17. Hanson and Gordon, *Antarctic Environments and Resources,* 198–99.

18. Clancy, Manning, and Brolsma, *Mapping Antarctica*, 96–97.

19. Ibid.

20. Ibid., 98.

21. Harrison, *Forgotten Footprints*, 49–50.

22. Hanson and Gordon, *Antarctic Environments and Resources,* 178–79.

23. Friend, "Sustainability and Habitation in Antarctica," 11.

24. See Hanson and Gordon, *Antarctic Environments and Resources*, 178–79; and Johan Nicolay Tonnessen and Arne Odd Johnsen, *The History of Modern Whaling* (London: C. Hurst & Co., 1982).

25. Ibid., 202.

26. Ibid.

27. Ibid., 205.

28. Clancy, Manning, and Brolsma, *Mapping Antarctica*, 99.

29. Harrison, *Forgotten Footprints*, 323.

30. Hanson and Gordon, *Antarctic Environments and Resources,* 204.

31. Clancy, Manning, and Brolsma, *Mapping Antarctica*, 99.

32. International Court of Justice, "Whaling in the Antarctic (Australia vs. Japan: New Zealand intervening)," judgment of March 31, 2014.

33. "Japanese Whaling Ships Depart for Antarctic Hunt," British Broadcasting Corporation (BBC), December 1, 2015, http://www.bbc.com/news/world-asia-34971637.

34. Hanson and Gordon, *Antarctic Environments and Resources,* 46–47.

35. Friend, "Sustainability and Habitation in Antarctica," 8.

36. Ibid., 1.

37. Chapter 4, "Potential Mineral Resources in Antarctica," *Polar Prospects: A Minerals Treaty for Antarctica*, U.S. Congress, Office of Technology Assessment, Washington, DC: U.S. Government Printing Office, September 1989, 96; https://www.princeton.edu/~ota/disk1/1989/8926/892606.PDF.

38. Friend, "Sustainability and Habitation in Antarctica," 1.

39. Ibid., 2–3.

40. Ibid.

41. Friend, "Sustainability and Habitation in Antarctica," 6–8. The Russian Vostock Station is located inland in Princess Elizabeth Land and was originally founded in 1957.

42. Hanson and Gordon, *Antarctic Environments and Resources,* 46–47.

43. Ibid., 45.

44. Friend, "Sustainability and Habitation in Antarctica," 6–8.

45. John Turner et al., "Antarctic Climate Change During the Last 50 Years," *International Journal of Climatology* 25 (2005): 279–94.

46. John Turner et al., eds. *Antarctic Climate Change and the Environment: A Contribution to the International Polar Year 2007–2008.* Scientific Committee on Antarctic Research, November 25, 2009, http://www.scar.org/scar_media/documents/publications/ACCE_25_Nov_2009.pdf.

The preparation of ACCE was first discussed by the SCAR Executive Committee at its meeting in Bremen in July 2004, which agreed that SCAR carry out an Antarctic climate assessment as a contribution to the Antarctic Treaty Consultative Mechanism and IPCC deliberations. The project was coordinated by a nine-person steering committee and involved a total of 97 scientists from 12 countries. For a history of the ACCE report, see the SCAR website at http://acce.scar.org/wiki/Key_points_-_2013_version.

47. Ibid., xvi. On Antarctica's changing climate, also see R. Winkelmann et al., "Combustion of Available Fossil Fuel Resources Sufficient to Eliminate the Antarctic Ice Sheet," *Science Advances* 1 (2015): 1–5; and N. R. Golledge et al., "The Multi-Millennial Antarctic Commitment to Future Sea-Level Rise," *Nature* 526 (2015): 421–25.

48. John Turner et al., "Antarctic Climate Change and the Environment: An Update," *Polar Record* 50 (2014): 238.

49. Ibid., 242.

50. Ibid., 249. The update states that in order to assess Antarctica's climate over the next 100 years, "The climate models used for assessing future climate change vary in complexity. Most of the current generation of climate models are 'climate simulators,' in the sense that they attempt to represent, as realistically as is practical, known processes that are important in influencing climate change. They are the most realistic means we have currently of providing projections of future environmental behavior, albeit crudely and at coarse resolution."

51. Ibid., 244.

52. Ibid., 246–47.

53. IPCC, "Climate Change 2014: Synthesis Report Summary for Policymakers," 2014, 4; https://www.ipcc.ch/pdf/assessment-report/ar5/syr/AR5_SYR_FINAL_SPM.pdf.

54. SCAR, "Antarctic Climate Change and the Environment—2015 Update," May 4, 2015, 3; http://www.scar.org/scar_media/documents/science/antera/ATCM38_ip092_e.pdf.

55. U.S. Congress, Office of Technology Assessment, *Polar Prospects*, 96.

56. Darcy Broughton, "The Potential for Mineral Exploration and Extraction in Antarctica," *PCAS 2011/2012,* Gateway Antarctica, University of Canterbury, 2; http://www.anta.canterbury.ac.nz/documents/PCAS_14/ProjectReports/Broughton%20-The%20potential%20for%20mineral%20exploration%20and%20extraction%20in%20Antarctica.pdf.

57. David B. Doan, "The Mineral Industry of Antarctica," U.S. Geological Survey—Mineral Information, 2; http://minerals.usgs.gov/minerals/pubs/country/1995/9501095.pdf.

58. The many prohibiting factors for mineral exploitation in Antarctica are described on p. 26 of the joint U.S. Department of Interior and National Science Foundation Geological Survey Circular 909, "Petroleum and Mineral Resources of Antarctica," 1983, http://pubs.usgs.gov/circ/1983/0909/report.pdf.

59. In Chapter Four of this book, the full history of the ban on minerals and the many reasons for it are discussed. Few countries truly believed Antarctica to be economically exploitable at the time.

60. U.S. Department of the Interior, Geological Survey Circular 705, "Mineral Resources of Antarctica," N. A. Wright and P. L. Williams, eds., 1974, http://pubs.usgs.gov/circ/1974/0705/report.pdf.

61. See, for example, Joseph P. Riva and James E. Mielke, Congressional Research Service, "Polar Energy Resources Potential," U.S. House of Representatives, Ninety-Fourth Congress, Second Session, 1976; and Robert J. Tingey, *The Geology of Antarctica* (Oxford: Clarendon Press, 1991).

62. Maarten J. de Wit, *Minerals and Mining in Antarctica: Science and Technology, Economics and Politics* (Oxford: Clarendon Press, 1985), 6.

63., "The Potential for Mineral Exploration and Extraction in Antarctica," 3–4. On exposed rock, see also U.S. Congress, Office of Technology Assessment, *Polar Prospects*, 93.

64. Egil Bergsager, "Basic Conditions for the Exploration and Exploitation of Mineral Resources in Antarctica: Options and Precedents," in Francisco Orrego Vicuna, ed., *Antarctic Resources Policy: Scientific, Legal, and Political Issues* (Cambridge, UK: Cambridge University Press, 1983), 173.

65. U.S. Congress, Office of Technology Assessment, *Polar Prospects*, 120, 122.

66. Lauro Crispini et al., "The Dorn Gold Deposit in Northern Victoria Land, Antarctica: Structure, Hydrothermal Alteration, and Implications for the Gondwana Pacific Margin," *Gondwana Research* 19 (2011).

67. Tore Gjelsvik, "The Mineral Resources of Antarctica: Progress in Their Identification," in *Antarctic Resources Policy*, 62–63 (1983).

68. De Wit, *Minerals and Mining in Antarctica,* 7.

69. Gjelsvik, "The Mineral Resources of Antarctica," 62–63.

70. Bergsager, "Basic Conditions for the Exploration and Exploitation of Mineral Resources in Antarctica," 173.

71. De Wit, *Minerals and Mining in Antarctica,* 15.

72. Gjelsvik, "The Mineral Resources of Antarctica," 63.

73. U.S. Department of the Interior and U.S. Geological Survey, "Explanatory Notes for the Mineral-Resources Map of the Circum-Pacific Region

of the Antarctic Ice Sheet," 1998, 17–19, http://pubs.usgs.gov/cp/42/report.pdf.

74. U.S. Department of the Interior, Geological Survey Circular 705, "Mineral Resources of Antarctica," 12–13, 15.

75. De Wit, *Minerals and Mining in Antarctica,* 17.

76. The six platinum group metals are ruthenium, rhodium, palladium, osmium, iridium, and platinum. They are generally used in applications that exploit their unique properties, such as their inertness, oxidation-resistance, biocompatibility, high melting temperature, good conductivity, and other electronic and catalytic characteristics.

77. Broughton, "The Potential for Mineral Exploration and Extraction in Antarctica," 9–10.

78. Ibid., 10.

79. De Wit, *Minerals and Mining in Antarctica,* 34–35.

80. Ibid.

81. Ibid., viii.

82. The Statistics Portal, "Distribution of the Global Population 2015, by Continent" and "Forecast of the World Population for 2100, by Continent," last accessed March 15, 2016, http://www.statista.com/statistics/237584/distribution-of-the-world-population-by-continent/.

83. United Nations Population Fund (UNFPA), "State of World Population 2014: The Power of 1.8 Billion Adolescents, Youth, and the Transformation of the Future," 2014, 115; https://www.unfpa.org/sites/default/files/pub-pdf/EN-SWOP14-Report_FINAL-web.pdf.

84. The Statistics Portal, "Distribution of the Global Population 2015, by Continent" and "Forecast of the World Population for 2100, by Continent."

85. IOM, "World Migration Report 2010: The Future of Migration: Building Capacities for Change," 2010, 121; http://publications.iom.int/system/files/pdf/wmr_2010_english.pdf. The IOM proposes the following definition of an environmental migrant: "Environmental migrants are persons or groups of persons who, for compelling reasons of sudden or progressive changes in the environment that adversely affect their lives and living conditions, are obliged to leave their habitual homes, or choose to do so, either temporarily or permanently, and who move either within their country or abroad." See IOM, "Migration and the Environment," last accessed March 15, 2016, https://www.iom.int/jahia/webdav/site/myjahiasite/shared/shared/mainsite/microsites/IDM/workshops/evolving_global_economy_2728112007/presentations/presentation_migration_environment.pdf.

86. IOM Migration Portal, "Migration, Climate Change, and the Environment: A Complex Nexus," https://www.iom.int/complex-nexus#estimates.

87. IOM, "World Migration Report 2010," 3.

88. Ibid., 4.

89. IPCC, "Climate Change 2014," 16.

90. Ibid., 13–15.

91. See the Nansen Initiative at https://www.nanseninitiative.org.

92. Climate Action, "Pacific Island Nations Warn of Large-Scale Climate Migration," October 14, 2015, http://www.climateactionprogramme.org/news/pacific_island_nations_warn_of_large_scale_climate_migration. The leaders of Fiji, Kiribati, Tuvalu, and Tokelau participated in the summit.

93. "Press Conference: Human Mobility in the UNFCC: Paris COP21 Draft Text 10th February, 2015, Palais des Nations, Geneva," Environmental Migration Portal, February 10, 2015, http://environmentalmigration.iom.int/press-conference-human-mobility-unfcc-paris-cop21-draft-text-10th-february-2015-palais-des-nations.

94. Recommendations from the Advisory Group on Climate Change and Human Mobility, "Human Mobility in the Context of Climate Change, UNFCCC-Paris COP-21," November 2015, http://www.internal-displacement.org/assets/publications/2015/201511-human-mobility-in-the-context-of-climate-change-unfccc-Paris-COP21.pdf.

95. Center for International Climate and Environmental Research et al., "The Nansen Conference: Climate Change and Displacement in the 21st Century," Norwegian Refugee Council, June 5–7, 2011, 3, http://www.unhcr.org/4ea969729.pdf.

96. Recommendations from the Advisory Group on Climate Change and Human Mobility, "Human Mobility in the Context of Climate Change, UNFCCC-Paris COP-21."

97. See UNFCCC, Conference of the Parties, "Adoption of the Paris Accord," FCCC/CP/2015/L.9/Rev.1, December 12, 2015, section on "Loss and Damage," 7; https://unfccc.int/resource/docs/2015/cop21/eng/l09r01.pdf.

98. S. Kreft et al., "Global Climate Risk Index 2015: Who Suffers Most from Extreme Weather Events? Weather-Related Loss Events in 2013 and 1994 to 2013," Germanwatch, November 2014, https://germanwatch.org/en/download/10333.pdf.

99. See ND-GAIN Index at http://index.gain.org.

100. IOM, "World Migration Report 2010," 83.

101. Ibid.

102. "Maldives Government Highlights the Impact of Climate Change . . . by Meeting Underwater," *DailyMail.com*, October 20, 2009, http://www.dailymail.co.uk/news/article-1221021/Maldives-underwater-cabinet-meeting-held-highlight-impact-climate-change.html.

103. Laurence Caramel, "Besieged by the Rising Tides of Climate Change, Kiribati Buys Land in Fiji," *The Guardian*, July 1, 2014.

104. Kenneth R. Weiss, "The Making of a Climate Refugee," *Foreign Policy*, 2015, http://foreignpolicy.com/2015/01/28/the-making-of-a-climate-refugee-kiribati-tarawa-teitiota/.

105. See Joyce A. Quinn and Susan L. Woodward, *Earth's Landscape: An Encyclopedia of the World's Geographic Features* (Santa Barbara: ABC-CLIO, 2015).

106. Ibid.

107. Stefansson Arctic Institute, *Arctic Human Development Report: Regional Processes and Global Linkages* (Copenhagen: Nordic Council of Ministers, Denmark, 2014), 109, http://norden.diva-portal.org/smash/get/diva2:788965/FULLTEXT03.pdf.

108. Central Intelligence Agency (CIA), "North America: Greenland," in *The World Factbook*, https://www.cia.gov/library/publications/the-world-factbook/geos/gl.html (last updated June 7, 2016).

109. "Greenland," in the *New World Encyclopedia*, last accessed March 15, 2016, http://www.newworldencyclopedia.org/entry/Greenland.

110. Ibid.; also see Stefansson Arctic Institute, *Arctic Human Development Report*, 82–87.

111. Stefansson Arctic Institute, *Arctic Human Development Report*, 96–97.

112. Mark Jenkins, "True Colors: The Changing Face of Greenland," *National Geographic*, special issue entitled "Greenland: Ground Zero for Global Warming" (June 2010): 46–47.

113. Hanson and Gordon, *Antarctic Environments and Resources,* 52.

114. Portland State University, "Larsen Ice Shelf 2002: Warmest Summer on Record Leads to Disintegration," 2002, http://web.pdx.edu/~chulbe/science/Larsen/larsen2002.html.

115. "NASA Map Shows What Antarctica Would Look Like Without Ice," *CBC News*, June 7, 2013, http://www.cbc.ca/news/technology/nasa-map-shows-what-antarctica-would-look-like-without-ice-1.1304997.

116. Hanson and Gordon, *Antarctic Environments and Resources,* 52.

117. Stefansson Arctic Institute, *Arctic Human Development Report*, 109–110 and 200–201. See also CIA, "North America: Greenland."

118. Priit J. Vesilind, "Hunters of the Lost Spirit," *National Geographic*, special issue entitled "People of the Arctic" (February), 163:2, 155 (1983).

119. Ibid., 191.

120. Stefansson Arctic Institute, *Arctic Human Development Report*, 274–77.

121. Ibid.

122. ABB Communications, "Clean Sustainable Energy for Greenland," December 10, 2012, http://www.abb.com/cawp/seitp202/b08ea3b92dc74ac8c1257aaf0047543c.aspx.

123. Stefansson Arctic Institute, *Arctic Human Development Report*, 307, 333–36.

124. Ibid., 21–23.

125. For example, see the statistics on the cost of energy transportation to Antarctica in Tina Tin et al., "Energy Efficiency and Renewable Energy Under Extreme Conditions: Case Studies from Antarctica," *Renewable Energy* 35 (2010): 1715–23.

126. CIA, "North America: Greenland.

127. See, for example, the recent NSF $3 million cross-disciplinary research grant that is enabling researchers from George Washington University to study Arctic cities and develop an index to measure the sustainability of communities. As the project's principal investigator said, "Understanding how people survive in the Arctic might give us insight into how people live in hot places, too. It will be significant in terms of looking at life in extreme circumstances." Julyssa Lopez, "How to Survive in the Arctic: $3 Million Grant from the National Science Foundation Will Expand Research on Urban Sustainability in Extreme Climate Conditions," *GW Today*, November 30, 2015, http://gwtoday.gwu.edu/how-survive-arctic.

128. Arthur K. Britton, *Extreme Jobs in Extreme Places: Life at a Polar Research Station* (New York: Gareth Stevens Publishing, 2013), 6–7.

129. For a history of McMurdo, see Andrew G. Klein et al., "The Historical Development of McMurdo Station, Antarctica: An Environmental Perspective," *Polar Geography* 31:3–4 (2008): 119–144.

130. Rear Admiral David Tyree, "New Era in the Loneliest Continent," *National Geographic*, special edition entitled "Burma" (February 1963), 123:2.

131. Ibid., 268.

132. Ibid., 281.

133. Ibid., 266, 270, 273.

134. Ibid., 296.

135. See "Three Women Manage Antarctic Stations: Go South Young Woman," written by Janet Phillips (once manager at the U.S. Palmer Station on the Antarctic Peninsula). The article first appeared on the website that her husband, Juan Reyes, published during his 1999–2000 summer at the South Pole; last accessed March 15, 2016, http://www.southpolestation.com/trivia/90s/janet.html.

136. For a history of McMurdo's nuclear reactor, see Tyler Reid, "Nuclear Power at McMurdo Station, Antarctica," submitted as coursework for PH241, Stanford University (winter 2014), http://large.stanford.edu/courses/2014/ph241/reid2/.

137. Klein et al., "The Historical Development of McMurdo Station," 121.

138. Ibid., 131.

139. Ibid., 136, 138, 142.

140. Kris Kuenning, "Green Antarctica: Station Greenhouses Produce Fresh Food, Feel-Good Environments," NSF press release, February 25, 2004, http://www.spaceref.com/news/viewpr.html?pid=13724.

141. Carolyn Gramling, "Overhaul in the Works for Aging U.S. Antarctic Station," *Science,* December 8, 2015, http://news.sciencemag.org/climate/2015/12/overhaul-works-aging-u-s-antarctic-station.

142. David Hill, "A New Master Plan for the McMurdo Research Station in Antarctica: OZ Architecture Attempts to Reinvent a Utilitarian Science Outpost in the Wilds of Antarctica," *Journal of the American Institute of Architects* (October 9, 2013), http://www.architectmagazine.com/design/buildings/a-new-master-plan-for-the-mcmurdo-research-station-in-antarctica_o.

143. Friend, "Sustainability and Habitation in Antarctica," 57.

144. Maijinn Chen, "Antarctic Architecture," *Architecture Week*, February 16, 2005, http://www.architectureweek.com/2005/0216/building_1-1.html.

145. Friend, "Sustainability and Habitation in Antarctica," 65.

146. Hanne Nielsen, "From Shelter to Showpiece: The Evolution of Scientific Antarctic Stations," University of Canterbury Summer Research Project, November 2013–January 2014, 6.

147. Isabel Mino, Carlos Naranjo, and Andres Montero, "Evaluation of Energy Efficiency Measures for the Ecuadorian Antarctic Station Pedro Vicente Maldonado Through a Calibrated Building Energy Model," *Revista Ingenieria de Construccion* 30:2 (2015): 127.

148. Stations that have been rebuilt include Amundsen-Scott South Pole Station (United States), Halley Station (United Kingdom), Neumayer Station (Germany), and SANAE Station (South Africa). Stations that have been renovated include Mawson Station (Australia) and Scott Base (New Zealand). Stations that have been repurposed include the original SANAE Station (South Africa), Comandante Ferraz Station (Brazil), and Neumayer II (Germany). Some of the newest stations include Bharati Station (India), Princess Elisabeth Station (Belgium), and Jang Bogo Station (South Korea). See Nielsen, "From Shelter to Showpiece," 4.

149. Ibid., 5.

150. Oliver Wainwright, "Antarctic Research Stations: Exposed to the Elements and Now to the Public," *The Guardian*, July 25, 2013.

151. Ibid.

152. Ibid.

153. Nielsen, "From Shelter to Showpiece," 15.

154. William Brooks, "Elevated Station Design for the South Pole Redevelopment Project at Amundsen-Scott South Pole Station," a paper presented at the 10th International Cold Regions Conference, August 16–19, 1999, 6.

155. Nielsen, "From Shelter to Showpiece," 6–7.

156. Ibid., 35.

157. Friend, "Sustainability and Habitation in Antarctica," 28–34.

158. Ibid., 34–39.

159. Mino, Naranjo, and Montero, "Evaluation of Energy Efficiency Measures," 127.

160. Tiago Malavazi et al., "Design and Analysis of Hybrid Energy Systems: The Brazilian Antarctic Station Case," *Renewable Energy* 88 (2016): 236–46.

161. Nicole Calder-Steele, "Electricity in Isolation: The Progress of Power Generation in Antarctica," *PCAS 15 (2012/2013), Critical Literature Review (ANTA602),* 6–7, http://www.anta.canterbury.ac.nz/documents/PCAS_15/Literature%20Review/PCAS_15_Lit_Nicole_Calder-Steele.pdf.

162. Tin et al., "Energy Efficiency and Renewable Energy," 1718.

163. Rowena Ryan, "Boeing 757 Commercial Plane Lands on Blue-Ice Runway in Antarctica for the First Time," *news.com.au*, December 2, 2015, http://www.news.com.au/travel/world-travel/antarctica/boeing-757-commercial-plane-lands-on-blueice-runway-in-antarctica-for-the-first-time/news-story/da2b1c0f07bad8dc074d39f800dbdee6.

164. Donald Acosta, "Antarctic Mission for C-17A Globemaster, a Success," *International Business Times*, November 23, 2015, http://www.ibtimes.com.au/antarctic-mission-c-17a-globemaster-success-1485549.

165. Huang Jin, "China's First Polar Fixed-Wing Aircraft Arrives at Research Station in Antarctica," *People's Daily Online Exclusives*, December 3, 2015, http://english.cas.cn/newsroom/china_research/201512/t20151203_156975.shtml.

# THREE

# Who Owns Antarctica?

As the case for exploiting Antarctica grows, questions will be raised as to who "owns" the continent. Various unilateral territorial sovereignty claims have been made in Antarctica through a fairly chaotic and disorganized process in the first half of the 20th century—claims that received no international recognition. The world came to understand the complex web of territorial claims and counterclaims being made only when developing nations pushed the secretary general of the United Nations (UN) to intervene to clarify the so-called Question of Antarctica. By then, a small set of countries had already come to monopolize the continent through a treaty that gave them supremacy. While the Antarctic Treaty froze these territorial claims and devoted the continent to the causes of peace and science, the question of sovereignty has remained very much alive, permeating virtually all spheres of Antarctic policymaking.

## CLAIMS OF TERRITORIAL SOVEREIGNTY

Following the continent's heroic age of discovery by 19th-century explorers, a number of countries rushed to claim sovereignty over various parts of Antarctica. Territorial claims were not made within the context of an orderly or organized process to divide up the land; rather, they took the form of unilateral declarations in national legislation or international statements by high-level governmental officials. The first comprehensive overview of these claims to be placed before the international community happened only in 1984, when the UN secretary general prepared a report for the UN General Assembly on the Question of Antarctica.[1]

While various justifications had been advanced in support of territorial claims, their legal grounds included a combination of the following principles: discovery, geographical proximity, territorial contiguity, geological affinity, inherited rights occupation, formal acts of possession, performance of administrative acts, and the "sector principle" of using the meridians to demarcate boundaries.[2]

Since the beginning of the 20th century, seven nations had made independent and unilateral territorial claims: Argentina, Australia, Chile, France, New Zealand, Norway, and the United Kingdom.[3] The geographical coverage of these claims later assisted the parties to the Antarctic Treaty in defining the boundaries of the continent, which came to be set as all the land and ice shelves south of 60°S latitude. The claimants included Australia, Argentina, Chile, France, New Zealand, Norway, and the United Kingdom. All claims were made between 1904 (the first claim by Argentina) and 1940 (the last claim by Chile).[4] They all represented wedgelike shapes, as can be seen in Figure 3.1. Australia claimed the lion's share of the continent—approximately two-fifths of its surface area. On the other hand, about 15 percent of the continent was left unclaimed.

The Australian segment represented two different areas, separated in the middle by a relatively small portion claimed by France.[5] The French claim, Terre Adélie, which was discovered by Dumon d'Urville in 1840, represented the smallest territorial claim made in Antarctica.[6] While New Zealand lay claim to a well-defined area, the Norwegian claim was less precise: an area known as the Droning Maud Land, which left open the limits of its claim to the north and to the south.[7] However, the most problematic set of claims consisted of overlapping ones made by the United Kingdom, Argentina, and Chile, on what is now known to be the fastest-warming part of the continent—namely, the Antarctic Peninsula, which sticks out of the continent to the left, giving Antarctica the shape of an elephant's head.[8] The United Kingdom's claim covers the entire area claimed by Argentina, a large part of the area claimed by Chile, and other neighboring portions of land.[9]

These seven territorial claims were by no means accepted by all nations, nor did they receive any form of international recognition. Notable, of course, is the fact that of the five countries in closest geographical proximity to Antarctica (Argentina, Chile, South Africa, Australia, and New Zealand), South Africa was alone in not making an official territorial claim, primarily because it was not involved in any of the early expeditions to the continent. Equally notable is that countries as far away as Norway did submit claims because of their early presence in the area.

In fact, explicit rejections of territorial claims were made by a number of governments, but most importantly by the United States and the Soviet

**Figure 3.1 Antarctic Territorial Claims**

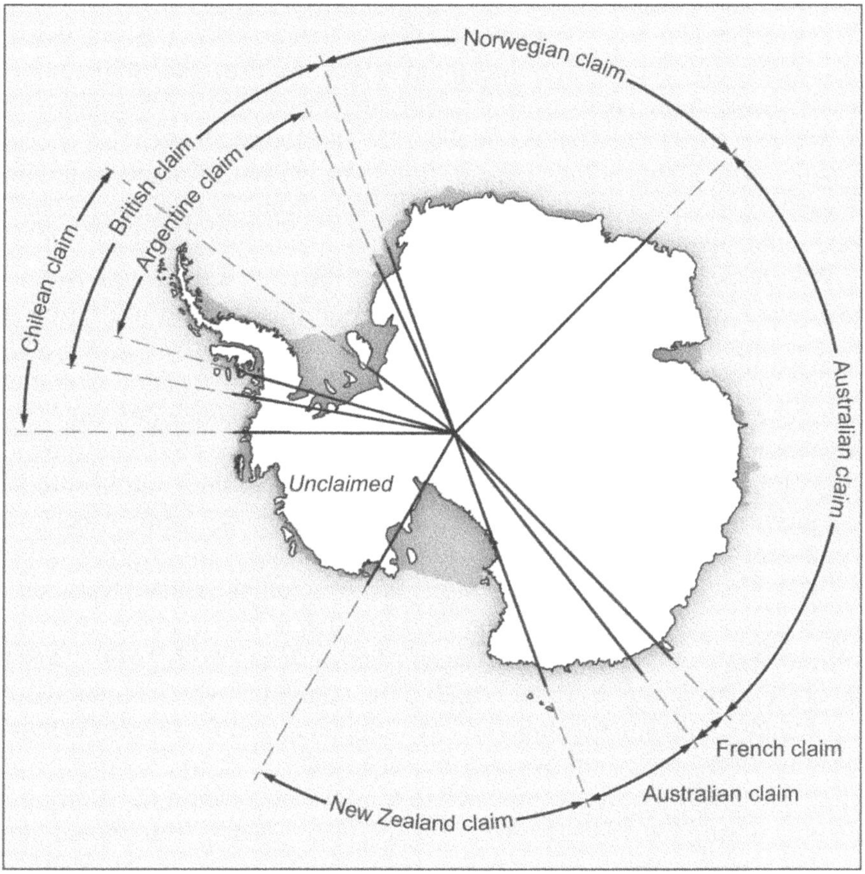


(Joseph Hill)

Union. In 1939, President Franklin D. Roosevelt delivered the following instructions to the commanding officer of the U.S. Antarctic Service: "The United States has never recognized any claims to sovereignty over territory in the Antarctic regions asserted by any foreign state. No member of the United States Antarctic Service shall take any action or make any statements tending to compromise this position."[10]

In a similar vein, the Soviet Union stated, "As to the question of territorial claims advanced by certain states in the Antarctic, the Soviet Government considers it necessary to state once again that it has not recognized and cannot recognize as legitimate any kind of separate solution for the problem of territorial positions in the Antarctic."[11] Both the United States and the Soviet Union stated not once, but multiple times, that they could (and perhaps would) table their own claims,[12] but they subsequently refrained from doing so, for reasons that will be discussed later in this chapter. Japan, once considered a potential claimant to Antarctica, was made to renounce all its claims in the Treaty of Peace with Japan, signed in San Francisco on September 8, 1951, after World War II. The treaty stated that "Japan renounces all claims to any right or title to or interest in connection with any part of the Antarctic area, whether deriving from the activities of Japanese nationals or otherwise."[13]

With respect to the overlapping claims of the United Kingdom, Argentina, and Chile, it is clear that all three countries saw the overlap as a potential time bomb and sought to defuse the tension early. For example, on March 4, 1948, Argentina and Chile issued a Joint Declaration on the Antarctic, stating that "until a friendly agreement is concluded concerning the common boundary line of the Antarctic territories of Chile and the Argentine Republic," both governments "will act in common accord in the juridical protection and defense of their rights in the South American Antarctic."[14]

In addition, to avoid military conflict in the region, the United Kingdom, Argentina, and Chile agreed soon after to a Tripartite Naval Declaration, which was reported on by the Foreign Office on January 18, 1949: "Being anxious to avoid any misunderstanding in Antarctica which might affect the friendly relations between the UK, Argentina, and Chile, the Governments of these three countries have informed each other that, in the present circumstances, they foresee no need to send warships south of latitude 60 degrees during the 1948–49 Antarctic season."[15] Despite this accord, the United Kingdom would bring its territorial dispute against the two nations to the International Court of Justice (ICJ) as soon as 1955. As can be seen, therefore, Antarctica was beset by a patchwork of territorial claims made before World War I and all way to the start of the Second. This situation made the search for a stable Antarctic regime imperative.

## SKIRMISHES LEADING TO A TREATY

A number of serious incidents during and soon after World War II came to reinforce the belief among territorial claimants, as well as other states with a serious interest in the continent, that a formal international arrangement of some sort was becoming essential. Some of these incidents were documented in the report of the UN secretary general on the Question of Antarctica. The most serious confrontations tended to involve the three overlapping claimants.

The secretary general reported that the battle over the continent had led some claimants to tear down each other's flags, destroy each other's scientific stations, and send rival expeditions to the same locations. In a famous incident, the British warship H.M.S. *Caranvon Castle* removed the commemorative bronze plaques installed in 1942 by an Argentinian expedition to the Deception and Wenckie Islands by the vessel *Primero di Mayo.*[16] In 1952, Argentinian forces in Hope Bay (south of the Orkney Islands) opened fire on a British scientific expedition attempting to reoccupy a station. The goal was to ensure its withdrawal.[17] Soon after, in 1953, the British dismantled certain constructions by Chilean and Argentinian nationals on Deception Island and arrested two Argentinians.[18] The scuffles continued after that, with matters coming to a head in 1955, when the United Kingdom decided to summon Argentina and Chile before the ICJ. An ICJ press release of 6 May 1955 stated:[19]

> The United Kingdom of Great Britain and Northern Ireland has filed in the Registry of the Court two Applications, one against Argentina and the other against Chile, relative to certain Antarctic territories. The United Kingdom contends that the Government of Argentina on the one hand, and the Government of Chile on the other, have encroached upon certain territories which are situated South of the 50th parallel of the South latitude and which are under British sovereignty. It requests the Court to recognize the validity of its titles to sovereignty and to declare that the pretensions of Argentina and Chile, as well as their encroachments in those territories, are contrary to international law.

The cases were discontinued in 1956, as both Argentina and Chile refused to recognize the jurisdiction of the court in this matter. As a result, this dispute has remained unresolved for decades, and continues so to this day.[20] The amount of territory in dispute is by no means minor. The British Antarctic Territory (BAT) extends over an area of 1.7 million km$^2$ (in other words, approximately three times the size of the United Kingdom). The area claimed by Chile encompasses 1.2 million km$^2$, and by Argentina 965,000 km$^2$.[21]

In 1948, Chilean president Gabriel Videla specifically visited Antarctica to, in his words, "defend the sovereignty and unity of our country, from Africa to the South Pole." In the meantime, the creation in 1962 of the BAT had taken place while the United Kingdom was consciously divesting itself of its colonies in Africa and stepping up its presence in the Antarctic. Its presence had been amplified in the lead-up to the 1957–1958 IGY through mapping, surveillance work, and the construction of scientific bases in the continent.[22]

A further serious contributor to instability in Antarctica was its unclaimed territorial portion, which kept large powers guessing as to whether, when, and how someone would step up to claim it. The area 90–150°W, commonly known as Marie Byrd Land and Ellsworth Land, has been closely associated with the United States due to various scientific expeditions that it undertook there. While claims by the United States over the area were asserted, they were never formalized.[23] The following question, therefore, continued to trouble the international community: would either the United States or the Soviet Union soon make claims in this area, and what would the impact be of a territorial claim by such large powers?

## THE UNITED STATES IN THE DRIVER'S SEAT

The negotiating history of the Antarctic Treaty demonstrates that the United States, which had emerged as one of two superpowers at the end of World War II, was very much in the driver's seat in the creation of a new polar order. Proposals by the U.S. Department of State, as well as the numerous conversations that it prompted with other interested parties, had placed goals such as the internationalization of Antarctica and the search for a final solution for the continent at the center of international affairs by the 1950s.

In a 1951 Department of State policy statement on U.S. interests in the polar regions, the United States described its Antarctic interests.[24] It made clear that the existing information available to it did not support the view that the Antarctic would acquire any commercial significance in the near future as a source of minerals, or that it would likely become an important communications link or a decisive location for military control of the sea routes around South America. However, Antarctica was important for the United States because of the efforts made by U.S. explorers to discover the continent and the many scientific expeditions that the United States had fielded to the area. These had not only contributed to a greater understanding of the Antarctic, but also evoked an emotional response in the minds of the U.S. public and placed the United States at the forefront of geographic and scientific discovery. This contributed directly to the prestige of the United States in the international arena.

In addition, the United States could acquire scientific and meteorological data in Antarctica that was unavailable elsewhere on the planet, with the continent also providing a valuable location for cold-weather training of its military personnel and testing of its Arctic equipment.[25] Internal U.S. government documents show that the United States felt it necessary to maintain a foothold in Antarctica in case further technological progress would one day unveil the existence of valuable minerals or other resources. The U.S.

National Security Council spoke of Antarctica's potential resources as follows: "Although they cannot yet be appraised, it would be imprudent for the U.S. to relinquish potential claims attaching to so vast a region of unforeseen possibilities."[26]

Starting from that premise, the United States characterized its policy in the Antarctic as a strategic precaution designed primarily to oppose participation in the control of the region by what it described as "unfriendly powers." To meet its defense needs, it argued that any solution to the "territorial problem" would have to fulfill three criteria: (1) denying its enemies participation in the control of any or all parts of Antarctica; (2) ensuring that such control would only be exercised by friendly powers; and (3) not establishing a precedent that would adversely affect U.S. interests in the Arctic. Only a solution of this nature, it believed, would ensure continued freedom for scientific discovery and exploration in the continent.[27]

A more detailed review of the U.S. negotiating stance reveals that it also harbored some very specific concerns with regard to the issue of territorial claims. First, it feared that the world could embark on a costly "claims race" that would require a continuous investment of funds to defend U.S. interests.[28] This was based on the understanding that to maintain a serious and credible foothold in the Antarctic, the United States would have to complement the efforts made by its early explorers with the establishment of scientific research stations and the creation of a serious research agenda. Otherwise, what would prevent any other nation—with no relationship to the continent—from also making claims? Table 3.1 demonstrates how the U.S. government had even started to tabulate the actions that would be necessary to maintain that credible claim, indicating that "No U.S. government agency has funds appropriated to carry out this policy."

Second, the United States appeared to fear for some of the research stations that it had already established on the continent. The McMurdo

**Table 3.1 Costing of U.S. Policy in Antarctica***

| **Estimated Cost of the Proposed Policy** | |
|---|---|
| Expenditures by Programs FY 1955–FY 1957 | |
| Expeditions | $10,000,000 |
| Permanent Station Maintenance | $2,000,000 |
| Organization | $150,000 |
| **Total** | **$12,150,000** |

*Ibid.

Station had been built in the New Zealand segment and the Palmer Station in the area of the three overlapping claims in the Antarctic Peninsula. The records of a 1958 Joint Chiefs of Staff meeting in the State Department illustrate this concern: "Admiral Burke referred to the great interest of the Navy in McMurdo Bay and to New Zealand claims of sovereignty. . . . Admiral Burke noted that the U.S. Navy was spending a great deal of money in McMurdo Bay and elsewhere, and expressed concern that the U.S. might be 'building a house on someone else's lot'"[29]

Third, the United States feared that the Soviet Union could stake claims in areas not only of interest to the United States, but in locations that could prompt an entirely new set of claimants. The United States had on numerous occasions, both in diplomatic notes and publicly, expressed its policy of reserving all its rights in the Antarctic, and the Soviet Union had officially asserted the right to participate in any territorial settlement. Declaration of a claim by the United States, the Soviet Union, or other powers could precipitate additional announcements of claims by countries such as South Africa. According to the United States, South Africa had begun to show an increasing interest in the area.[30]

Fourth, the United States was concerned that the longer this unsatisfactory territorial arrangement remained, the louder the calls would be for the United Nations to step in to administer the continent. At the 1956 and 1958 sessions of the UN General Assembly, the Indian delegation sought to have the question of Antarctica placed on the agenda but later withdrew its proposal. The Indian government was concerned that the Antarctic may become grounds of further international contention and thus contribute to a continuation of the Cold War.[31] However, for the United States, the United Nations was not the entity best placed to govern the continent, particularly in the absence of an agreement among the states directly concerned.

As a result, the United States debated for many years how best to shape its policy in Antarctica. Would it be wiser for it to table a national claim or to refrain from doing so, and what would the consequences be of making such a claim? In addition, what would an international arrangement need to look like to best protect U.S. interests in the region? The United States explained the basis for its decision to neither table a territorial claim nor to recognize the claims of others as being grounded in the difference between "discovery" and "exploration," on the one hand, and "effective control" and "occupation" on the other:[32]

> The failure thus far of the US to announce an official claim and our refusal to recognize the Antarctic claims of other countries rest on the traditional and orthodox view that *discovery* and *exploration* of new

> territories, which are *not followed by effective control or occupation*, do not give valid title. Discovery alone is said to give merely an inchoate title which must later be perfected by occupation. Because of the impossibility of occupying certain polar regions, however, it may in the future be generally recognized that title to such regions may be acquired by some form of control short of actual settlement (italics mine).

Internal communications also make clear that the United States had reflected long and hard about the size of any potential claim that it would make. Its claim would have to comprise all the areas first explored and mapped by U.S. nationals, which it knew "would conflict with all existing claims."[33] In other words, the United States was fully aware that were it to make a claim based on discovery and mapping—concepts used to support most, if not all, other claims—then there would be significant encroachment on the territories claimed by others. This situation would certainly exacerbate international tensions.

A 1954 report of the U.S. National Security Council shows how the U.S. thinking on the issue gradually matured. The report acknowledged the U.S. right to table claims "to major areas in all of the explored parts of Antarctica." However, it aired the concern that to maintain a claim, considerable U.S. investment would be required to justify it. This investment would take the form of expeditions and scientific bases, and would be needed in order for "our potential Antarctic rights to be preserved." It came to a conclusion that represented a radical shift in U.S. Antarctic policy. Rather than tabling its own claim, it would be wiser to support a freezing of all other claims:[34]

> In the absence of U.S. activities in all or certain portions of the region, the only feasible way of protecting our potential claims from *relative deterioration* would be a standstill agreement among the U.S. and the present claimants. Such an agreement might well be very difficult to negotiate (italics added).

The report added that were the United States to make a claim, then any announcement of its specific area of interest would be sure to spark more Soviet investment in that particular location. A subsequent 1957 report by the U.S. National Security Council pointed to the fears aroused by the multiple Soviet scientific expeditions to the continent made in connection with the IGY—ones that appeared to have a longer-term horizon than just that year.[35] But it was in a 1959 U.S. National Security Council report that the United States sounded the loudest alarm bells yet with regard to Soviet engagement, strengthening the momentum to reach a so-called final solution for the continent:[36]

> In addition to extensive participation in the IGY, the Soviet program in Antarctica includes such activities as ground-controlled aerial photography, mapping, hydrographic charting, and basic geology and biology, which have been only incidentally a part of U.S. activities. The USSR's announcement of plans to expand the Soviet network of stations in Antarctica and to penetrate into the Unclaimed Sector, where U.S. rights are now stronger than those of other nations, the ambitiousness of the traverse plans of the Soviet Union, and its initiative in pressing for an international program for mapping the continent could represent a governmental decision to make the Soviet Union the leading scientific nation in the Antarctic. Whether or not all of these plans are actually carried out, the already impressive Soviet scientific results could eventually be increased to a point where they might overshadow the present pre-eminence of the United States in Antarctic affairs with serious implications to U.S. prestige in the Antarctic.

## A "CONDOMINIUM" OR A "TRUSTEESHIP"? THE ESCUDERO PROPOSAL

As of 1948, the United States became central to the search for an orderly transition toward a stable territorial arrangement in the Antarctic. Two main approaches to resolving the territorial question were tabled by the United States and underwent various iterations based on the reactions of key players. These approaches can be said to have been absolutely central to arriving at a final solution for the continent (in other words, to concluding the Antarctic Treaty itself).[37]

The first of these approaches was to place Antarctica under some sort of a condominium arrangement that would allow it to be governed by all territorial claimants plus the United States, and the second was to place the continent under a UN Trusteeship, broadening the basis for its future governance.[38] The goals of both proposals, as communicated by the United States to its allies, were to ensure a peaceful solution to the territorial problem of Antarctica, improving the basis for the establishment of sovereignty over the area by the United States and other friendly powers; to ensure the freedom of exploration and authentic scientific investigation throughout the continent; and to guarantee access by the United States and its allies to the natural resources of the region.[39]

The proposal to establish a condominium[40] had as its core the merger of all territorial claims. The Antarctic would be governed by a "joint administration" comprising the United States and the other seven territorial claimants. This would occur under the terms of a special legal regime that would also ensure cooperation with the appropriate organs and specialized agencies of the United Nations. The continent would be governed with the purpose of maintaining international peace and security in accordance with Article 84 of the UN Charter.

The proposal to place the continent under a UN Trusteeship,[41] on the other hand, would be crafted on the basis of the rules of the UN Charter for the establishment of such arrangements. There would be two main options to consider: that the Trusteeship be administered by the United Nations itself, or that it be administered by the main powers. While the idea of establishing UN Trusteeships had been primarily created to ensure the development of people in inhabited lands, it was noted that from a legal perspective, nothing would exclude its application to the uninhabited Antarctic.

The United States postulated that three main articles of the charter would need to be taken into account:[42] Article 77(c), which states that the Trusteeship system could apply to territories placed under the system by "states responsible for their administration"; Article 79, which stipulates that the terms of the Trusteeship be agreed upon by the states directly concerned; and Article 81, which calls upon the Trusteeship arrangement to designate the authority that would exercise the administration of the Trust. In 1948, the United States explored with the seven claimants the idea of a "joint trusteeship administered by the foregoing powers who were to merge and join their respective claims to and interests in the area under a special regime."

In the aforementioned 1957 U.S. National Security Council report, the United States set out the advantages and disadvantages of each of the two options.[43] The condominium proposal would have the advantage of obviating the need to negotiate rival territorial claims. The arrangement would place the United States and the seven powers in a position to maintain their de jure control of the Antarctic. As a joint sovereign over the areas where the Soviets are located, the condominium arrangement would also better position the United States to control the Soviet Union. At the same time, the agreement to cooperate with various UN institutions would blunt the concern that the United States and its allies wished to use the continent for other than peaceful purposes. However, from a U.S. perspective, the most important advantage of the condominium would be that the United States would not have to table a separate a territorial claim. The merger of claims would allow Antarctica to be administered as a single unit for peaceful, scientific, purposes.

The Trusteeship Proposal, on the other hand, could work to the favor of the United States and its allies if Article 77(c) were read as conferring the administration of the continent to the territorial claimants. In other words, if the term "states responsible for their administration" could be read to mean the states that consider themselves sovereign within the areas of their claims. It would also serve U.S. interests if the terms of the Trusteeship between the United States and its allies could provide for the designation of one or more of their number as the administering authority. A key difference between this proposal and that of the condominium was that the Trusteeship could necessitate the tabling of a separate U.S. territorial claim to place the United States on an equal footing with other powers in the administration of the continent.

The disadvantages of both proposals, as described by the U.S. National Security Council, tended to be similar and to revolve around two main issues. The first was how the Soviet Union would react to its exclusion and maneuver to be let in; and the second was how U.S. allies would react to the idea of merging their claims, with Argentina and Chile known to be particularly nationalistic about the continent.[44]

The negotiating history makes clear that as the United States sought to resolve the Antarctica question, it very much felt that it was in a race against time. The U.S. National Security Council proposed that, if prior to achieving the U.S. strategy of an international settlement, "the USSR makes a claim, or if UN action or other developments make it necessary or desirable," then the United States should immediately "assert a U.S. claim to the unclaimed sectors of Antarctica, and to such of the presently-claimed areas regarding which arrangements have been made with the Free World claimants concerned, and a reservation of U.S. rights within other presently claimed sectors."[45]

In other words, its strategy was to negotiate a solution with its allies while keeping open the possibility of tabling an independent territorial claim, were the Soviets to do so or were the United Nations to begin to interfere in a heavy-handed manner at the request of India or any other nation. As it happened, many of the U.S. fears materialized, with key allies objecting to the merger of their claims and the Soviet Union expressing its dissatisfaction, in no uncertain terms, about the secret, closed-door negotiations. A surprising development, however, was that certain quarters of the United Nations that heard of the Trusteeship Proposal quickly seized it and gave it their own flavor.

In September 1948, almost immediately after the United States had floated its proposal to internationalize Antarctica, Dr. Julian Huxley, the first director of the United Nations Educational, Scientific, and Cultural

Organization (UNESCO), suggested the creation of an International Antarctic Research Institute under the initiative of UNESCO for the internationalization of scientific research in the continent. He proposed that either the eight nations (seven claimants and the United States) take ownership of the institute or that the institute be placed under the United Nations.[46] Needless to say, the idea met with resistance by the concerned parties. The proposal can be said to be have reflected a fundamental misunderstanding of the forces truly driving scientific research, expeditions, and the establishment of scientific bases in the continent. The latter were all designed to buttress different territorial claims, and the suggestion that they be folded into a global research program was seen as naïve.

It did not take long for the Soviets to voice their objections to the secret negotiations taking place among the eight powers. On February 10, 1949, in Leningrad, the president of the Geographical Society of the USSR submitted a report entitled "Russian Discoveries in the Antarctic and Present-Day Interest in the Area."[47] The report detailed one of the early Russian voyages to the region.[48] The meeting of the Geographical Society criticized the reports cited in the foreign press about the attempt by certain states to decide Antarctica's fate to the exclusion of the Soviet Union. It adopted a resolution stating, "No decisions on the question of the regime of the Antarctic without the participation of the Soviet Union can have legal force, and the USSR has every reason not to recognize such a solution."[49]

Furthermore, in a June 9, 1950, memorandum by the USSR to the governments of Australia, France, Norway, New Zealand, the United Kingdom, and the United States, the Soviet Union asserted its right to be included in the discussions initiated by the United States:[50]

> The Soviet Government considers it necessary to state that in accordance with international practice, all interested countries should be entitled for the discussion of the regime for any area of international significance. The Soviet Government holds that this international practice should also be observed in settling the question of Antarctica.

A subsequent 1951 Department of State policy statement made reference to this memo, taking note of its attempt to justify Soviet interest on the basis of a Russian naval voyage in 1819–1821 that resulted in the successful circumnavigation of the continent by Fabian Gottlieb von Bellingshausen and Mikhail Lazarev. The Department of State acknowledged that while "they sighted no part of the mainland[,] the Russians can be said to be the first to have discovered land south of the Antarctic Circle (Peter I and Alexander I Islands, in the area opposite South America)."[51]

Later internal communications of the U.S. government demonstrate that by 1958, the United States had started to seriously consider including the Soviet Union in the settlement of the Antarctic question—a matter that had become unavoidable. In fact, U.S. allies had themselves initiated conversations with the Soviets. By 1958, the Department of State was calling for dialogue with U.S. allies on whether the USSR would be better controlled within a regime or without, suggesting that Japan too be brought into the fold.[52]

Meanwhile, the Chileans and Argentinians both rejected the U.S. proposal to merge all territorial claims but continued to engage in the search for a constructive international solution to the Antarctic problem. In a 1948 Note Verbale from the Chilean Ministry of Foreign Affairs to the U.S. embassy in Santiago, Chile rejected the uniting of claims but called for further study of the matter, in particular with a view to avoiding "incidents of an international character" that could affect cordial relations between states.[53]

It was not be long before Chile proposed what came to be known as the "Escudero Proposal"[54]—which formed the basis for the final settlement of the Antarctic question. Chile suggested that, as a first step toward the resolution of the Antarctic problem, governments could subscribe to a declaration affirming that the establishment of new scientific bases, the undertaking of research expeditions, and the exercise of associated activities would not prejudice their rights of sovereignty and could not be invoked to establish dominion. It proposed that the declaration apply to all territory south of 60°S latitude and that it remain in effect for five years.[55] With the understanding that all territorial claims previously announced could be preserved, many more countries were willing to come to the international negotiating table. The idea also appealed to the United States—which would then take it further to suggest a freezing of new claims as well, since it had always been concerned about an expensive and unnecessary claims race.

On May 3, 1958, President Dwight Eisenhower announced that he had sent 11 countries,[56] including the Soviet Union, an invitation to confer on the question of Antarctica, particularly on how to keep the continent "open to all nations to conduct scientific or other peaceful activities." Arriving at an effective governance arrangement, the note stated, would avoid the uneconomic expenditure of funds to defend national interests and undesirable international diplomatic incidents.[57] The note described the legal solution that had been found to overcome the defeat of the U.S. original 1948 suggestion—the solution inspired by the Escudero Proposal: "It is believed that such a treaty can be concluded without requiring any participating nation to renounce whatever basic historic rights it may have in Antarctica, or whatever claims of sovereignty it may have asserted."[58]

It then added a corollary concept that would freeze the legal status of Antarctica during the duration of the treaty and therefore prevent new territorial claims from being made: "It could be specifically provided that . . . . no new rights would be acquired and no new claims made by any country during the duration of the treaty."[59] This statement quickly formed the basis of Article IV of the Antarctic Treaty, loosely referred to as the provision on "Territorial Sovereignty," and concluded what was considered a historic agreement. As is clear from this discussion, the issue of territorial sovereignty was the main hurdle that bedeviled the negotiation of an Antarctic treaty. This hurdle was overcome by kicking the can down the road. All claims were frozen in what was an icy continent, with no new claims to be authorized.

## SOVEREIGNTY, AS CAPTURED IN THE ATS

The Antarctic Treaty is the main building block of the Antarctic Treaty System (ATS), a much wider group of agreements governing the Antarctic. It is complemented by the Conservation of Antarctic Seals (CCAS), concluded in 1972; the Convention on the Conservation of Antarctic Marine Living Resources (CCAMLR) concluded in 1980; and the 1991 Environment Protocol. Following the successful resolution of the issue of territorial sovereignty, Article IV of the Antarctic Treaty came to form the backbone of the agreement. It was subsequently reproduced in all other treaties that form part of the ATS to safeguard the rights of past and future territorial claimants.[60] It states that:[61]

### Article IV of the Antarctic Treaty

Nothing contained in the present treaty shall be interpreted as:

(a) a renunciation by any Contracting Party of previously asserted rights of or claims to territorial sovereignty in Antarctica;

(b) a renunciation or diminution by any Contracting Party of any basis of claim to territorial sovereignty in Antarctica which it may have whether as a result of its activities or those of its nationals in Antarctica, or otherwise;

(c) prejudicing the position of any Contracting Party as regards its recognition or non-recognition of any other States right of or claim or basis of claim to territorial sovereignty in Antarctica.

No acts or activities taking place while the present treaty is in force shall constitute a basis for asserting, supporting or denying a claim to territorial sovereignty in Antarctica or create any rights of sovereignty in Antarctica. No new claim, or enlargement of an existing claim, to territorial sovereignty in Antarctica shall be asserted while the present treaty is in force.

As can be seen from this language, every conceivable precaution was taken in the drafting of this article to ensure that no past territorial claims [Article 4(a)], present or future [(Article 4(b)], would be prejudiced by the entry into force of this new accord, and nor would the issue of the recognition or nonrecognition of claims made by others [Article 4(c)]. Parties to the Antarctic Treaty later ensured that the spirit of Article IV was captured in all other instruments of the ATS. In 1978, for example, Article IV was cross-referenced in the CCAS. Article 1 of the convention provides that "This Convention applies to the seas south of 60° South Latitude, in respect of which the Contracting Parties affirm the provisions of Article IV of the Antarctic Treaty."[62]

In 1982, Article IV was also reflected in CCAMLR, which applies to the entire Antarctic marine ecosystem.[63] This was achieved using very similar language to Article IV, and with the relevant provision in CCAMLR given the same number as in the original treaty—the Roman numeral IV. However, since CCAMLR is open to states and entities that are not parties to the Antarctic Treaty (such as the European Union, for example), Article IV of the convention begins by asserting that all parties to the convention are also bound by the Antarctic Treaty's Article IV. This is designed to make Article IV, as well as the long-negotiated solution to sovereignty, binding for all. Two other important variations can be seen in the CCAMLR language relative to the Antarctic Treaty's Article IV, which will be discussed next.[64]

### Select Portion of Article IV of the CCAMLR

1. With respect to the Antarctic Treaty area, all Contracting Parties, whether or not they are Parties to the Antarctic Treaty, are bound by Articles VI and VI of the Antarctic Treaty in their relations with each other.
2. Nothing in this Convention and no acts or activities taking place while the present Convention is in force shall:
   (a) constitute a basis for asserting, supporting or denying a claim to territorial sovereignty in the *Antarctic Treaty area or create any rights of sovereignty in the Antarctic Treaty area*;

(b) be interpreted as a renunciation or diminution by any Contracting Party of, or as prejudicing, any right or claim or basis of claim *to exercise coastal state jurisdiction under international law within the area to which this Convention applies* (italics added) . . .

In covering the entire Antarctic marine ecosystem, CCAMLR applies both to areas that have the legal status of the high seas and various Exclusive Economic Zones (EEZs) in the Southern Ocean. It also applies to areas that third parties have rights to, and interests and obligations in, under the United Nations Law of the Sea (UNCLOS). With this in mind, it was important for CCAMLR to shield all claims of sovereignty and jurisdiction over these waters from any potential prejudice that the convention could bring.[65]

The change made to Article IV (a)—i.e., its reference to an "Antarctic Treaty area"—was designed to capture the full scope of CCAMLR's area of application. Similarly, the addition introduced to Article IV (b)—i.e., the reference to "coastal state jurisdiction"—was designed to shield from prejudice the claims of any coastal state in Southern Ocean waters covered by CCAMLR.

In 1988, Article IV of the Antarctic Treaty also found reflection in Article 9 of Convention on the Regulation of Antarctic Mineral Resource Activities (CRAMRA), entitled the "Protection of Legal Positions under the Antarctic Treaty."[66] The article was aimed at extending the balance reflected in Article IV of the treaty to new activities related to mineral resources, using the same legal format as CCAMLR. CRAMRA, a legal regime that was later abandoned, will be discussed in subsequent sections of this book. With the careful incorporation of the spirit of Article IV into all facets of the ATS, parties to the original treaty clearly sought to seal off any open door to the delicate and very divisive question of sovereignty.

## JURISDICTION IN THE SOUTHERN OCEAN

With the issue of territorial sovereignty settled—or rather frozen—a host of related issues started to emerge. The territorial claimants could in theory also claim that a 200-nautical-mile EEZ was their territorial claim to be recognized. Within an EEZ, coastal nations have the sole exploitation right over natural resources. Claiming a 200-nautical-mile EEZ in the Antarctic would mean that the freedom of the high seas would extend only to the limit of that EEZ. If the EEZ were unrecognized (like the issue of territorial sovereignty itself), then the freedom of the high seas would extend to the edge of the continental shelf or land mass.

All seven territorial claimants in Antarctica also happen to be parties to UNCLOS.[67] Under this convention, a coastal nation may request the establishment of a continental shelf beyond the 200 nautical miles. It would have to submit data in support of its request to the Commission on the Limits of the Continental Shelf (CLCS) for examination and an eventual recommendation.[68] The limits of the shelf established by a coastal state on the basis of a CLCS recommendation are final and binding. The rules of procedure of the CLCS make clear that it cannot consider areas of a shelf subject to dispute unless all states that are parties to the dispute consent.[69]

Despite territorial claims being frozen under the Antarctic Treaty, the issue of jurisdiction over the territorial sea, the establishment of EEZs, and the extension of the continent shelf became hotly contested after the treaty was concluded. In other words, the issue of sovereignty remained very much alive. The report by the UN secretary general on the question of Antarctica detailed the diametrically opposed views that prevailed on the subject: those of the territorial claimants, the nonclaimants, and the nonparties to the Antarctic Treaty.[70]

One set of views held that Antarctic territorial claimants can act in full accord with their claims of territorial sovereignty in any matter not expressly addressed by the treaty.[71] Since the Antarctic Treaty is silent on the territorial sea, continental shelf, or jurisdiction in maritime areas adjacent to Antarctica, then territorial claimants are entitled to claim them too. The report explained that all the claimants had indeed indicated that, within their maritime areas, they would only accept restrictions to which they had specifically agreed. Nonclaimant states, party to the Antarctic Treaty, held a different view.[72] According to them, all waters adjacent to Antarctica must be considered high seas. In these waters and their corresponding seabed areas, there are no territorial seas, 200-nautical-mile EEZs, or continental shelves, and they are free to conduct any activity not prohibited by the treaty within those areas.[73]

Finally, the report detailed the views of the nonparties to the Antarctic Treaty,[74] which held that the status of the maritime areas surrounding Antarctica are as unresolved as the status of the Antarctic land mass itself. In the UN General Assembly, many members of the Group of 77 (G-77), a coalition of developing countries, in fact put the legitimacy of the entire ATS into question.

Adding a further complication to the discussion was Article VI of the Antarctic Treaty, which states that "The provisions of the present treaty shall apply to the area south of 60° South Latitude, *including all ice shelves, but nothing in the present treaty shall prejudice or in any way affect the rights, or the exercise of the rights, of any State under international law with regard to the high seas within that area*" (italics added).[75] In other words, the negotiators of the Antarctic Treaty wished to ensure that the treaty would not prejudice their rights under any other body of international law governing the high seas. Born in 1982, UNCLOS was the principal international legal instrument that came to occupy that space.[76] However, as the report of the UN secretary general pointed out, Article VI raised the problem of whether the "ice shelves" to which it referred would fall under the legal regimes governing land or those governing water. "The juridical nature of shelf ice, pack ice, and underlying waters is unclear and appears to be a very difficult issue to resolve," it wrote.[77]

While the report of the secretary general provided an overview of the situation as it stood back in 1984, what does state practice to date tell us with respect to the EEZ and the continent shelf? In fact, it tells us that significant tension continues to exist on the issue of sovereignty. State practice in Antarctica has varied with respect to the declaration of EEZs. Argentina, Chile, and France had all applied a legislative claim of a 200-nautical-mile EEZ to their Antarctic territories prior to the entry into force of UNCLOS. New Zealand, on the other hand, had excluded the Ross Dependency from its EEZ declaration in 1977. Australia had initially done the same, excluding the Australian Antarctic Territory (AAT) from the EEZ declaration of 1979,[78] only to reverse itself through the entry into force on August 1, 1994, of the its Maritime Legislation Amendment Act.[79] Despite this variation in state practice, all claimants—New Zealand included—assert that they have a right to an EEZ around their Antarctic territories.

Antarctic claimants also tested the issue of sovereignty through their continental shelf extension claims.[80] While all seven claimants have made assertions to Antarctic continental shelves, state practice has varied here too. In accordance with decision SPLOS/72, all Antarctic claimants were required to make a submission to the commission by May 13, 2009, and all of them obliged.[81] In their submissions to the CLCS, claimants took three different approaches. One approach has been to make unconditional assertions to the shelf; a second has been to make conditional assertions; and a third has been to make no assertion, but to reserve the right to do so in the future.[82]

Unsurprisingly, Argentina falls into the first category and has been the most active claimant. In its 2009 submission to the CLCS, it included data covering the Argentine Antarctic Sector, as well as the Islas Malvinas, Georgias del Sur, and Sandwich del Sur, making no reference to the contestability of its Antarctic claims.[83] Given that the CLCS cannot consider areas of shelf subject to a dispute unless all states that are parties to the dispute consent, it went without saying that the CLCS would not be opining on the Argentinian request.

Australia, Norway, and Chile took a more cautious approach, falling into the second category. They included Antarctic continental shelf data in their submissions but explicitly requested the commission not to act upon it. The three remaining claimants—New Zealand, the United Kingdom, and France—refrained from including any Antarctic data in their submissions, reserving their rights to do so later.[84] Of note is that Australia asserted one of the largest claims to the continental shelves of the Southern Ocean, with its claim including areas around several sub-Antarctic islands as well. It was, in fact, the first claimant to have covered the Antarctic so comprehensively in its submission and to have included a vast amount of supporting data.[85] Aware that few states recognized its Antarctic claim, however, it requested the CLCS not to consider it, with the CLCS obliging.

However, in light of the breadth of its claim, six parties to the Antarctic Treaty felt obliged to lodge an objection with the CLCS, indicating that they did not recognize any claim to the Antarctic, nor to its continental shelf. These countries included the United States, the Russian Federation, Germany, Japan, India, and the Netherlands.[86] CLCS records also demonstrate that a number of countries had chosen to systematically object to any Antarctic continental shelf claim. India and the Netherlands appear to have been among the most active respondents to such claims. They questioned their validity, reminding the claimants of the Antarctic Treaty's Article IV, arguing that there would have to be a commonsense extension of its logic to the continental shelves.[87]

While it can be argued that Antarctic claimants had no option but to table submissions to the CLCS in view of SPLOS/72 and its 2009 cutoff date,[88] the strength with which some claims were made (the assertiveness of the Argentinian claim and breadth of the Australian claim), as well as the equally strong reaction to them, tend to indicate that the issue of sovereignty—whether at land or at sea—will not go away easily.

## THE QUESTION OF ANTARCTICA AT THE UNITED NATIONS

It was as early as 1956 that India sought to initiate a discussion in the United Nations on the Question of Antarctica. In both 1956 and 1958, Indian attempts were stymied by other nations seeking to arrive at an accord to internationalize Antarctica. The conclusion of the Antarctic Treaty in 1959 and its subsequent entry into force in 1961, however, did little to alleviate the concern of nonparties with respect to the preservation of their own rights in the continent.

It was 1982 that the Question of Antarctica was presented to the UN General Assembly by the prime minister of Malaysia, Dr. Mahathir Bin Mohamad. In his address, the prime minister called upon the international community to turn its attention to Antarctica now that it had successfully completed the UNCLOS. Like the seas, he argued, Antarctica belonged to the international community and should either be turned over to the United Nations to administer or its "present occupants" should act as trustees for the nations of the world.[89] His statement had clearly been anchored in the early 1948 U.S. proposal to place Antarctica under a UN Trusteeship and demonstrated the extent to which a large portion of the international community had lagged in its understanding of the developments that had taken place in the continent.

However, the timing of the Malaysian proposal and its presentation to the United Nations in 1982 was also driven by the fact that negotiations had commenced that same year on an Antarctic minerals regime under the auspices of the Antarctic Treaty. While CRAMRA collapsed in 1988, the mere fact that parties to the Antarctic Treaty had started discussing the continent's mineral wealth clearly unnerved the rest of the international community.

The Malaysian proposal resulted in the placement of the Question of Antarctica on the UN agenda for years. The question was treated annually from 1983–1995, then every three years from 1996–2005, and was eventually discontinued in 2005.[90] It was in the context of this process that the UN secretary general was requested to prepare a comprehensive report on the Question of Antarctica, which he did on the basis of extensive consultations with member-states and which was eventually delivered in 1984. He was repeatedly asked to update his report until this report too was eventually discontinued.

The UN process allowed a very large portion of its membership that had never been consulted on the question of Antarctica to air its concerns, and obliged parties to the Antarctic Treaty to engage in greater outreach toward the rest of the international community and to gradually broaden the membership of the treaty.[91] It also led them to cooperate on a more transparent and stable basis with the United Nations.

In 1982, when the Question of Antarctica first made its way to the United Nations, the number of parties to the Antarctic Treaty was still fairly limited. Many UN members argued that Antarctica formed part of the "common heritage of mankind" principle. It was only in 1985, with China's entry into the Antarctic Treaty with a Consultative Party status (i.e., an ability to participate in decision making) that all permanent members of the UN Security Council became formal members of the Antarctic Treaty.[92] This placed the treaty on a more stable footing.

The gradual inclusion of key emerging powers in the Antarctic Treaty can also be said to have done much to diffuse the tension that had started to build in the United Nations. Other developing countries that acceded to the treaty on the basis of full Consultative Party status were India and Brazil in 1983 and South Korea in 1989.[93] It was only when Malaysia itself was invited to participate in meetings of the Antarctic Treaty, however, that the Question of Antarctica at the United Nations wound down.[94] Malaysia became a non–Consultative Party in 2011.

While still alive at the United Nations, the Question of Antarctica led countries, and in particular the G-77 nations, to express concern about a wide variety of issues. The first was the price of entry into the Antarctic Treaty, which was considered exorbitant.[95] According to Article IX.2 of the Antarctic Treaty, countries are entitled to participate in the Consultative Meetings and to accede to the treaty during such times as they demonstrate their interest in Antarctica by "conducting substantial research activity there."[96] Only then can their status be elevated to that of Consultative Parties.

Many countries believed this to be an excessively high entrance fee, particularly when they lacked the technological capability to participate in a meaningful manner in scientific research or expeditions to the continent. In his report, the UN secretary general added, "It is to be noted that the requirement does not apply to the original parties to the Treaty, not all of which have continuing research programs in Antarctica at this time."[97] While he left undefined the particular countries that he was referring to, South Africa was most likely the object of this comment, since its inclusion in the Antarctic Treaty was resented by other African nations.

The second issue at stake was the negotiation of a minerals regime that several G-77 members argued was being conducted in a secretive manner. In fact, the General Assembly passed Resolution A/Res/41/88 in 1986 to call "upon the Antarctic Treaty Consultative Parties to impose a moratorium on the negotiations to establish a minerals regime until such time as all members of the international community can participate fully in such negotiations."[98]

The third was the United Nation's lack of involvement in the Antarctic Treaty. Repeated UN General Assembly resolutions called upon the parties to the Antarctic Treaty to invite the UN secretary general to participate in their deliberations and to share more regular information with the United Nations.[99] In fact, in 1990, a resolution urged the United Nations to establish its own scientific research station on the continent—but the proposal never materialized due to lack of funding.[100] It was only in 2007 that Ban Ki-moon visited Antarctica, becoming the first sitting UN secretary general to do so.[101] In response to these concerns, however, parties to the Antarctic Treaty invited the head of the United Nations Environment Programme (UNEP) to participate on a regular basis in their deliberations.[102]

The fourth was the participation of South Africa as the only African nation in the Antarctic Treaty. Resolution after resolution called upon parties to the Antarctic Treaty to terminate South Africa's membership because of its apartheid regime.[103] When that regime fell in 1994, this matter died down as well. What seems clear from all this is that discussions in the United Nations were triggered by an extreme fear by nonparties of being excluded from the new minerals regime. The collapse of that regime—a topic that will be dealt with in the next chapter—did much to alleviate that concern, as did the gradual expansion of the Antarctic Treaty's membership.

The conclusion to draw from this discussion is that more than a decade's debate about the Question of Antarctica at the United Nations laid a strong marker by the international community on its need to be consulted on the fate of the seventh continent. While the expansion of the Antarctic Treaty's membership diffused a time bomb, the debate—with all of its underlying tension—had simply shifted to another forum: the United Nations.

## IS SOVEREIGNTY HERE TO STAY?

Clearly, the issue of territorial sovereignty in Antarctica was not only the main driving force for an international settlement, but also the main obstacle to be overcome. The ingenious Escudero Proposal allowed this hurdle to be crossed by setting the question aside, opening the door to peaceful scientific investigation of the continent. The Escudero Proposal was made possible by the fact that no mineral resources were known to exist on the continent, with U.S. Department of State memos confirming that there could no imminent commercial exploitation of Antarctica.

Article IV essentially allowed parties to freeze their territorial claims over the continent's land mass and its mineral wealth until such a time when the situation would need to be reassessed. The freezing of territorial claims was facilitated in no mean way by the actual climatic freeze on the continent, which rendered it virtually inaccessible. Despite the Article IV compromise, territorial claimants continue to assert their sovereignty over the continent in other international fora. This is visible in their declarations of EEZs and their assertions of continental shelves under the UNCLOS. Similarly, the remainder of the international community continue to assert its rights as well. Many countries have made it a point to systematically object to continental shelf claims under UNCLOS, and the G-77 has declared its right to become a partner in deciding the fate of the continent through the lengthy discussion in the United Nations of the Question of Antarctica.

In fact, it can be argued that the signature of the Antarctic Treaty made the quest for sovereignty even more pronounced. From the first minute, the treaty appears to have set off aggressive nationalist instincts, which took the form of a battle over the creation and hosting on its Secretariat. In 1961, South Africa offered to host a Secretariat, but the idea was immediately rejected by both Argentina and Chile, who feared an excessive internationalization of Antarctica. The question remained dormant for several years. With time, and as the need for a Secretariat became more evident, Argentina offered to host it. However, the Argentinian proposal was met by rival claims by the United Kingdom and Chile, and it was only in 2004 that a Secretariat was finally established and placed in Buenos Aires. The United Kingdom compromised to ease tensions with Argentina resulting from the Falklands War.[104]

However, since then, the quest for sovereignty in Antarctica has continued to be demonstrated every day. In addition to the growth of scientific programs and a spiral in the number of research stations, countries have been flying their political leaders (and now pregnant women) to the continent, issuing Antarctic stamps, producing maps, and carefully recording the details of their every voyages of discovery.[105] In short, the question of territorial sovereignty has not gone away.

## NOTES

1. UN General Assembly, *Question of Antarctica*, a study requested by General Assembly Resolution 38/77, report of the Secretary General, A/39/538, October 31, 1984. http://www.unep.org/dewa/Portals/67/pdf/A-39-583-Part_I.pdf.

2. For a discussion of these general principles, see Y. Z. Blum, *Historical Titles in International Law* (The Hague: Martinus Nihoff, 1965); R. Y. Jennings, *The Acquisition of Territory in International Law* (Manchester, UK: Manchester University Press, 1963); I. Brownlie, *Principles of Public International Law*, 8th ed., (Oxford, UK: Oxford University Press, 2012); and D. P. O'Connell, *International Law*, 2nd ed. (London: Stevens & Sons Ltd., 1970). See, for example, the statement made by Argentina on February 20, 1953, in which it asserted that its request was based on "the historical antecedents of its titles—maintained firmly in all circumstances by the

Argentine Government and spiritually identified with the feeling of the nation's entire population; in the insuperable geographical portion of the Republic; in the geological continuity of its land with the Antarctic lands; in the climatological influence that neighboring polar zones exercise over its territories; in the right of the first occupancy; in the pertinent diplomatic measures; and finally, in uninterrupted activity in the same Antarctic terrain" [as quoted in C. Joyner, "The Exclusive Economic Zone and Antarctica: The Dilemmas of Non-Sovereign Jurisdiction," *Ocean Development and International Law* 19 (1988): 705].

3. The exact scope of these claims, which are expanded on in the next footnote, was detailed in UN General Assembly, *Question of Antarctica*, para. 13–20.

4. The Argentinian claim was made in 1904; the United Kingdom's in 1908; New Zealand's in 1923; France's in 1924; Norway's partly in 1929, and then 1939; Australia's in 1933; and Chile's in 1940. Five territorial claimants have cross-recognized each other's claims; they include the United Kingdom, New Zealand, Australia, France, and Norway, leaving out only Chile and Argentina, whose claims overlap with the United Kingdom's. On this point, see Alan D. Hemmings, "Security Beyond Claims," in *Antarctic Security in the Twenty-First Century: Legal and Policy Perspectives,* Alan D. Hemmings, Donald R. Rothwell, and Karen N. Scott, eds. (New York: Routledge, 2012), 87.

5. The two claims lie south of the 60°S latitude. The first claim is located between 45° and 136°E longitude and the second, between 142° and 160°E longitude.

6. Its geographical coordinates are 136° and 142°E longitude south of the 60° parallel.

7. New Zealand claims the area between 160°E longitude and 150°W longitude south of the 60° parallel. Norway claims the area situated between 20°W longitude and 45°E longitude, leaving the limits of the claim to the north and south undefined.

8. The Argentinian claim encompasses the area between 25° and 74°W longitude, extending from the South Pole to the 60° parallel. Embodied in Presidential Decree No. 1747 of November 6, 1940, the Chilean claim extends to all lands "and their respective territorial sea" lying between 53° and 90°W longitude.

9. More specifically, the United Kingdom claim encompasses all territory between longitudes 20° and 50°W, and south of latitude 50°S. It also covers territories between 50° and 80°W and south of latitude 58°S.

10. Marjorie M. Whiteman, ed., *Digest of International Law* 1 (Washington, DC: U.S. Government Printing Office, 1963), 1248.

11. UN General Assembly, *Question of Antarctica*, para. 33. The report refers to an article published in *Pravda,* June 4, 1958, containing the Soviet position.

12. The Soviet Union, for instance, stated that it "reserves to itself all rights based on discoveries and explorations of Russian navigators and scientists, including the right to present corresponding territorial claims in the Antarctic," also in *Pravda,* June 4, 1958. Similarly, a note dated May 2, 1958, from the government of the United States proposing an international conference on Antarctica, stated, "In view of the activities of the US and its nationals . . . my Government reserves all of the rights of the US with respect to the Antarctic region, including the right to assert a territorial claim or claims" (as quoted in Whiteman, *Digest of International Law*, 1248).

13. Ibid., 1261.

14. Ibid., 1259.

15. Ibid., 1238.

16. UN General Assembly, *Question of Antarctica*, para. 36–37.

17. Ibid., para. 42.

18. Ibid., para. 43.

19. ICJ, communique no. 55/26, May, 6, 1955, http://www.icj-cij.org/docket/files/27/12325.pdf.

20. Donald R. Rothwell, "The Antarctic Treaty as a Security Construct," in *Antarctic Security in the Twenty-First Century*, 35.

21. Klaus Dodds, "The Antarctic Peninsula: Territory, Sovereignty Watch, and the 'Antarctic Problem,'" ibid., 96–97.

22. Ibid., 96–98.

23. Rothwell, "The Antarctic Treaty," 36.

24. U.S. Department of State policy statement, "Polar Regions," doc. 762, July 1, 1951, https://history.state.gov/historicaldocuments/frus1951v01/d762.

25. Ibid.

26. National Security Council Planning Board, report to the National Security Council, doc. 301, June 28, 1954, https://history.state.gov/historicaldocuments/frus1952-54v01p2/d301.

27. U.S. Department of State, "Polar Regions."

28. National Security Council Planning Board, doc. 301.

29. U.S. Department of State–Joint Chiefs of Staff meeting, memorandum of discussion, doc. 264, January 10, 1958, https://history.state.gov/historicaldocuments/frus1958-60v02/d264.

30. National Security Council, "Statement of U.S. Policy on Antarctica," doc. 286, April 7, 1959, https://history.state.gov/historicaldocuments/frus1958-60v02/d286.

31. Ibid.

32. U.S. Department of State, "Polar Regions."

33. Ibid.

34. National Security Council Planning Board, doc. 301.

35. National Security Council, "Note by the Executive Secretary to the National Security Council on Antarctica," doc. 333, June 29, 1957, http://history.state.gov/historicaldocuments/frus1955-57v11/d333.

36. National Security Council, "Statement of U.S. Policy on Antarctica," doc. 286.

37. For a comprehensive record of the negotiating history of the Antarctic Treaty, see John Hanesian, "The Antarctic Treaty 1959," *International and Comparative Law Quarterly* 9 (1960): 449; Paul A. Berkman, "President Eisenhower, the Antarctic Treaty, and the Origin of International Space," in *Science Diplomacy: Antarctica, Science, and the Governance of International Spaces,* Paul Arthur Berkman et al., eds. (Washington, DC: Smithsonian Institution Scholarly Press, 2011); and Klaus Dodds, *The Antarctic: A Very Short Introduction* (Oxford, UK: Oxford University Press, 2012).

38. National Security Council, "Note by the Executive Secretary," doc. 333.

39. Ibid.

40. Ibid.

41. Ibid.

42. Ibid.

43. Ibid.

44. Ibid.

45. Ibid.

46. Hanesian, "The Antarctic Treaty 1959," 449.

47. Ibid., 443.

48. In reference to Bellingshausen's voyage of 1819–1821.

49. Hanesian, "The Antarctic Treaty 1959," 443–444.

50. Ibid., 446.

51. U.S. Department of State, "Polar Regions."

52. U.S. Department of State–Joint Chiefs of Staff meeting, doc. 264.

53. Hanesian, "The Antarctic Treaty 1959," 441.

54. The proposal was named after Chilean professor Julio Escudero. The Chilean Antarctic research base located on King George Island was subsequently also named after him.

55. Hanesian, "The Antarctic Treaty 1959," 441.

56. The 11 countries included all seven territorial claimants, plus the Soviet Union, South Africa (due to its geographic proximity to the continent), and Belgium and Japan because of their levels of engagement in the IGY and their other ties to the Antarctic. Belgian nationals had played an

important role in the discovery of Antarctica, and Belgium had stepped up its presence in the continent in the run-up to the IGY. It established King Bedouin Base in 1958 in an expedition led by Gaston de Gerlache, the son of Adrien de Gerlache, who had led the 1897–1899 Belgian Antarctic expedition. In the run-up to the IGY, Japan established a foothold on the continent. The first Japanese Antarctic observation corps, commanded by Takeshi Nagata, left Japan in 1956, arriving in Antarctica on January 29, 1957. Showa Station was the first Japanese observation base in Antarctica, and it was set up on the same day.

57. Hanesian, "The Antarctic Treaty 1959," 457.

58. Ibid., 458.

59. Ibid.

60. For a discussion of how the Antarctic Treaty's Article IV was generalized across the ATS, see Vladimir Golitsyn, "Balancing Sovereign Interests Beyond National Jurisdictions," in *Science Diplomacy*, 53–54.

61. Article IV of the Antarctic Treaty, Annex I.

62. Article 1 of CCAS in Secretariat of the Antarctic Treaty, *Compilation of Key Documents of the Antarctic Treaty System*, 2nd ed., Buenos Aires, 2004. All subsequent references in this discussion to provisions of the CCAS can be found in this source.

63. Article IV of CCAMLR in Secretariat of the Antarctic Treaty, *Compilation of Key Documents of the Antarctic Treaty System.*

64. Ibid.

65. Golitsyn, "Balancing Sovereign Interests," 53–54.

66. Article 9 of CRAMRA, http://sedac.ciesin.columbia.edu/entri/texts/acrc/cramra.txt.html.

67. For the membership of UNCLOS, see http://www.un.org/depts/los/reference_files/chronological_lists_of_ratifications.htm.

68. The outer limits of the shelf beyond 200 nautical miles can be defined by various formulas relating to the foot of the continental slope, sediment depth, and specific distance measurements, but must not exceed the 350 nautical miles from the baselines from which the territorial sea is measured.

69. For an understanding of UNCLOS, see "UNCLOS at 30," http://www.un.org/depts/los/convention_agreements/pamphlet_unclos_at_30.pdf.

70. UN General Assembly, *Question of Antarctica*, para. 94–95.

71. Ibid., para. 94.

72. Ibid., para. 95.

73. Of note with respect to this set of positions is that the United States, which is not a territorial claimant and which has not recognized any territorial claim made by another state, is not a party to UNCLOS.

74. UN General Assembly, *Question of Antarctica*, para. 96.

75. The Antarctic Treaty, Annex I.

76. UNCLOS replaced four other treaties: the 1962 Convention on the High Seas; the 1964 Convention on the Territorial Sea and Contiguous Zone; the 1964 Convention on the Continental Shelf; and the 1966 Convention on Fishing and Conservation of Living Resources of the High Seas.

77. UN General Assembly, *Question of Antarctica*, para. 104.

78. Gillian D. Triggs, "The Antarctic Treaty System: Some Jurisdictional Problems," in *The Antarctic Treaty Regime: Law, Environment, and Resources*, Gillian D. Triggs, ed. (Cambridge, UK: Cambridge University Press, 1987), 91.

79. Alan D. Hemmings, "Problems Posed by Attempts to Apply a Claimant's Domestic Legislation Beyond Its Own Nationals in Antarctica," *Asia Pacific Journal of Commonwealth Law* 2 (2008): 216.

80. For a comprehensive discussion of Antarctic continental shelf claims, see Mel Weber, "Delimitation of the Continental Shelves in the Antarctic Area: Lessons for Regime, Resource, and Environmental Security," in *Antarctic Security*, 172–196; and Tim Stephens, "Polar Continental Shelves: Australian and Canadian Challenges and Opportunities," in *Polar Oceans Governance in an Era of Environmental Change*, Tim Stephens et al., eds. (Northampton: Edward Elgar, 2014), 146–165.

81. Australia tabled its submission in 2004, New Zealand in 2006, the United Kingdom, Norway, and France in various stages between 2006 to 2009, and Argentina and Chile in 2009.

82. Stephens, "Polar Continental Shelves," 149.

83. Ibid.

84. Ibid., 151.

85. Ibid., 155.

86. Ibid., 151.

87. Weber, "Delimitation of the Continental Shelves," 180–181 and 185–187.

88. Stephens, "Polar Continental Shelves," 152.

89. Moritaka Hayashi, "The Antarctica Question in the UN," *Cornell International Law Journal* 19 (1996): 275–276.

90. Peter J. Beck, "Twenty Years On: The UN and the 'Question of Antarctica' 1983–2003," *Polar Record* 40 (2004): 206.

91. With the entry of India and China, almost 90 percent of the world population today forms part of the Antarctic Treaty.

92. See the parties to the Antarctic Treaty and their date of accession, http://www.ats.aq/devas/ats_parties.aspx?lang=e.

93. Ibid.

94. Beck, "Twenty Years On," 209.

95. UN General Assembly, *Question of Antarctica*, para. 107.

96. The Antarctic Treaty, Annex I.

97. UN General Assembly, *Question of Antarctica*, para. 107.

98. UN General Assembly, "The Question of Antarctica," General Assembly Resolution, A/RES/41/88, December 4, 1986.

99. UN General Assembly, "The Question of Antarctica," General Assembly Resolution, A/RES/46/41, December 6, 1991.

100. Beck, "Twenty Years On," 208.

101. "Antarctica Shows Need for Action on Climate Change, Ban Ki-moon Says," *UN News Center*, http://www.un.org/apps/news/story.asp?NewsID=24613#.VX7ehuvdKfQ.

102. The General Assembly would then welcome "the invitations to the Executive Director of the UNEP to attend Antarctic Treaty Consultative Meetings in order to assist such meetings in their substantive work, and urges the parties to continue to extend such invitations for future consultative meetings"; "The Question of Antarctica," General Assembly Resolution, A/RES/51/56, December 10, 1996.

103. See the General Assembly's statement, "Noting with regret that the racist apartheid regime of South Africa, which has been suspended from participation in the General Assembly of the UN, has continued to participate in the meetings of the Antarctic Treaty Consultative Parties," in "The Question of Antarctica," General Assembly Resolution, A/RES/41/88, December 4, 1986.

104. Dodds, "The Antarctic Peninsula," 110–11.

105. Ibid., 107.

# FOUR

# The Conquest: Science and Minerals

Despite the ingenious solution of freezing all territorial claims in Antarctica, the quest for sovereignty has continued. Science has come to symbolize political power in Antarctica, with countries dotting the continent with scientific research stations. The scientific occupation of Antarctica is now well under way, with science increasingly used to uncover the mineral potential of the continent.

Aware of the need for an orderly division of the continent's mineral wealth, parties to the Antarctic Treaty attempted in the 1980s to negotiate a regime that would govern mining. Negotiated in secrecy, the regime collapsed because of its interference with territorial sovereignty and strong objections by countries excluded from the negotiations. Cast as a victory for the environment, the collapse of the mining regime was followed by an Environment Protocol that banned mining. In this protocol, however, parties would embed a "walkout" clause that would essentially allow them to revisit the ban some 50 years down the line. With the 2048 rendezvous approaching, could another battle for Antarctica be about to begin?

## SCIENCE IN THE ANTARCTIC TREATY

The Arctic and the Antarctic have always been considered "natural laboratories" for scientific research, with the North and South Poles holding the Earth's heat budget in balance. However, scientific research has played a much greater role in the Antarctic since it could proceed unhindered by claims of territorial sovereignty.[1] After all, the Antarctic Treaty in 1959 had specifically dedicated the continent to the pursuit of peace and science.[2]

## Antarctic Treaty

### Article II

Freedom of scientific investigation in Antarctica and cooperation toward that end, as applied during the International Geophysical Year, shall continue, subject to the provisions of the present treaty.

### Article III

1. In order to promote international cooperation in scientific investigation in Antarctica, as provided for in Article II of the present treaty, the Contracting Parties agree that, to the greatest extent feasible and practicable:
   (a) information regarding plans for scientific programs in Antarctica shall be exchanged to permit maximum economy and efficiency of operations;
   (b) scientific personnel shall be exchanged in Antarctica between expeditions and stations;
   (c) scientific observations and results from Antarctica shall be exchanged and made freely available.

Antarctica has always been of particular importance to the advancement of science since it plays a critical role in maintaining the climatic, physical, and biological stability of the rest of the planet. Antarctica contains 90 percent of the world's ice and approximately 70 percent of its freshwater, which would be enough to raise sea levels dramatically were the continent to warm to a serious degree.[3] Its geographic position has made it a unique vantage point for the study of the interaction between the Earth's magnetic field and the charged particles from the Sun. It also offers multiple advantages for astronomers, with its clear skies, high plateau, low humidity, and stable atmosphere. Astronomical research conducted on the continent continues to shed light on issues as complex as the location of the first stars and galaxies, the nature of dark matter, and gravity.[4] In addition, the Antarctic ozone hole has been one of the most significant scientific discoveries of the last century, which has spurred the creation of a new international treaty to replenish the ozone layer,[5] as well as the promulgation of new environmental policies across the globe. For the last 30 years, the ozone hole has shielded the continent from the impact of global warming.[6]

The Antarctic Ocean is one of the richest biological reservoirs on the planet, supporting a few marine species but with very large, commercially significant populations. In providing a meeting point between the Indian,

Atlantic, and Pacific Oceans and a mixing of their hot and cold waters, it is the birthplace of the special regime known as the *Antarctic Convergence*.[7] The convergence is literally the curve that encircles the continent right at the meeting point of its cold and hot waters, which has its own unique properties. It has no equivalent in the Arctic due to the vast amount of land encircling the northern polar region.

Scientific research in Antarctica has never been driven exclusively by environmental or pure research motives. Economic, political, and security questions have been among the key drivers of scientific inquiry on the continent.[8] The economic drivers have consisted of a desire to exploit the continent's biological and mineral wealth.[9] Geological research in Antarctica, in particular, has often been criticized as being driven by commercial interests. Within the context of the Antarctic Treaty, interest in minerals was first expressed at the Antarctic Treaty Consultative Meeting (ATCM) in Tokyo in 1970 in an attempt to convince Consultative Parties to craft a minerals regime. The timing proved to be fortuitous since the Organization of Petroleum Exporting Countries (OPEC) restricted world supplies of hydrocarbons soon afterward, in 1973, which drove up oil prices.[10] Albeit somewhat unrealistically at that stage, Antarctica became one of the options considered to counter the power of OPEC.

In fact, as a result of the 1971–1972 drilling operations of the *Glomar Challenger* in the Ross Sea,[11] which found traces of hydrocarbons, governments and industry alike started to speculate about the existence of vast oil reserves on the continent.[12] This led major oil companies to encourage seismic surveys of the Antarctic seabed, with expeditions conducted under various national flags to collect seabed data in the Ross, Weddell, and Bellingshausen Seas, and off the Antarctic Peninsula. The political drivers, on the other hand, have consisted of using research to establish a national presence and to gain international visibility, all of which reinforce territorial claims.[13] In fact, subsequent sections of this chapter will illustrate how the creation of scientific research stations has largely been driven by the desire to establish a solid territorial foothold on the continent.

Noteworthy in this regard is that science, and the capacity to conduct serious and significant scientific research, were used as a benchmark in the Antarctic Treaty right from the start as to who could become a Consultative Party. In other words, science was set up to be the distinguishing variable between the would-be decision makers on Antarctic affairs and the observers. Under Article IX of the Antarctic Treaty, science was essentially used to confer power and influence over Antarctic matters—hence the political importance from a national standpoint of conducting regular scientific research on the continent.[14]

**Antarctic Treaty**
**Article IX**

> Each Contracting Party which has become a party to the present treaty by accession under Article XIII shall be entitled to appoint representatives to participate in the meetings referred to in paragraph 1 of the present Article, during such time as that Contracting Party demonstrates its interest in Antarctica by conducting substantial scientific research activity there, such as the establishment of a scientific station or the dispatch of a scientific expedition.

Article IX establishes "substantial research activity" as the key criterion for acceding to the status of a Consultative Party, with scientific stations or expeditions used for substantiation. This requirement was much criticized by the developing world in the United Nations (UN) as creating a rich man's club for the governance of the continent. Elzinger describes the role of science particularly well:[15]

> In the Antarctic, because of the treaty, which suspends territorial claims and makes science the ticket into the club of decision-makers, research represents a form of symbolic capital. There is a special kind of trade-off with politicians, whereby scientists are provided with funds to do research, but in doing this research they also perform a political task, advancing national interests of their own country in a geopolitical arena . . . Crudely put, one might say that politicians don't need to worry so much about the kind of work their scientists do, as long as they are there in Antarctica.

Finally, the security drivers have consisted of ensuring that no armed conflicts take place on the continent, and that the continent does not become a playground for terrorists—a concern visible in the negotiating history of part of the Antarctic Treaty System (ATS).[16] In short, scientific research in Antarctica has never been a purely investigative exercise. While much science is indeed conducted to enhance world knowledge of various scientific phenomena, science in Antarctica is very much a political enterprise that is used to assert and confirm territorial claims. Much of it is also economically driven, with the goal of better understanding the resource prospects of the continent. Security questions also played an important role since any serious security problem could jeopardize both political and economic interests.

## THE DELIBERATE FRAGMENTATION OF SCIENCE PLATFORMS

Established in 1958, the Scientific Committee on Antarctic Research (SCAR) had its origins in the Special Committee on Antarctic Research of the International Geophysical Year (IGY). It was the first body to be established and entrusted with research at the conclusion of the IGY. Its creation, therefore, predated the Antarctic Treaty.[17] It is a nongovernmental organization, albeit funded through national government contributions, and forms part of the International Council for Science.[18] Initially, SCAR's main role was to develop and coordinate international scientific research and to provide advice to Antarctic Treaty parties through national governments. SCAR's role later evolved into providing direct advice to ATCMs, as well as to submitting information and working papers to these meetings. It was eventually granted observer status under the Antarctic Treaty in 1987.[19]

The first meeting of SCAR took place in the Hague in 1958 and had as its primary purpose to agree on the nature of the scientific research that would take place on the continent after the conclusion of the IGY.[20] It was attended by the 12 countries considered to be the most active in the Antarctic affairs, as well as representatives of five scientific unions.[21] The meeting was marked by serious divisions as to who would continue working in Antarctica and what the scope of their work would be. While several governments were eager to contain the expensive scientific research that had been conducted during the course of the IGY, a proposal from the Soviet Union to expand its scientific engagement on the continent forced others to reconsider their position. If the Soviets stayed in Antarctica, others would too.[22]

Of course, the conclusion of the Antarctic Treaty soon followed, and the modalities of scientific engagement on the continent was clarified. With the birth of the Antarctic Treaty, SCAR was quick to receive official recognition of its leading role in research and in the counseling of Antarctic Treaty parties. At the first ATCM held in 1961, parties clarified and welcomed the role of SCAR:[23]

> The Representatives agree without prejudice to the rights of Governments, to make such arrangements as they deem necessary to further the objectives of scientific co-operation set forth in the Treaty:
>
> 1. That the free exchange of information and views among scientists participating in SCAR, and the recommendations concerning scientific programs and cooperation formulated by this body constitute a most valuable contribution to international scientific co-operation in Antarctica,

2. That since these activities of SCAR constitute the kind of activity templated in Article II of the Treaty, SCAR should be encouraged to continue its advisory work which has so effectively facilitated international co- operation in scientific investigations.

SCAR's advisory role led to the formulation of the Agreed Measures for the Conservation of Antarctic Fauna and Flora of 1964, and for the first 40 years of the Antarctic Treaty, SCAR provided advice on protected areas and other environmental issues. SCAR also played an important role in bringing about the Conservation of Antarctic Seals (CCAS) and the Conservation of Antarctic Marine Living Resources (CCAMLR).[24] Soon, however, SCAR was to find itself weakened by governments, in particular the United States and Australia, in search of greater autonomy in the conduct of their scientific operations. Edward Todd, the director of the Office of Polar Programs of the U.S. National Science Foundation (NSF) was the first to express reservations with respect to SCAR.[25]

In 1983, he complained that the Logistics Working Group of SCAR was starting to interfere dangerously with the running of the U.S. science program and its associated logistics. He wrote: "some SCAR participants forget that commitments to SCAR are not governmental commitments by most SCAR participants who have no such charter; this confusion has led SCAR to assume management direction of research activities to which governments are not committed, and to unwarranted criticism of governments that have declined the presumed commitment of resources necessary to implement them."[26] Visible in this statement is that the United States felt that SCAR had started to dictate to governments how they should run their research programs, as well as the scale of the resources that they should commit.

The appointment of Peter Wilkness as the next head of the Office of Polar Programs exacerbated matters further, as he too started to question why governments should take instruction from a nongovernmental entity such as SCAR. Jim Bleasel the director of the Australian Antarctic Division and chair of the SCAR Working Group on Logistics sided with the United States, and together they began to persuade others of the need to establish a separate and independent network of Antarctic program managers.[27] On September 15, 1988, the Council of Managers of National Antarctic Programs (COMNAP) was formed. While COMNAP was described as being "federated" to SCAR as a face-saving solution, in reality the Antarctic program managers would entirely break free of SCAR's control.[28]

Various other developments gradually contributed to a weakening of SCAR, some of which predated the establishment of COMNAP. With the conclusion of CCAMLR in 1982 and the creation of a special advisory organ within it, SCAR had already started to see its role reduced. However, its role was diminished further when the 1991 Environment Protocol—also known as the Madrid Protocol—established a Committee on Environmental Protection.[29] The committee was seen as displacing SCAR on environmental issues. These developments, together with the diminishing budget of SCAR, led some policymakers to call for its restructuring, arguing also that it had mostly become reactive and was no longer able to take the policy lead.[30]

Overall, these developments led to the fragmentation of science and of scientific platforms in Antarctica. The creation of COMNAP, in particular, allowed countries to retain much greater autonomy in the running of their research programs. This fragmentation undoubtedly was intended to prevent SCAR from monopolizing the scientific field and becoming the exclusive source of scientific counsel. With SCAR's status reduced, countries were to enjoy greater freedom in shaping their scientific agendas around the political, economic, and other objectives that truly mattered to them.

## RESEARCH STATIONS THAT DOT THE CONTINENT

Based on the most recent data from COMNAP,[31] 104 facilities have operated in Antarctica in 2014. These facilities include research stations; camps, which are less permanent installations; refuges, which are usually very small or basic installations; airfields; and depots for food or other supplies. Of these 104 facilities, 82 are scientific research stations. The stations belong to a total of 29 different Consultative Parties, 18 of which have multiple stations.[32]

Many stations are built near the coast or on nearby islands to facilitate the movement of people and supplies. In some instances, station buildings are built on poles that raise them above snow and ice, allowing them to be moved to other locations to escape being buried by snow. In fact, heavy snow had previously crushed the United Kingdom's Halley Research Station, forcing engineers to be as creative as possible in rendering the buildings and facilities resistant to extreme weather. Three of the Halley Research Station buildings had rested on platforms with steel legs, which were raised annually to keep them above the snow, and two had rested on skis to be systematically towed to new positions. When the Station was rebuilt, it was reinforced and now has a total of seven buildings.[33]

The biggest research station in Antarctica by far is the U.S. McMurdo Station, which has 85 buildings and a peak population of 1,000 personnel.

Of those, only about 250 spend the winter there. "MacTown," as it is called by its residents, is built on a rock on Ross Island. The buildings include restaurants and a radio station, and there is also a harbor, an airport, and a landing pad for helicopters.[34]

In fact, only nine stations in Antarctica have a peak population of over 100 persons, including the U.S. McMurdo station (1,000), followed by U.S. Amundsen-Scott South Pole Station (250), Russia's Mirny Station (169), Argentina's Marambio Station (150), the United Kingdom's Rothera Station (130), Chile's Julio Escudero Station (120), Japan's Syowa Station (110), Argentina's Carlini Station (100), and France's Dumont d'Urville (100).[35] As can be seen from this list, the U.S. presence on the continent is unrivaled, despite the United States not making any formal territorial claim.

Of the 29 Consultative Parties with research stations in Antarctica, 10 are developing or emerging countries, including Argentina, Brazil, Chile, China, Ecuador, India, South Korea, Peru, South Africa, and Uruguay. In fact, some of the smallest stations in Antarctica, as measured by peak population, belong to these countries. For example, two of the Chilean stations have a peak population of 15, one of the Argentinian stations has 14, and a Brazilian station has 12.

In terms of having the highest number of separate stations, Argentina, Chile, and Russia are the leaders. Argentina has 13 stations, and Chile and Russia have 12.[36] The goal of a number of countries, in particular Argentina and Chile, has been to spread out as widely as possible—in particular across the areas where they have made territorial claims—even if that means becoming too thin in the process, as is evidenced by the low headcount in some of their facilities.

The oldest stations in Antarctica were constructed in the late 1940s and early 1950s and belong to the original Antarctic Treaty Consultative Parties. The newest stations, on the other hand, tend to belong to developing countries, with the most recent stations being China's Taishan Station (2014), Korea's Jang Bogo Station (2014), and India's Bharati Station (2012). In total, today, China has four research stations on the continent, India three, and Korea two.[37] This is a clear sign of the developing world's growing appetite for the continent. In the beginning of 2016, Turkey sent its first scientific expedition to Antarctica and may be on its way to joining the ranks of China, India, Korea, and the rest of the emerging nations.[38] Figure 4.1 shows the spread of research stations across Antarctica.

Many questions have been raised about the efficacy of the Antarctic science agenda, with research stations being this numerous. In fact, the United Nations had expressed concern in Resolution A/RES/46/41 of 1991 over this matter and asked the Consultative Parties "to increase the level of

**Figure 4.1 Map of Major Research Stations in Antarctica**

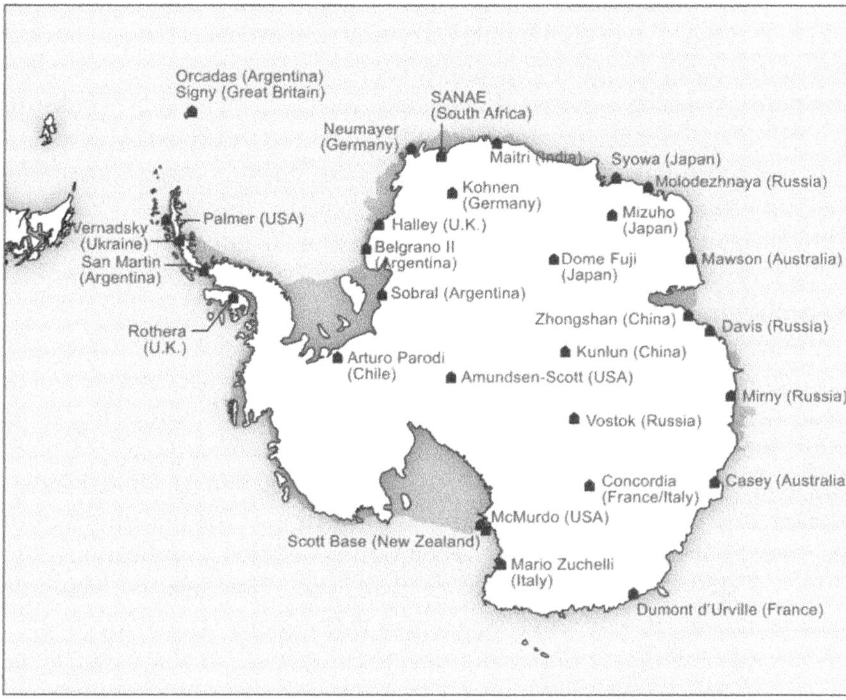


(Joseph Hill)

cooperation and collaboration with a view to reducing the number of scientific stations in Antarctica."[39] After all, a careful reading of Article III(a) of the Antarctic Treaty suggests that countries were expected to share information and to aim for efficiency and economy in their scientific research endeavors.

The siting of research stations, sometimes in very close proximity to each other, has fueled speculation about the true degree of commitment to the scientific cooperation called for the treaty. Much has in fact been written about the unnecessarily high number of stations and the political rather than scientific motivation behind their establishment, with some considering the continent to have been ravaged in the name of science.[40]

In 2011, Alan Hemmings, a well-known Antarctic scholar, published an article entitled "Why Did We Get an International Space Station Before an International Antarctic Station?" with his main thesis being that research stations in Antarctica have been a tool for asserting territorial sovereignty.[41] Despite the exhortations in the Antarctic Treaty for scientific collaboration among parties, only two research stations are joint, he noted. These are the

Concordia, operated by France and Italy, and the Law-Racovita, operated by Australia and Romania.

In fact, a massive overcrowding of stations is particularly visible in the area of overlapping territorial claims—the Antarctic Peninsula—where the siting of stations has a clear political goal. Today, the peninsula is home to six Chilean, five Argentinian, and three U.K. stations (operated by the countries with territorial claims in that specific area). In addition, because of its proximity to South America, it has also become the location of choice for any Consultative Party wishing to establish a quick and easy foothold on the continent. Figure 4.2 below shows the density of research outposts on the peninsula.

Another particularly crowded area is King George Island,[42] which sits right across from the Antarctic Peninsula to the north, and is even closer to South America. It is separated from the peninsula by only the Bransfield Strait. It houses research stations from Argentina, Brazil, Chile, China, Ecuador, South Korea, Peru, Poland, Russia, Uruguay, and the United States. It raises the exact same questions as does the peninsula on the efficiency of research and the real purpose of the stations. In addition, it is clear that several of the stations in Antarctica are outdated and of little use, but that little attention is being paid to removing them or to combining them with others.[43]

The spread of research stations across the continent has also triggered its own set of environmental concerns. For example, in 1988, the Environmental Defense Fund issued a scathing report entitled "The Failure of the National Science Foundation (NSF) to Protect Antarctica," in which it illustrated the heavy environmental footprint of the U.S. presence on the continent.[44] The report accused the NSF of failing to take all reasonable steps to reduce the pollution of Antarctic waters, citing violations relating to the discharge of waste and toxic chemicals, in addition to the burning of combustible solid wastes in open pits.[45]

In response to these concerns, Congressman Walter Jones, chair of the House Committee on Merchant Marine and Fisheries at the time, together with other political actors, introduced legislation to obligate the NSF to address these issues. The outcome was the Antarctic Environmental Protection, Clean-Up, and Liability Act of 1990, which addressed problems such as waste management practices at U.S. bases, as well as of marine pollution and garbage disposal practices from ships visiting the continent.[46]

It is important to point out that U.S. bases were by no means alone in these damaging practices. The French base in Dumont D'Urville, for instance, is notorious for its establishment of an airstrip, which environmentalists warned would have serious negative ecological consequences. In fact, the creation of that airstrip led to the establishment of a Greenpeace

**Figure 4.2 Map of Stations in the Antarctic Peninsula**

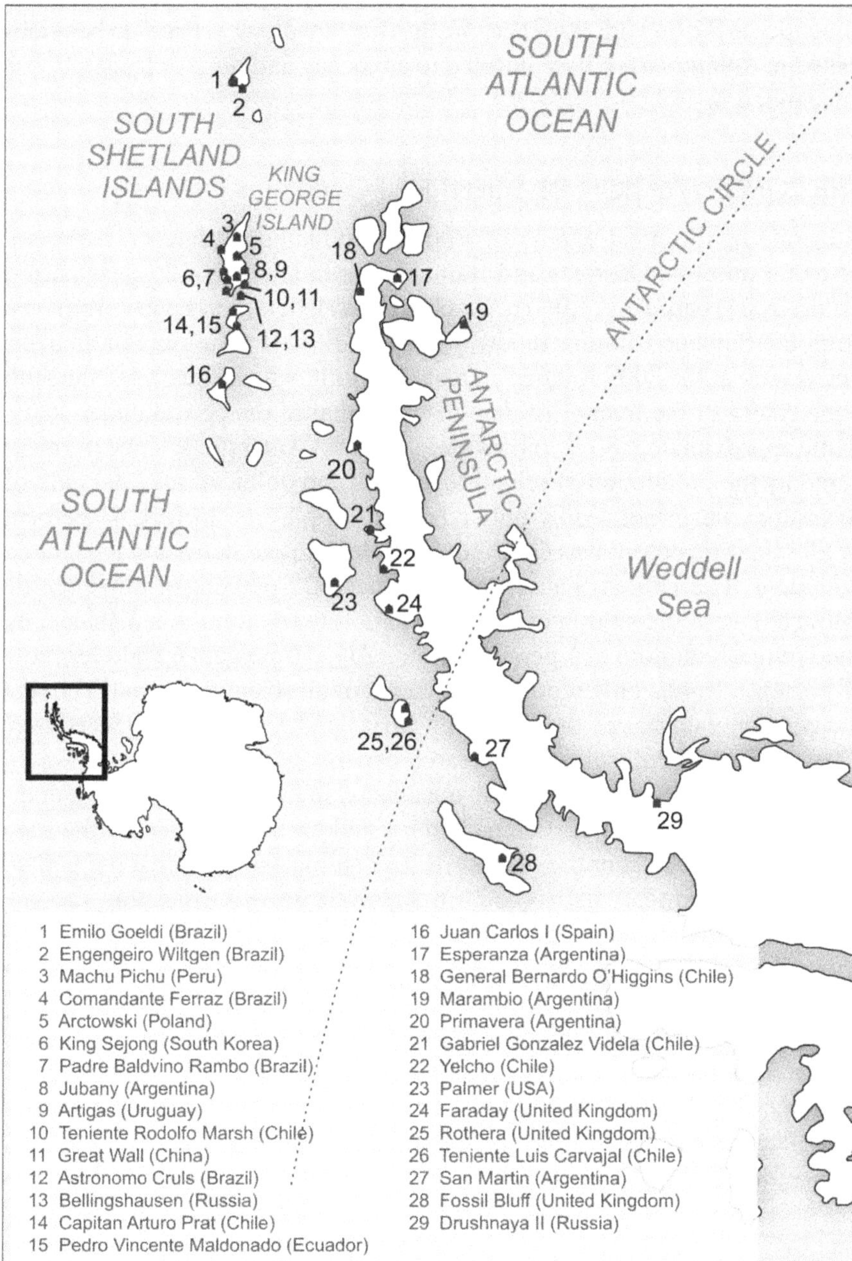


(Joseph Hill)

World Park Base on Ross Island in 1987 as a sign of protest, with Greenpeace calling upon policymakers to halt the pillaging of the continent.[47] The Greenpeace World Park Base became the first "ideological base" to be established in Antarctica—a base devoted to advocacy and policy change. It was dismantled in 1992.

## THE CONCENTRATION OF SCIENTIFIC POWER

In one of the most extensive studies conducted to date on Antarctic research outputs, Dastidar and Ramachandran analyze a 25-year period (1980–2004) based on 11,000 research articles, letters, and reviews retrieved from the Science Citation Index.[48] Their work was the object of analysis and review at the 2009 Antarctic Treaty Summit, which celebrated the 50th anniversary of the treaty.[49] It demonstrates that in the period under study, while approximately 80 countries were active in Antarctic science (as measured by the production of scientific articles and publications), the bulk of the output was produced by only a handful of players. In fact, 41 percent of all scientific output was contributed by the United States and the United Kingdom. The United States alone produced 27 percent of all scientific publications in the study period.[50] It was followed, at quite a distance, by the United Kingdom (14 percent), Australia (10 percent), and Germany (8 percent). Russia only produced 3 percent of all scientific output. The first developing countries on the list, South Africa and Argentina, produced only 2 percent each of all scientific research papers.

This is, of course, an astonishing conclusion in light of the very high number of scientific research stations that existed on the continent at the time and the number of countries involved. The study also finds that the United States was the most "networked" in Antarctic science, meaning the one conducting the largest amount of collaborative research with other countries. The findings demonstrate that every third organization participating in Antarctic science belonged to the United States, with 75 percent of these being universities.[51] In addition, 40 percent of all scientific papers published on the Antarctic were disseminated in only 16 journals (i.e., 1.6 percent of all scientific journals worldwide). The most cited journals were *Nature* and *Science,* published from the United Kingdom and the United States, respectively.[52]

The authors attribute their findings to the following factors: first, the fact that the United States has the largest research complex in Antarctica and by far the largest number of personnel on the continent; second, that the U.S. program hosts visiting scientists from other Antarctic Treaty parties; and third, that the United States spends more per capita on research and development (R&D) than many other nations.[53] While the study has not been brought up to date to 2015, and while it is reasonable to anticipate a growing involvement by the emerging developing world in light of the new research stations that they have opened, the United States clearly continues to dominate. Particularly astonishing is that Australia—which claims 42 percent of Antarctica—produced only 10 percent of its scientific output.

In fact, much has been written about the lack of strategic focus of Australia's Antarctic science program. It was as far back as 1993 that Bruce Davis, then the deputy director of the Institute of Antarctic and Southern Ocean Studies at the University of Tasmania, warned about the situation. "During the past decade, attempts have been made to improve priority identification and resource allocation within Australian Antarctic Science, but doubt remains as to the degree the program is effectively integrated and managed," he wrote.[54] More recently, in June 2013, Anthony Bergin, deputy director of the Australian Strategic Policy Institute and Anthony Press, chief executive of the Antarctic Climate and Ecosystems Co-operative Research Centre in Australia, wrote:[55]

> We're running our program on the smell of an oily rag: for this coming financial year, it's about $169 million, an 8 per cent cut from last year. A continued downward trajectory in budget allocations will lead to closure or mothballing of stations and reduced science, and will diminish our standing in Antarctic affairs . . . our present capability means that we can't match what others are doing in our territory, let alone lead. If we're going to regain our leading position, we'll need to ensure that critical Antarctic science is adequately funded and supported.

To understand the nature of scientific enterprise in Antarctica, it is equally important to zoom in on the nature of the collaborative effort that has taken and is taking place. Much collaboration over the years has occurred between U.S. and Soviet scientists. For example, the Soviet icebreaker *Mikhail Somov* was used in 1981 as a research platform by 13 Soviet and 13 U.S. scientists, managing to produce the first comprehensive data set on winter ice. A decade later, the Russian icebreaker *Akedemik Federov* and the U.S. icebreaker *Nathaniel Palmer* collaborated in the same area to establish a drifting camp on frozen waters.[56]

The Antarctic Gamburstev Province Project has involved the United States, the United Kingdom, Russia, Germany, China, and Australia and led to the discovery of river valleys in the Gamburstev Mountains under the Antarctic ice sheet. This allowed scientists to form an opinion of what the continent looked like prior to its freeze. The involvement of China in this collaborative effort is of course notable.[57] A large number of countries also participate in the Polar Earth Observing Network to map the Antarctic crust, although the United States plays a central role in funding the network.[58] Furthermore, Germany, Italy, New Zealand, the United Kingdom, and the United States have contributed to the Antarctic Geological Drilling Program, succeeding in obtaining sediment core.[59] Several other examples of collaborative efforts exist.

Notable in all of this information is the centrality of the developed world to science in Antarctica, whether individual or collaborative, and the particularly significant role played by the United States. If science under the Antarctic Treaty is power, then the United States clearly emerges as the strongest force.

## WHY HAVE A MINERALS REGIME FOR ANTARCTICA?

Negotiated in 1959, the Antarctic Treaty was silent on the question of minerals exploitation on the continent. At the time, little was known about the continent's resource prospects, which is precisely why countries wished to dedicate Antarctica to science. However, scientific investigation from the 1950s to the 1980s caused a continent of unknown potential—a *terra incognita* of sorts—to become less of a geological mystery. In addition, technological progress had rendered the question of the access to and exploitation of Antarctic minerals slightly less theoretical.[60]

While up until the 1970s, there remained tremendous uncertainty about the mineral deposits of the Antarctic, the 1971–1972 oil drilling operations of the *Glomar Challenger* had started to whet the appetite of governments and industry alike. As early as 1969, several Consultative Party governments began to receive requests from companies interested in mineral prospecting on the continent, particularly in offshore oil. However, governments were fearful of upsetting the stability of the Antarctic Treaty by allowing such activities to go forward.[61] With respect to hard minerals, the Antarctic Peninsula appeared to present the best prospects on the continent—especially for base metals and precious minerals—because of its greater proportion of exposed rock.[62]

In 1977, parties adopted a recommendation urging voluntary restraint in exploration and exploitation, conditional on progress toward a minerals regime. From 1972 onward, mineral resources became a regular item of discussion at ATCMs, and it was at their 11th meeting in Buenos Aires in 1981 that the Consultative Parties formally decided to negotiate a minerals regime for the continent.[63] A combination of motives appear to have been behind this decision. The first was the fear that a regime would best be negotiated before the discovery of major mineral deposits.[64] In other words, there would be less political tension were a regime to be crafted in the absence of precise information on the mineral wealth of the continent than if such information were available and itself become a part of the negotiations.

The second was the desire to prevent disorganized exploitation from taking place, whether by Antarctic Treaty parties or by third parties with the technological capacity to do so.[65] After all, a gold rush could seriously destabilize the delicate balance struck in the Antarctic Treaty on the question of sovereignty. In the same vein, there appeared to be the fear that the absence of a minerals regime linked to the Antarctic Treaty could prompt third parties to settle the issue outside.

Furthermore, in the absence of a regime, no steps could be taken to prevent developers from monopolizing either a particular set of resources or particular sections of the Antarctic.[66] Australia, the claimer of 42 percent of the continent, appeared eager to ensure that any resource exploitation by Australians would not be seen as a unilateral act, and equally keen to prevent disorganized exploitation from taking place within what it considered to be Australian Antarctic Territory (AAT). Bergin described the Australian position as follows:[67]

> The knowledge of resources within the AAT is not great and only four other states expressly recognize Australia's territorial claim to Antarctica.[68] Other states thus might not accept any unilateral mining within the AAT. A treaty regulating Antarctic minerals extraction would minimize any opprobrium that may arise if Australia were to extract minerals outside of an agreed treaty. Since most states would not accept Australia denying to other states access to minerals within the AAT an international agreement would provide the possibility of Australian companies having guaranteed access and security of tenure.

Similarly, the U.S. motivation, like that of several other negotiating powers, was to ensure access to Antarctic minerals on a "non-discriminatory basis."[69] However, since there appeared to be no confirmation of the existence of mineral deposits in any significant quantity at the time of the negotiation, the imperatives for the development of a minerals regime can be said to have been more political than economic.

Parties to the Antarctic Treaty soon established a set principles for the negotiation of the minerals regime.[70] Their main tenets were that the

Antarctic Treaty be maintained in its entirety; that Article IV of the treaty be preserved; that the interests of "all of mankind" be taken into account; that any regime be acceptable and without prejudice to claimant states or those states which neither recognize such claims or assert any such right (meaning the United States and Russia); that any state adhering to the regime that is not a member of the Antarctic Treaty be bound by the basic principles of the Treaty; that the regime apply to all resource activities on the continent and its adjacent offshore areas; that the regime cover mineral resource activities at every stage (exploration, development, and production); and that due regard be paid to the impact of minerals exploitation on the Antarctic environment.

Particularly interesting was the emphasis on the need to protect the Antarctic Treaty, and in particular its provisions on territorial claims (Article IV), and the desire to bind nonparties to the treaty to those main provisions were they accede to the new regime. Equally interesting was the use of the words *claimant states* and the requirement for their approval, demonstrating the extent to which sovereignty remained at the core of Antarctic affairs. Finally, it is noteworthy how cognizant the negotiators were of the need to take the interests of all humanity into account—in other words, that they not allow themselves to get away with a regime that catered to their economic interests alone.

In fact, the 1984 Report of the UN Secretary General on the "Question of Antarctica" described the anxiety of various members of the United Nations with respect to the negotiating process.[71] They saw it as secretive, urging that it protect the interests of all people,[72] and that it take place under the more global UN umbrella. The United Nations nevertheless felt that a minerals regime was required since there was concern that "the Treaty would fail to keep the peace in Antarctica once the exploitation of mineral resources became feasible."[73]

## STRUCTURE OF THE MINERALS REGIME

On June 2, 1988, Christopher Beeby, the deputy secretary of the New Zealand Ministry of Foreign Affairs, and chair of the CRAMRA negotiations, announced the adoption of a new minerals regime:[74]

> This is an historic occasion which I believe will go down in Antarctic history as the most important political development regarding the regulation of Antarctica since the Antarctic Treaty itself was adopted in 1959.

What Beeby had succeeded in shepherding countries toward was one of the most complicated international accords; an agreement 98 pages long, consisting of 67 separate articles and an annex of another 12 articles.[75] The convention covered all nonliving, nonrenewable, natural resources, including fossil fuels and metallic and nonmetallic minerals, but it excluded ice—a source of fresh water—which was to become the subject of further negotiations after the adoption of the agreement.[76]

The convention appeared in the four official Antarctic Treaty languages—English, French, Russian, and Spanish—with the question of using Chinese becoming an issue at the final negotiating session.[77] Although the Chinese delegation denied any desire to add Chinese to the official ATS languages, it nevertheless succeeded in obtaining Article 67 of CRAMRA, which stated: "This Convention, of which the Chinese, English, French, Russian, and Spanish texts are equally authentic, shall be deposited with the Government of New Zealand." This was a strong signal from China of its desire to take part in the regime and become a full partner in its implementation and evolution.

One particularly complex aspect of this negotiation was the extent to which it required "internal accommodation" among the parties to the Antarctic Treaty, "external accommodation" vis-à-vis the wider international community,[78] and special attention to interests of developing countries so they would not feel excluded.[79] The internal accommodation consisted of finding an *entente* among claimant states, among nonclaimant states, and across both sets of players. In addition, the negotiation had to balance economic interests against environmental concerns, ensuring at all times that it was not opening up the continent for pillaging by resource developers. For all these reasons, the outcome was an accord of great complexity.

Offering an account in 1982 of the main challenges likely to be confronted in the negotiations, Beeby wrote that sovereignty issues were likely to loom large, hence the need to establish strong linkages between the new legal regime and the balance already struck in the Antarctic Treaty.[80] A prominent Chilean lawyer and expert in Antarctic affairs, Francesco Orrego Vicuna, added at the time that claimant states would have to move toward a system that would not void their claims, while nonclaimant states would have to search for an approach that would not recognize the claims of others.[81]

Beeby stressed that finding "external accommodation" would also be key, since the principle of the Common Heritage of Mankind that the developing world was advocating could not be ignored. If it was, nothing could prevent the rest of the international community from creating a rival regime.

Because the developing world did not have the financial resources or technological capacity to participate immediately in mining, he believed that special mechanisms would need to be devised to allow it to benefit from the opening of this new continent anyway. He gave the transfer of funds to the developing world as an example.[82]

Of particular concern to Beeby, as chair, was how the negotiation would take into account the interests of those "not at the negotiating table."[83] This was the case due to the fact that CRAMRA was being negotiated only by parties to the Antarctic Treaty, with serious accusations of secrecy being made. The CRAMRA negotiations nevertheless went forward. They resulted in a minerals regime that addresses the entire life cycle of mineral activities on the continent; all the way from prospecting to exploitation and development.[84] The convention started with a series of general provisions, which are important to understand because they provide context for the subsequent, more detailed modalities of the accord.

The convention began by asserting that "operators" could engage in Antarctic mineral resource activities when they had a "sponsoring state." It defined an operator as "an agency or instrumentality of a Party," a "juridical person established under the law of a Party," or a "joint venture consisting exclusively of any combination of the foregoing."[85] This essentially opened up the continent to the entry of both private and public entrepreneurs, provided that they could find a sponsoring state.

In Article 3, the convention prohibited any mineral resource exploitation from being conducted outside its framework. The intention was to prevent any rival regimes from emerging or any third parties with sufficient technological capacity from stepping into the continent and acting unilaterally. However, the extent to which this provision would have been respected by nonparties is questionable.

Each party to CRAMRA was made responsible for enforcing the convention upon its own nationals,[86] with a Protocol on Liability to be developed, in the absence of which no exploration or development could take place.[87] In other words, when CRAMRA was concluded, it was understood that minerals activities could not be initiated until such time as its limited rules on liability could be developed further.

The general provisions of CRAMRA restated the language of Article IV of the Antarctic Treaty to evacuate the issue of territorial sovereignty.[88] The relevant provision were entitled "Protection of Legal Positions Under the Antarctic Treaty." In a later provision, CRAMRA established that it would be open to accession only by contracting parties to the Antarctic Treaty.[89] The purpose of these two provisions, taken together, was to ensure that nothing under CRAMRA would rock the delicate balance on sovereignty struck within the Antarctic Treaty. This represented a crucial element of "internal accommodation"—essentially providing comfort to the parties to the Treaty (arguably at the expense of nonparties).

Under its general provisions, CRAMRA established a "non-discrimination" principle against any party or its operators.[90] This had always been one of the key demands of the United States and one of its key motivations for the negotiation of CRAMRA. The convention also called for the protection of information with commercial value, which could continue to be retained by the operator.[91] In terms of institutional structure, CRAMRA created a system that would manage the entire life cycle of minerals activities on the continent. Its most powerful organ was to be the "Commission," whose role was to identify possible areas for exploration and development.[92] Conversely, it could designate areas prohibited to mining and adopt measures for the protection of the environment.

The Commission was open to all Consultative Parties, any non–Consultative Party actively engaged in research, and to parties sponsoring a minerals activity. Any decision on areas to be opened up for exploration or development required consensus among the members of the Commission.[93] Therefore, another critical element of "internal accommodation" lay in the fact that no decision could be taken on minerals activities without the agreement of at least all Consultative Parties, including both claimants and nonclaimants. This also meant that any Consultative Party could veto minerals development in particular sections of the continent.

However, it is important to note that the veto power also could have worked to the benefit of nonclaimants. Wolfrum explains this well in the following statement:[94]

> Each single claimant state may forestall any resource activity taking place within its claimed sector. This right to object, however, works in the other direction as well, which may have made it acceptable for other participants of the negotiations. It empowers every single member of the Commission to object to the identification of an area on, for example, environmental grounds, which represents a counterweight to the claimant state's right to object. To that extent, the introduction of the consensus principle is bifocal as far as the positions of the claimant and non-claimant states are concerned.

CRAMRA created extremely important "Regulatory Committees," whose functions were to consider the specific applications made for exploration and development in the sections of the continent already authorized by the Commission, as well as to approve management schemes and to issue exploration and development permits. These committees were also made responsible for monitoring minerals activities once underway.[95]

The Regulatory Committees had very precise, but also very complicated, rules on composition.[96] They were to include 10 members consisting of 4 claimants (the United States and the Soviet Union, who assert a "basis for a claim" but not an actual claim, would have had to be among them), and 6 nonclaimants. In the claimants category, the member (or members) that assert a claim, or basis of a claim, in the "identified area" were to be included. In addition, the chair of a committee would have been under obligation to ensure at all times "adequate and equitable representation of developing country members of the Commission, having regard to the overall balance between developed and developing country members of the Commission."[97]

Decisions by the Regulatory Committees to approve a management scheme and issue a development permit were to be taken by a two-thirds majority. However, within that majority, a simple majority had to be achieved across claimants (and those asserting a basis for a claim), and another simple majority achieved across the nonclaimants.[98] Therefore, with respect to the Regulatory Committees, elements of "internal and external accommodation" could be found. The claimants would play a very strong role, with the United States and the Soviet Union always present, but so would the six nonclaimants, who would also have a say.

The Regulatory Committees were to play another critical role under CRAMRA—that of deciding between competing applications for the same minerals development activity.[99] A committee could either invite the competing applicants to resolve the matter, or, if the matter was not resolved, to take it upon itself to do so. The latter had to take place with priority given to "the application with the broadest participation among interested Parties which are Antarctic Treaty Consultative Parties, in particular, developing countries in either category."[100] In other words, the Consultative Parties had to be given priority over the non–Consultative Parties when there were competing applications (more evidence of "internal accommodation").

Finally, CRAMRA established an "Advisory Committee" open to all parties to the Antarctic Treaty, and to which parties could be represented by scientists or other individuals with technical or environmental competence.[101] The committee advised the Commission and the Regulatory Committees on the scientific, technical, and environmental aspects of any mining activity.[102]

A number of other provisions of CRAMRA are important to mention. The first is the fact that minerals prospecting did not require any authorization by the institutions of the convention, with the term *prospecting,* used to refer to activities aimed at identifying locations with minerals potential.[103] As will be demonstrated later in this chapter, this became a major source of concern for environmentalists. The second is the fact that the convention strongly encouraged "joint ventures" to allow developing countries to participate in minerals exploitation.[104] This became a clear sign of "external accommodation," intended to signal inclusion for the developing world. Finally, CRAMRA would enter into force only if all 16 original members of the Antarctic Treaty ratified it.[105]

On the whole, it can be stated that while CRAMRA attempted to protect the interests of claimant states throughout the convention, particularly by granting them veto power in the areas of their claims through the consensus principle at the commission level, it nevertheless made some effort to accommodate the broader international community. This was manifest in its inclusion of the developing world in the various bodies of CRAMRA and the encouragement of joint ventures, as well as attempts to give non-claimants power, particularly in the Regulatory Commissions.

As evidenced by the mere language of CRAMRA—its continuous use of the terms "claimant" and "non-claimant" states, and countries with a "basis for a claim"—the issue of sovereignty was clearly front and center during the negotiations. A particular balance was struck on the sovereignty question through CRAMRA's reiteration of Article IV of the Antarctic Treaty, its requirement that all parties to CRAMRA accede to the treaty, the institutions that it would create, and their composition and role in decision making. But how long would this delicate balance hold? As we will see, not long.

## THE COLLAPSE OF CRAMRA

Concluded on June 2, 1988, in Wellington, New Zealand, after six years and 12 formal negotiating sessions, CRAMRA collapsed almost immediately. Conventional wisdom has it that CRAMRA failed because of heavy lobbying against it by the environmental community.[106] While the environmental community indeed took up arms against CRAMRA, there were other reasons behind its quick demise.

On the environmental front, two major accidents in 1989 forced governments to reflect on the soundness of the accord. The first took place in February, when the Argentinian naval vessel *Bahia Paradiso* sank in Antarctic waters, releasing oil near Anvers Island and raising fears of possible damage to the Antarctic environment.[107] The *Sydney Morning Herald* reported on the incident as follows:[108]

> Diesel fuel, almost one million liters of it, is right now spreading along the frigid Antarctic coast. The suffocating black slick, stretching more than 3 km, is threatening the wildlife-rich area with disaster. Penguins are emerging from the icy depths covered in oil. Their natural oil glands clogged, they are unable to function and will die. Other animals such as seals will be affected by the destruction of krill, the crustacean at the center of the Antarctic's food chain.

A month later, an environmental accident occurred in the Arctic that trumped *Bahia Paraiso*. The *Exxon Valdez* ran aground in Prince William Sound in Alaska, considerably damaging marine life and associated ecosystems over a huge area.[109] These two accidents combined led environmentalists to campaign against CRAMRA, arguing that oil spills in cold Antarctic waters would result in exceptionally long amounts of time for the oil to degrade, as well as complicated cleanup operations. Mining in Antarctica would also require massive infrastructure compared to other parts of the globe, which would take an even heavier toll on the environment.[110]

Environmentalists criticized CRAMRA for not containing adequate environmental safeguards. They feared that horse trading would ensue in its Regulatory Committees, arguing that the Commission should have been given much greater power so that it could block off entire sections of the Antarctic to exploitation. In addition, they believed that the negotiation of a Protocol on Liability had been unnecessarily postponed and that the absence of nongovernmental organization (NGO) observership during the negotiations had been unacceptable.[111] They also criticized the regime for allowing minerals prospecting to go forward without any authorization from the institutions of CRAMRA, a situation that they saw as hazardous since it could subject Antarctica to heavy commercial pressure by developers.[112]

NGOs started advocating for the designation of Antarctica as a World Park, calling for a prohibition on mining and protection of Antarctica's wilderness. In January 1988, the Greenpeace World Park base in Antarctica was used as a platform to campaign against CRAMRA. An expedition led by Peter Wilkinson initiated the World Park campaign, in which he stated:[113] "We intend to take our message to . . . the continent. The public at large and those presently living on the continent must be made aware of the consequences of oil exploration in this most delicate and special place."

Almost in parallel, French naval officer and well-known conservationist and explorer Captain Jacques-Yves Cousteau weighed in against CRAMRA, arguing that it would open up the seventh continent to industrial exploitation like a Pandora's box.[114] In a congressional hearing before the U.S. Committee on Foreign Affairs, he testified that he had not even been aware that CRAMRA was in the making because of the level of secrecy that shrouded the negotiations. Had it not been for three lines at the bottom of a *New York Times* article, he said, he would not have known that it had been concluded.[115]

Australia became the first country to express serious hesitations about CRAMRA. It announced in May 1989 that it was opposed to the convention, expressing its support for the declaration of Antarctica as a World Park or a Wilderness Reserve.[116] France soon followed, swayed by Cousteau's resistance to CRAMRA.[117] French prime minister Michel Rocard, under intense lobbying from the Cousteau Foundation, started to publicly criticize the convention. Soon enough, like Australia, France indicated that it would not ratify CRAMRA.[118] It then fielded Cousteau across the globe to convince others of the need to dismantle the convention. In the interview in Box 4.1, Rocard provides his personal account of the events leading to CRAMRA's collapse.

Having convinced France of its position, Australia then turned to convincing others through diplomatic channels,[119] even if the refusal by one of the original parties to the Antarctic Treaty to ratify would have sufficed to keep CRAMRA from entering into force.[120] However, Australia did not want to be singled out for scuppering a treaty that had taken so long to negotiate. In the end, Australia and France succeeded in swaying other players, and CRAMRA was shelved.

While environmental concerns were certainly partly responsible for the collapse of CRAMRA, there were other deeper reasons in play. These included (1) intense opposition by the United Nations to CRAMRA, (2) serious concern among claimant states about the consequences of CRAMRA on sovereignty, and (3) fear that the absence of proper rules on competition policy within CRAMRA would disadvantage certain countries.

The United Nations had been uncomfortable with the ATS right from the start, seeing it as an exclusive millionaire's club, so it is no surprise that it was equally—if not more—uncomfortable with its new minerals regime, which began to divide up the mineral wealth of the continent. In fact, a call was made in a 1986 UN General Assembly Resolution for a moratorium on the negotiation of CRAMRA until the entire international community could participate. It stated that the General Assembly:[121]

1. Reaffirms that any exploitation of the resources of Antarctica should ensure the maintenance of international peace and security in Antarctica, the protection of its environment, the non-appropriation and conservation of its resources and the international management and equitable sharing of the benefits of such exploitation;

### Box 4.1 Interview of Former French Prime Minister Michel Rocard[a]

1. *To what extent is the issue of territorial sovereignty still alive in Antarctica?*
The ingenious treaty of 1959, the result of the will of President Dwight Eisenhower, did not recognize any of the territorial claims that had been made. It prohibited the presentation of new territorial claims, the introduction of national borders, passports, or photos. The continent was demilitarized, and this was a miracle, as we were right in the middle of the Cold War. We wanted to do without a Hot War. Nobody imagined that it would be possible to demilitarize Antarctica in the middle of the Cold War. But it was a reality and it was done. I do not see the sovereignty question raising its head again, but there may be a desire to exploit the mineral resources of the continent. A clause to revisit the prohibition on the ban on mineral resources is to be revisited in 2048, as per the date set in the Madrid Protocol, and there will be pressure to lift it. However, environmentalists would be likely to object, and in my opinion, the prohibition may be saved.

2. *Why did CRAMRA collapse in 1988?*
Prime Minister [Bob] Hawke of Australia and I were the ones to kill it. The main factor was the orientation of the Australian parliament. The convention was signed in 1988 after 6–7 years of negotiation at New Zealand's initiative. It reflected New Zealand's belief that the continent was fragile, and that if its ice were to melt, it would lead to a planetary catastrophe. So they initiated a negotiation that would aim to authorize minerals development only if the best mining techniques were used, and the greatest environmental protection applied through a peer review process—a process that would be exercised by a college of nations.

However, when it came up for signature, Australian prime minister Bob Hawke saw that his government had become increasingly green, and that environmentalist did not want minerals exploitation in Antarctica. For a full year, Hawke could not get the convention ratified. On 17 June 1989, he visited Paris. I met him, we were friends, and we had a coffee together at the Jardin de Matignon. He told me that his parliament was completely refusing to sign the convention. The president of the French Republic at the time did not like environmentalists, and some of the biggest nations had already signed the convention like the Russia and the U.S. So what I had before me was a challenge, but both of us would soon announce that neither France nor Australia would sign the convention nor send it Parliament. We had obtained the accord of the French President.

We received diplomatic messages from governments such as Belgium and Italy imploring us to sign. They said this was no way to work. We tried to mobilize support for our position. Paris was an air hub and this helped me. We contacted Russia and China, and the U.S. was very angry with us. The U.S. then became busy with national elections and stopped responding to diplomatic messages. In the meantime, I decided to invite Jacques Yves Cousteau—a well-known environmentalist who had a bad character and hated government and public affairs, to spend three months playing his documentary *The Silence of the Sea* around the globe—it was on the oceans and the need to protect Antarctica. The French embassy in Washington also went to work and [started] lobbying U.S. senators to oppose the treaty. We had excellent young staff at the embassy at the time and they were extremely active. Every time we would receive a message that Senator X or Senator Y had changed their minds after talking with Cousteau, we would cross their names out and lobby the next person. We finally won the battle, and new negotiations commenced in Madrid.

[a]Interview conducted by the author on October 5, 2015.

2. Calls upon the Antarctic Treaty Consultative Parties to impose a moratorium on the negotiations to establish a minerals regime until such time as all members of the international community can participate fully in such negotiations.

In addition, the impact of CRAMRA on territorial sovereignty was an issue that grew after the convention was concluded, and it became heavily and openly debated. For Australia in particular—the claimant of 42 percent of the Antarctic—this issue was paramount.[122] In November 1988, Australian treasurer Paul Keating and resource minister Peter Cook argued that ratifying CRAMRA would be tantamount to admitting that Australia did not own its own AAT, particularly since the convention did not guarantee that royalties would be paid to claimant states.[123] In the negotiations, Australia had in fact stubbornly argued that claimant states should have the automatic right to derive significant revenue through royalties and taxes from mining. However, for obvious reasons, this was vehemently opposed by the nonclaimant states.[124]

There is evidence that the U.S. side also was keenly aware of the implications of CRAMRA on sovereignty. In its explanation of CRAMRA to the U.S. Congress in 1989, the Office of Technological Assessment wrote that ratification of this accord could mean that claimants recognize that they do not have exclusive mineral rights in their territories.[125] However, it went on to reassure Congress that the United States would be well represented within the institutions of the convention, so it would be an important decision maker, urging Congress to approve it anyway.[126] Nevertheless, Congress chose not to ratify it shortly afterward.

In fact, while the convention was still being negotiated, an academic debate began on its implications on sovereignty. In 1984, for instance, F. M. Auburn wrote:[127]

> The regime would be likely to mark the practical end to claims to sovereignty. . . . Even the most rudimentary form of licensing and decision-making must be regarded as having a considerable impact on sovereignty . . . once this has been achieved in a stable form continuing indefinitely, there will be no apparent difference in Antarctic system entitlements between claimants and non-claimants.

Finally, there was tremendous concern that CRAMRA did not include rules to ensure fair competition among resource developers (specifically, antisubsidy provisions). Keating and Cook argued that with no antisubsidy provisions, Australian mining interests on the continent would be adversely affected.[128] In addition, the Treasury Department expressed the concern that the mining of Antarctica would be impossible without heavy subsidies that Australia could ill afford. Therefore, it would be automatically disadvantaged were CRAMRA to enter into force.[129] In fact, it appeared that, at the time of the negotiations, the player that Australia feared the most was Japan, from its own neighborhood. It worried that resource-poor countries, such as Japan, would enter the continent with a heavy subsidy machinery behind them.[130]

Then, on November 21, 1988, a confidential letter from Keating to other senior ministers was leaked by the *Sydney Morning Herald*, revealing the much deeper concerns that some politicians harbored. The *Herald* stated:[131]

> The Treasurer, Mr. Keating, has warned other senior ministers that Australia risks losing sovereignty over its Antarctic territories if it

> ratifies an international agreement on minerals exploitation, open for signature this week. In a confidential letter to several colleagues, including the Ministry of Foreign Affairs, Senator Evans, Mr. Keating says the signing of the Antarctic Minerals Convention 'would concede our economic claims over Antarctica for virtually nothing' . . . Mr. Keating wrote that claimant nations, like Australia, might not obtain the revenue it believes it should earn from the harvest of resources. 'I believe we must be prepared to stand alone if necessary on this issue,' he writes. Secondly, Mr. Keating rejects as 'completely unsatisfactory' a loophole which could allow the superpowers to subsidize unprofitable mining operations for strategic purposes. There would be no recourse if the Australian mining industry was adversely affected.

As can be seen, therefore, the reasons for the collapse of CRAMRA went beyond the environmental sphere. Political and economic motivations were also a factor. Sovereignty issues, as well as the specific modalities for resource competition in the Antarctic, were at the heart of the matter. Of course, because CRAMRA was negotiated before any major discovery of minerals, the agreement became easy to shelve. The question now is: If confronted with major discoveries, what will the international community be able to do? CRAMRA sank when competition over specific minerals was a nonissue; it died on mere theoretical grounds. This can only mean that matters will be far more complicated once specific resources are found.

## THE BIRTH OF THE MADRID PROTOCOL AND ITS "WALKOUT" CLAUSE

As it started to become clear that CRAMRA would not be accepted by the international community, two crucial proposals were made by France and Australia.[132] France suggested that a special meeting be convened to elaborate an environmental regime for Antarctica, and Australia called for a new environmental convention that would designate Antarctica as a Wilderness Reserve. From these two proposals, the Protocol on Environmental Protection to the Antarctic Treaty, otherwise known as the Madrid Protocol, emerged on June 21, 1991. The protocol entered into force on January 14, 1998.[133]

The Madrid Protocol designates Antarctica as a "natural reserve," establishes environmental principles for the conduct of all activities on the continent, subjects all activities to environmental impact assessments, establishes a Committee for Environmental Protection to advise the ATCM, requires the development of contingency plans to address environmental emergencies, and calls for the elaboration of rules on liability for environmental damage. The protocol was complemented with several annexes covering a range of different issues, such the establishment of protected areas and waste management.[134]

Article 7 of the protocol bans mining on the continent and constitutes the thrust of the new accord. It states that "Any activity relating to mineral resources, other than scientific research, shall be prohibited."[135] In addition, Article 25 establishes procedures for the modification and amendment of the protocol.[136] It allows countries to propose a review of the protocol only 50 years after its entry into force—which makes the review year 2048. No amendment can enter into force without the ratification, acceptance, or approval of all states that were Antarctic Treaty Consultative Parties at the time the protocol was adopted.

The protocol goes on to state that the ban on mining can be reversed through Article 25 only if an alternative legal regime is in place to govern mining on the continent, and if that regime were to fully safeguard the interests of all states referred to in Article IV of the Antarctic Treaty. In addition, if a modification or amendment emanating from a Review Meeting that is specific to mining does not enter into force within 3 years of the date of its adoption, any party would have the right to withdraw from the protocol—commonly known as the "walkout clause."[137]

In other words, what the Madrid Protocol effectively established was a 50-year moratorium on mining. The ban would remain in place after the 50-year period unless a party to the protocol explicitly called for its review (i.e., it would not automatically expire). The "walkout clause" resulted from a controversy that arose at the June 1991 ATCM meeting in Madrid, when the United States insisted on language that would allow the lifting of the moratorium at a future date. The meeting adopted all the provisions of the protocol except this one, prompting the United States to walk out—the only Antarctic Treaty party to do so. The United States continued to insist on its language, however, so it was eventually included. After that, the protocol was adopted by all parties at the end of 1991.[138]

Clearly, minerals exploitation of the seventh continent has been an extremely contentious issue, with countries struggling to devise a regime that would adequately address the concerns of parties to the Antarctic Treaty and the wider international community. The first attempt at such an agreement—the CRAMRA—collapsed, with the sovereignty issue proving to be a particularly high hurdle. The Madrid Protocol, which emerged from the ashes of CRAMRA, simply kicked the problem 50 years down the road, buying the parties additional time. In the meantime, scientific research stations continue to dot the continent, and the hunt for minerals—and territorial sovereignty in Antarctica—remains well under way.

## NOTES

1. Aant Elzinga, ed., *Changing Trends in Antarctic Research* (Boston: Kluwer Academic Publishers, 1993), 7.

2. Articles II and III of the Antarctic Treaty, Annex I.

3. Mahlon Kennicutt II, "New Frontiers and Future Directions of Antarctic Science," in *Science Diplomacy,* 155.

4. Ibid., 159.

5. That would be the Montreal Protocol on Substances That Deplete the Ozone Layer, which entered into force in 1989.

6. Kennicutt II, "New Frontiers and Future Directions," 155.

7. Prabir G. Dastidar and S. Ramachandran, "Intellectual Structure of Antarctic Science," *Scientometrics* 77:3 (2008): 390.

8. Elzinga, *Changing Trends*, 9.

9. Ibid.

10. David W. H. Walton, "The Scientific Committee on Antarctic Research and the Antarctic Treaty," in *Science Diplomacy,* 81.

11. The *Glomar Challenger* was a deep-sea research and scientific drilling vessel used for oceanography and marine geology studies, contracted to the NSF and the Scripps Institution of Oceanography.

12. Walton, "The Scientific Committee," in *Science Diplomacy,* 81.

13. Elzinga, *Changing Trends*, 9.

14. Article IX of the Antarctic Treaty, Annex I.

15. Elzinga, *Changing Trends*, 22.

16. See the discussion on how CRAMRA broke new ground in including a reference to acts of terrorism in Peter J. Beck, "Convention on the Regulation of Antarctic Mineral Resource Activities: A Major Addition to the Antarctic Treaty System," *Polar Record* 25 (152): 31. Beck writes: "Terrorism has rarely been discussed in an Antarctic context, but this reference represents a realistic appreciation of current possibilities."

17. Walton, "The Scientific Committee," in *Science Diplomacy,* 75.

18. See SCAR website, last accessed March 15, 2016, http://www.scar.org.

19. Walton, "The Scientific Committee," in *Science Diplomacy,* 75.

20. Ibid., 76.

21. Namely, the International Union of Geodesy and Geophysics, the International Geographical Union, the International Union of Biological

Sciences, the International Union of Pure and Applied Physics, and the Union Radio Scientifique Internationale.

22. Walton, "The Scientific Committee," in *Science Diplomacy,* 78.

23. Ibid.

24. Ibid., 75.

25. Ibid., 82.

26. Ibid.

27. Ibid.

28. Ibid.

29. Elzinga, *Changing Trends*, 38.

30. Ibid., 145–146. For a contrasting opinion, see how Jorn Thiede strikes a more charitable note on the role of SCAR and its ability to reinvent itself: "The SCAR has changed greatly over the past 10 years, from an almost closed shop into an open science organization. The introduction of the Open Science Conferences since 2004 has attracted many established as well as numerous young polar scientists. They have provided an important base for the large efforts of the Fourth IPY, which brought more than 50,000 scientists into the polar regions. In its role advising the ATS, SCAR has to strive to provide independent, high-quality advice, and it also has to critically evaluate the methods and ethics of the conduct of Antarctic research. To retain its position as the major scientific advisory body, SCAR has to express informed opinions on Antarctic research, be proactive in flagging important new developments for the ATS, and respond as far as possible to requests from the ATS; the recent publication of a major report on the impact of climate change in the Antarctic is an excellent example of this change in SCAR's attitude. The SCAR has succeeded in regaining the attention of the ATS and can look with confidence into the future in this area of science diplomacy. However, the future is not without dangers."Jorn Thiede, "Modern Research in Polar Regions," in *Science Diplomacy,* 163.

31. See COMNAP's Antarctic Facility List of February 13, 2014; https://www.comnap.aq/Information/SitePages/Home.aspx.

32. The only non–Consultative Party with a physical presence in Antarctica is Romania, which shares the operation and running of one of the Australian stations.

33. Britton, *Extreme Jobs in Extreme Places*, 6–7. Winter temperatures in Antarctica reach −94°F (−70°C), and in summer, the temperature is rarely above freezing. Winds can exceed 100 miles (160 kilometers) per hour.

34. Ibid., 8.

35. See COMNAP's Antarctic Facility List of February 13, 2014.

36. Ibid.

37. Ibid.

38. "Turkish Scientists Head for Maiden Antarctica Mission," *Hurriyet Daily News,* March 29, 2016. The article states: "A crew of Turkish academics and researchers set off on a journey to Antarctica on March 29, for the country's first scientific mission on the continent. The 14-person team, which has two female members and includes academics from Istanbul Technical University (İTÜ), Istanbul University, Erciyes University, Çanakkale Onsekiz Mart University, Marmara University, Kocaeli University, Çukurova University, and the Scientific and Technological Research Council of Turkey (TÜBİTAK), will conduct research in areas such as climate change and marine pollution for three weeks."

39. UN General Assembly, *The Question of Antarctica*, General Assembly Resolution 46/41, December 6, 1991.

40. Elzinga, *Changing Trends*, 119, 130.

41. Alan Hemmings, "Why Did We Get an International Space Station Before an International Antarctic Station?" *Polar Journal* 1:1 (2011): 5–7.

42. Thiede, "Modern Research in Polar Regions," in *Science Diplomacy,* 163.

43. Ibid.

44. Joan Bondareff, "Congress Acts to Protect Antarctica," *Territorial Sea Journal* 223 (1990–1991): 239–240.

45. Ibid.

46. Ibid.

47. Marcus Haward and Tom Griffiths, eds., *Australia and the Antarctic Treaty System: 50 Years of Influence* (Sydney: University of New South Wales Ltd., 2011), 245.

48. Dastidar and Ramachandran, "Intellectual Structure of Antarctic Science," 389–414.

49. Antarctic Treaty Summit 2009, http://www.atsummit50.org/session/index.html.

50. Dastidar and Ramachandran, "Intellectual Structure of Antarctic Science," 396.

51. Ibid., 401–402.

52. Ibid., 409.

53. Ibid., 412.

54. Elzinga, *Changing Trends*, 126.

55. Anthony Bergin and Anthony Press, "Funding Cuts Threaten Our Standing in Antarctica," *The Australian*, June 8, 2013; http://www.theaustralian.com.au/news/inquirer/funding-cuts-threaten-our-standing-in-antarctica/story-e6frg6z6-1226659489671.

56. Karl A. Erb, "International Collaboration in the Antarctic for Global Science," in *Science Diplomacy,* 266.

57. Ibid., 267.

58. Ibid., 268.

59. Ibid., 269.

60. U.S. Congress, Office of Technology Assessment, *Polar Prospects: A Minerals Treaty for Antarctica, OTA-O-428* (Washington, DC: U.S. Government Printing Office, Sept. 1989), 8–9.

61. Ibid.

62. Ibid., 16.

63. Ibid., 8–9.

64. James Hansom and John Gordon, *Antarctic Environment and Resources: A Geographical Perspective* (New York: Addison Wesley Longman Ltd., 1998), 283.

65. G. D. Triggs, "Negotiation of a Minerals Regime," in *The Antarctic Treaty Regime*, 190.

66. Rüdiger Wolfrum, Convention on the Regulation of Antarctic Mineral Resource Activities: An Attempt to Break New Ground (Berlin: Springer-Verlag, 1991), 12, 20.

67. Anthony Bergin, "The Politics of Antarctic Minerals: The Greening of White Australia," *Australian Journal of Political Science* 26 (1991): 220.

68. These are the United Kingdom, Norway, France, and New Zealand, due to the absence of any overlap in their claims.

69. U.S. Congress, Office of Technology Assessment, *Polar Prospects*, 17.

70. Triggs, "Negotiation of a Minerals Regime," 184.

71. UN General Assembly, *Question of Antarctica*, study requested under General Assembly Resolution 38/77, report of the Secretary General, A/39/538, October 31, 1984, para. 151, 157, 271–74.

72. Some called for the application of the principle of the Common Heritage of Mankind to Antarctic minerals.

73. UN General Assembly, *Question of Antarctica*, October 31, 1984, para. 151.

74. Beck, "Regulation of Antarctic Mineral Resource Activities," 19.

75. The full text of CRAMRA can be found at http://sedac.ciesin.org/entri/texts/acrc/cramra.txt.html. All subsequent references to CRAMRA can be found at this source.

76. Beck, "Regulation of Antarctic Mineral Resource Activities," 21. The relevant provision in CRAMRA is Article 1(6).

77. Wolfrum, Regulation of Antarctic Mineral Resource Activities, 97.

78. See Wolfrum's discussion of internal and external accommodation, ibid., 13–15.

79. Vicuna, ed., Antarctic Resources Policy, 196.
80. Ibid., 192.
81. Ibid., 206.
82. Ibid., 196.
83. Ibid.
84. See full text of CRAMRA.
85. Article 1(11) of CRAMRA on "Definitions."
86. Article 7 of CRAMRA on "Compliance with This Convention."
87. Article 8 of CRAMRA on "Response Action and Liability."
88. Article 9 of CRAMRA on "Protection of Legal Positions Under the Antarctic Treaty."
89. Article 61 of CRAMRA on "Ratification, Acceptance, Approval, or Accession."
90. Article 14 of CRAMRA on "Non-Discrimination."
91. Article 16 of CRAMRA on "Availability of Confidential Data and Information."
92. Article 18 of CRAMRA on "Commission" and Article 21 on "Functions of the Commission."
93. Article 22 of CRAMRA on "Decision Making in the Commission"; in particular, see 22(2)(c).
94. Wolfrum, Regulation of Antarctic Mineral Resource Activities, 50.
95. Article 31 of CRAMRA on "Functions of the Regulatory Committee."
96. Article 29 of CRAMRA on "Regulatory Committees."
97. Ibid.
98. Article 32 of CRAMRA on "Decision Making in Regulatory Committees."
99. Article 45(4) of CRAMRA on "Examination of Applications."
100. Article 43(e) of CRAMRA on "Preparatory Work by Regulatory Committees."
101. Article 23 of CRAMRA on "Advisory Committee."
102. Article 26 of CRAMRA on "Functions of the Advisory Committee."
103. Article 37 of CRAMRA on "Prospecting."
104. Article 41(d) of CRAMRA on "Action by the Commission."
105. Article 62 of CRAMRA on "Entry into Force."
106. Bergin, "The Politics of Antarctic Minerals," 222.
107. S. K. N. Blay and B. M. Tsamenyi, "Australia and the Convention for the Regulation of Antarctic Mineral Resource Activities (CRAMRA)," *Polar Record* 26:158 (1990): 198.
108. Ibid.

109. Ibid.

110. See testimony by Dobie, in which the environmental perspective is reflected. *Antarctica Mining Prohibition Bill: Second Reading*, Parliament of Australia, House of Representatives (1991), http://parlinfo.aph.gov.au/parlInfo/search/display/display.w3p;db=CHAMBER;id=chamber%2Fhansardr%2F1991-03-13%2F0008;query=Id%3A%22chamber%2Fhansardr%2F1991-03-13%2F0059%22.

111. Beck, "Regulation of Antarctic Mineral Resource Activities," 25–26.

112. See testimony by Downer, in which the environmental perspective is reflected, *Antarctica Mining Prohibition Bill.*

113. Beck, "Regulation of Antarctic Mineral Resource Activities," 25.

114. Bondareff, "Congress Acts to Protect Antarctica," 233.

115. *Preserving Antarctica's Ecosystem: Hearing on Title II of H.R. 2984, H. CON Res. 287, and H. CON. Res. 227* Before the Subcommittee on Human Rights and International Organizations of the Committee on Foreign Affairs, House of Representatives, 101st Cong. (1990), http://babel.hathitrust.org/cgi/pt?id=pur1.32754074682067;view=1up;seq=1.

116. Blay and Tsamenyi, "Australia and the Convention," 195.

117. Bondareff, "Congress Acts to Protect Antarctica," 228.

118. Bergin, "The Politics of Antarctic Minerals," 226.

119. Haward and Griffiths, *Australia and the Antarctic Treaty System,* 255.

120. See Article 62 of CRAMRA on "Entry into Force."

121. UN General Assembly, *The Question of Antarctica*, Resolution A/RES/41/88, December 4, 1986.

122. Bergin, "The Politics of Antarctic Minerals," 225–26.

123. Ibid.

124. Ibid.

125. U.S. Congress, Office of Technology Assessment, *Polar Prospects*, 14.

126. See how the Office of Technological Assessment reassures Congress that the United States and the Soviet Union would be represented in the Commission and all Regulatory Committees, ibid., 11.

127. F. M. Auburn, "Antarctic Minerals and the Third World," *Journal of Polar Studies* 1:1 (1984): 201–23.

128. Bergin, "The Politics of Antarctic Minerals," 225–226.

129. Ibid., 229.

130. Beck, "Regulation of Antarctic Mineral Resource Activities," 27.

131. Mike Seccombe, "Keating Warns Against Signing Antarctic Treaty," *Sydney Morning Herald*, November 21, 1988.

132. Haward and Griffiths, *Australia and the Antarctic Treaty System,* 256.

133. Full text of the Madrid Protocol in Secretariat of the Antarctic Treaty, *Compilation of Key Documents of the Antarctic Treaty System.*

134. See all Annexes to the Madrid Protocol at Ibid.

135. Article 7 of the Madrid Protocol on "Prohibition of Mineral Resource Activities."

136. Article 25 of the Madrid Protocol on "Modification or Amendment."

137. Ibid.

138. Christopher C. Joyner, *Governing the Frozen Commons: The Antarctic Regime and Environmental Protection* (Columbia: University of South Carolina Press, 1998), 167.

# FIVE

# An Antarctic Economy: Lessons from the Melting Arctic

If the Antarctic continent were to become more accessible, and perhaps even more livable, as a result of climate change, what kind of economy would begin to emerge? A few decades ago, the economic exploitation of the region most often compared to Antarctica—namely, the Arctic—would have seemed unimaginable. Yet, today, the Arctic Ocean and its surrounding territories are increasingly coming into the fold of the global economy. What lessons, then, can we learn from the melting Arctic? More specifically, what can the Arctic tell us about the future of the Antarctic?

## RISING INTEREST IN THE ARCTIC—NEXT, THE ANTARCTIC

In 2004, the Stefansson Arctic Institute in Iceland launched the first Arctic Human Development Report—a report usually associated with the United Nations Development Programme (UNDP). This represented not only a significant and growing interest in the Arctic region, but the beginning of a new era in which the Arctic would be subject to the same metrics for the measurement of its human and economic development as other parts of the globe. A second Arctic Human Development Report was produced 10 years later, in 2014.[1]

Reflective of the growing interest in the Arctic, the World Economic Forum created a Global Agenda Council on the Arctic[2]—an interdisciplinary forum dedicated to promoting innovative thinking and business development in the region. In doing so, the council took the Arctic to the global stage, perhaps for the first time, going beyond the boundaries of customary regional talks among the main Arctic powers.

The trigger for this greater interest at the global level was, without a doubt, the unusual and unexpected pace of the melting of Arctic sea ice. This has significantly reduced the isolation of lands around the Arctic Circle and raised hopes for the Arctic Ocean's greater navigability. During the course of the past two to three decades, in fact, over 41 percent of the Arctic Ocean's hard, multiyear ice has vanished. This decomposition could lead the Arctic Ocean to soon become like the Baltic Sea—in other words, covered by only a thin layer of seasonal ice in winter and potentially navigable year round.[3]

The second Arctic Human Development Report wrote that the rapid retreat and thinning of Arctic sea ice at rates faster than those predicted by the Intergovernmental Panel on Climate Change (IPCC), was among the most dramatic changes to have occurred since the first report.[4] The rise in the mean temperature of the Arctic is now known to be taking place at double the global level.[5] As with the Antarctic, the main drivers of warming and environmental change come from outside the region, mainly from the combustion of fossil fuels by industrial and emerging economies and the emission of greenhouse gases.[6]

While the Arctic is mainly an ocean surrounded by land and Antarctica is a continent surrounded by an ocean, the two regions have much in common, making their comparison relevant. They both lie at the extreme ends of the globe and are geographical opposites. They share some of the Earth's harshest and coldest climate and are both undergoing climatic and environmental change at a much faster pace than the rest of the planet. In Antarctica, these changes are most pronounced on the Antarctic Peninsula. According to the British Antarctic Survey, the peninsula has become the fastest-warming part of the Southern Hemisphere, comparable to the rapidly warming regions of the Arctic.[7]

Finally, their warming will have an impact that goes much beyond their borders. The IPCC explains that changes at the poles will have both local and global implications. The retreat of sea ice and shrinking of ice sheets will cause global sea level rise, changes to ocean salinity, and alter feedback loops that will make both regions warm even faster.[8]

International attention has turned first to the Arctic, mainly because it is much warmer than the Antarctic and is becoming accessible at a faster pace. Antarctica is colder than the Arctic for three main reasons. First, much of the continent is more than 3,000 meters above sea level, with its temperature decreasing with elevation into the troposphere. Second, the Arctic Ocean's relative warmth is transferred through its icepack, preventing temperatures in the Arctic from reaching the extremes typical of the land surface of Antarctica. Third, the Earth is farthest from the Sun in the Antarctic winter, with this orbital distance contributing to a colder Antarctic climate.[9]

The environmental impact of the melting Arctic has been extremely significant. Polar bears have become a threatened species, new species of fish are now migrating north into the warming Arctic Ocean, and Greenland is experiencing a revolution in its farming, with crops such as potatoes and broccoli now being grown. Global warming is also making the region's minerals far more accessible. The freshwater emanating from its melting ice could also become a valuable resource fairly soon.[10]

Attention has turned first to the Arctic for another set of reasons as well: its geographical location and proximity to industrial centers, the fact that the Arctic littoral is the metropolitan territory of states (industrial states) with all the legal and administrative security that this brings, and the fact that minerals exploitation is less complicated and cheaper in the Arctic than it is in the Antarctic for geological and weather reasons.[11]

With respect to the latter, in the Arctic, the continental shelf makes up approximately 50 percent of the ocean floor and lies under shallow water. In contrast, in Antarctica, the continental shelf is generally narrower and occurs in much deeper water. The depth of the water in Antarctica requires a new generation of mining and production systems, as well as new types of floating structures to sustain mineral exploitation. Furthermore, oil and gas extraction in the Arctic depends heavily on pipeline technology, with current extraction mainly transported over land or the seabed by pipelines, as with the Trans-Alaska pipeline.[12]

If and when mining is allowed in Antarctica, its remoteness would necessitate that any minerals that are extracted in the continent be transported from it by ship. With the Southern Ocean being as perilous as it is known to be, this could represent a major hazard and would require a technological breakthrough.[13] In addition, the different weather patterns in Antarctica—the problems of wind, temperature, snow, and the movement of large icebergs—make exploitation more hazardous.[14] The ice in Antarctica is significantly weaker than that found in the Arctic, tending to form much larger and more mobile icebergs that represent a serious complication for navigation in the Southern Ocean.[15]

That said, there is no doubt that Antarctica's turn will come, and global attention will gradually begin to shift to the Earth's seventh continent as it too begins to thaw. Antarctica's attraction will lie not only in its potential offshore mineral wealth, but also in its onshore resources, and as previously discussed, its potential habitability. Its landmass may become as attractive as any oil, gas, or other mineral resource that it may harbor.

## STRUCTURE OF THE ARCTIC ECONOMY: A MIRROR FOR ANTARCTICA?

Five littoral or coastal states surround the Arctic Ocean and include the United States (Alaska), Canada, Russia, Denmark (Greenland), and Norway. Together with another three Arctic nations—Iceland, Sweden, and Finland—they constitute the eight members of the Arctic Council.[16] This is the loose forum created in 1996 for dialogue among the main Arctic powers. Figure 5.1 provides a map of the Arctic.

**Figure 5.1 The Arctic Ocean**

(Joseph Hill)

The size of the Arctic economy often comes as a surprise, being quite significant for a part of the world this remote. According to the most recent Arctic Human Development Report, the Arctic gross domestic product (GDP) in 2010 was estimated to be around USD443 billion, which the World Bank characterized as being equivalent to the combined size of the Malaysian and Colombian economies.[17]

In fact, the Arctic's share of global GDP was approximately four times its share of the world population. In 2010, it contributed 0.6 percent to global GDP while having only 0.15 percent of the world population. In fact, the region's GDP has risen exponentially over the years by about 42 percent from 2000 to 2010.[18] These figures, however, conceal a wide disparity across different Arctic nations, with some far wealthier than others. In fact, Russia represents approximately 71 percent of the Arctic GDP, followed by Alaska at 10.8 percent, Norway at 4.2 percent, Finland at 4.5 percent, Sweden at 4.6 percent, Iceland at 2.5 percent, Canada at 1.6 percent, and Greenland and the Faroe Islands at 0.4 percent.[19] Both geographically and economically, Russia is the Arctic's dominant power, with one-fifth of its landmass falling north of the Arctic Circle.[20]

As opposed to the virtually uninhabited Antarctic, approximately 4 million people lived in the Arctic in 2010, with this population projected to grow to 4.2 million by 2030.[21] Of these 4 million, around 1.5 million are indigenous people from 40 different ethnic groups, such as the Inuit, the Nanai, and the Chukchi.[22] Once again, Russia represents the bulk of the Arctic's population, with approximately 2 million of its citizens living within the Arctic Circle. It is followed by the approximately 700,000 residents of Alaska.[23]

While the size of the Arctic economy has started to be measured only recently, the Arctic is by no means an integrated region. Rather, it is composed of a set of regions that share similar climatic characteristics. The economy of the Arctic is divided between a local and international economy. A relatively small local economy exists to provide goods and services to the residents of the Arctic, encompassing traditional sectors such as fishing, hunting, herding, and gathering.[24]

The international economy, on the other hand, behaves like an entirely separate economy. It serves regional and international markets, providing a wide range of natural products such as oil and natural gas, diamonds, iron, zinc, fish, and timber. The Arctic's international economy largely depends on global demand for natural resources and the cost of production and transportation of resources to markets.[25] China's growing appetite for natural resources has been an important driver of the Arctic's international economy.[26]

The Arctic economy is also underpinned by a range of public services—such as transportation, postal, and health services—which are financed by transfer payments from central to regional governments. The existence of a very large international economy, but only a relatively small local one in the Arctic in financial terms, has meant that today, the Arctic exports enormous amounts of raw material for processing and imports massive amounts of finished products, consumer goods, and services for its residents.[27]

Some of the most sophisticated and mature Arctic economies are beginning to rely less heavily on their natural resource and public service sectors.[28] The Russian Arctic represents one extreme, with over 60 percent of its GDP being derived from natural resources. Iceland is at the other extreme, with approximately 50 percent of its GDP emanating from the services sector, which includes industries such as tourism. In general, the Arctic economies most heavily dependent on natural resources are Russia, North America, and the Danish Arctic, while the least dependent are Iceland, Finland, Norway, and Sweden.[29]

With Antarctica currently virtually uninhabited and closed to minerals exploitation, it is difficult to speculate on the exact form that its economy would take were it to become more accessible. However, one can easily imagine an economy gradually developing along the lines seen in the Arctic. A gradual habitation of Antarctica, as well as the development of its mineral wealth, could lead to the formation of both a local and an international economy.

The local economy may come to be based on industries such as fishing and the provision of services to local residents—postal, communication, education, and other public services. It may also include the provision of services to the operators of the international economy, such as the tourism industry, and logistical, construction, and other services for the mining industry. As in the Arctic, the international economy is likely to be heavily based on extractive resources, at least initially.

## ASIA IN THE ARCTIC—LESSONS FOR ANTARCTICA

Currently, the Arctic is loosely governed by the Arctic Council, a body established in 1996 as a forum for interstate cooperation among the eight Arctic nations. The governance of the Arctic differs markedly from that of the Antarctic, with the latter being much more sophisticated.

The Antarctic is primarily governed by the Antarctic Treaty System (ATS). However, in light of the fact that virtually all lands within the Arctic have their sovereignty clearly defined, it is national laws that prevail. The existence of the United Nations Law of the Sea (UNCLOS) has also meant that

Arctic nations have not felt any particular urgency in developing a specific legal regime applicable to the Arctic Ocean.[30]

The Arctic Council, therefore, has always represented a relatively loose coordination mechanism among the eight Arctic nations and was never vested with formal decision-making powers. As stated in the 2014 Arctic Human Development Report, the council uses "moral political suasion" rather than legislative authority to achieve its objectives.[31] Nevertheless, it is considered a particularly important forum, in that it not only includes the eight Arctic states as "Members," but also various indigenous people's groups as "Permanent Participants."[32] This allows these groups to express themselves and to articulate their interests before the key decision makers of the region.

While much of the polar governance debate among policymakers and academics has revolved around the replicability of the ATS in the Arctic,[33] there are nevertheless lessons to be drawn for Antarctica from some of the recent developments in the north. With the opening up of the Arctic and the gradual melting of its sea ice, a number of Asian states have started to take a keen interest in Arctic governance arrangements. This culminated in a number of them formally requesting observer status in the Arctic Council, and receiving it in the May 2013 meeting in Kiruna, Sweden. They included China, India, Japan, Singapore, and South Korea.[34]

Today, the largest Asian states all have research stations in Norway's Arctic archipelago.[35] As in Antarctica, the establishment of research outposts has not been exclusively driven by a scientific research agenda, but also by a desire to send a message that the country in question wants some sort of foothold in the region and a say in its governance. For example, when announcing a new Arctic research center in Cambridge Bay, Nunavut, at the end of 2012, Prime Minister Stephen Harper stated: "The Canadian High Arctic Research Station will also enhance Canada's visible presence in the Arctic. In this way, science and sovereignty are entwined."[36]

Asian interests in the Arctic are multiple, ranging from interest in all the new and emerging shipping routes that could significantly shorten the distances traveled by their merchandise across the globe, to interest in the growth of the shipping industry in the Arctic and the more accessible mineral wealth of the region.[37]

Asian countries are now among the world's top shipbuilding nations. In fact, 12 of the 20 largest liner shipping companies in the world are Asian.[38] China is one of the leading Asian shipbuilding nations, with its yards capable of constructing a range of different vessels. It owns the *Xuelong* icebreaker, which was built in Ukraine in 1993. It is building a second icebreaker, in which Chinese and other partners are involved: the Chinese Arctic and Antarctic Administration, the Polar Research Institute of China, and Finland's

Aker Arctic Technology, Inc. In addition, various Chinese shipyards, such as the Shanghai Shipyard and the Hudong-Zhonghua Shipbuilding Co. Ltd., are now constructing ice-class ships.[39]

Similarly, South Korea is another major shipbuilding nation, with three of its shipyards being the world's largest. They include Hyundai Heavy Industries, Samsung Heavy Industries, and Daewoo Shipbuilding and Marine Engineering. Currently, icebreaking transporters, Arctic drillships, and bidirectional tankers are being constructed in these yards. The latter is a new concept of ship that is designed to break ice to transport not only passengers, but also cargo. Japanese shipyards are equally renowned and have received an order for a polar ice-breaking bulk carrier.[40]

The negotiating history of the permanent observerships granted in the Arctic Council demonstrates that several Arctic players were particularly reluctant to allow Asia through the door, but finally opted to do so when the rules on observership were tightened and when they took a different view of their economic self-interest. Several Arctic powers have found Chinese capital in particular to be necessary for the development of the mineral wealth of the region and its new sea routes.

Before the May 2013 meeting of the Arctic Council that granted observerships to the aforementioned Asian countries, Russia's position had been to oppose China's entry.[41] It had long taken the stance that the main decision makers in the Arctic ought to be the five Arctic powers with territories within the Arctic Ocean (i.e., the littoral or coastal states). On this basis, it had even called for Finland, Sweden, and Iceland to play a lesser role. It was, of course, eventually overruled with respect to the latter point, with these three countries now forming part of the Arctic Council, but it wanted to at least put the brakes on China's growing interest in the region. The following U.S. diplomatic cable, which was posted on WikiLeaks, describes the Russian position:[42]

> Seeing the Arctic as its sphere of influence, Moscow favored a more exclusive and relevant Arctic Council reigned over by the Arctic Five, with a limited number of observers. Having numerous and large non-Arctic actors as observers, including China and the EU, was seen as possibly altering the balance of power and drawing unwelcome attention to the Arctic.

On China's observership, Russia changed its mind after an adjustment was introduced to the rules on observership in the Arctic Council. These made clear that applicants must respect the sovereign rights of the Arctic states and UNCLOS, and that their status would be reviewed every four years. Subsequent to these clarifications, the Russian scholar Lev Voronkov wrote: "The change of rules was decisive, and there became no danger in accepting the applicants. Only the Arctic eight take decisions, and there is no space for other states and international organizations to be involved. It should be called temporary rather than permanent observers, as memberships will be reviewed every fourth year."[43]

It appears that the United States itself was also wary of China's growing involvement in the Arctic at a certain stage. In fact, it was only in 2014 that the United States announced its plan to create the position of a U.S. special representative for the Arctic, devoting greater attention to this region, in part due to fear of being outcompeted.[44] In 2015, the United States took over as chair of the Arctic Council and was eager to affirm its interests in the region, despite being the only Arctic nonsignatory to UNCLOS. The United States is, in fact, known to have been the main Arctic player involved in trimming the powers of the Arctic Council, or as Borgenson (2008) states, "emasculating it at birth," by preventing it from addressing broader security interests.[45] It wanted the main Arctic powers to be the principal decision makers of the region—hence its hesitation when China applied for formal observership.

Concerns by Arctic states over China's growing role, as well as the growing interest of many other non-Arctic states, also explain why a broader Arctic governance regime was never developed. Charles Emmerson (2011) writes that in the Arctic, "different legal regimes apply to the land, the sea, and the seabed. . . . But none of the coastal states in the Arctic . . . wants to open Pandora's box which would result from trying to negotiate some new overarching deal for the Arctic. To do so would invite non-Arctic states to muscle in."[46]

In Antarctica, a large number of Asian powers are currently either Consultative Parties or non–Consultative Parties.[47] Of the five Asian countries that applied for observership in the Arctic Council, China, India, Japan, and South Korea are already Consultative Parties. Singapore, on the other hand, is neither a Consultative Party nor a non–Consultative Party. In fact, the Antarctic Treaty has embraced an even larger number of Asian players than has the Arctic Council, with Kazakhstan, North Korea, Malaysia, and Pakistan already enjoying non–Consultative Party status.

The lesson to be learned from the resource opening of the Arctic is the growing and significant role that these Asian powers will want to play as soon as the Antarctic continent opens up, with all the intra-Asian politics that this will bring. They will want to be as present and engaged as they currently are in the Arctic. Their presence in the Antarctic Treaty in such large numbers is already a strong signal in this direction.

But as Asian countries have started to assert their presence in the Arctic, another development has also been under way. Many Arctic powers have themselves invited Asian capital into the Arctic to assist with the development of the region's mineral wealth and sea routes. In addition, following its 2014 annexation of Crimea, Ukraine, Russia started looking beyond the United States and Europe for investment and partnership.

In fact, the largest Arctic deal involving an Asian company is the 2013 oil-for-cash agreement signed between Russia's largest oil company, Rosneft, and the state-owned China National Petroleum Corporation, an agreement expected to double the level of oil trade between the two countries.[48] In May 2014, Russia and China also signed a $400 billion agreement to supply 3.7 billion cubic feet of gas per day from Siberia to China for the next three decades, starting in 2018.[49]

In the 2013 Asia Pacific Economic Cooperation (APEC) Summit, Russian president Vladimir Putin specifically invited Asian countries to invest in the Arctic to help develop its minerals, as well as the thawing Northern Sea Route (NSR). In a press conference following the summit, he referred to the importance of "interconnectedness"—one of the summit's themes—and the role that the NSR could play in achieving this goal.[50] He also referred to bilateral meetings held with various Asian players, in which it is known that he requested assistance to develop the NSR.[51]

Russia was, of course, not alone in trying to attract Chinese investment capital into the Arctic. Iceland has publicly expressed its interest in developing stronger relations with China in resource development and scientific research. It granted the China National Offshore Oil Corporation a license to carry out oil and gas exploration in Iceland's Dreki region. Greenland has taken a similar position, with the country's minister of industry and mining visiting China in November 2013, and with the representatives of 10 major Chinese companies visiting Greenland the same year to explore investment opportunities. Iceland has also found a reliable partner in China, which came to its rescue in 2008 when its banking sector was on the edge of collapse, and which since then has signed a Free Trade Agreement (FTA) with Beijing in April 2013. It is eager to attract greater Chinese investment through this FTA.[52]

Needless to say, China has not been the only Asian investor in the Arctic. Other Asian powers have also been active, with Arctic players fully embracing the influx of their capital. Japanese and Korean companies are also actively engaged in exploration activities on the Norwegian continental shelf.[53] The demand and thirst for Asian capital is likely to find resonance in the Antarctic as well. While there will be mixed feelings about Asia's economic interests in the seventh continent, particularly in light of the fact that no Asian country has a territorial claim as yet, the capital and investment that it is likely to be bring will undoubtedly make it a strong force.

## A CONTESTED ARCTIC—LESSONS FOR ANTARCTICA

According to the United States Geological Survey (USGS), the Arctic contains approximately a quarter of the world's undiscovered oil and gas.[54] On September 7, 2015, however, the *International New York Times* published an article entitled "Melting Ice Isn't Opening Arctic to Oil Bonanza," arguing that the discovery of shale gas, and the resulting plummet in the price of oil, have made offshore oil exploration in the Arctic relatively unattractive.[55] The article quoted Scott D. Sheffield, the chief executive officer (CEO) of Pioneer Natural Resources, a Texas-based oil company, as saying, "The entire cost structure up there is three to five times more expensive than onshore" at lower latitudes. While there is significant controversy at the moment as to how accessible the offshore mineral resources of the Arctic are, there is no doubt that the competition for the contested Arctic seabed is in full swing.

In 2007, Russian parliamentarian and explorer Artur Chilingarov planted a Russian flag at the bottom of the Arctic Ocean, under the North Pole. The Russian government, as part of the third International Polar Year (IPY), had supported his expensive expedition to the Arctic.[56] While the move was widely seen as a publicity stunt, it reflected Putin's strong interest in the Arctic and his desire to assert control over its vast resources.

After all, it was in June 1997 that a senior government official pursuing a doctoral degree at the St. Petersburg Mining Institute submitted an 200-plus page dissertation on the role of mineral resources in the Russian economy.[57] This student, who was none other than Vladimir Putin, pushed for the development of the Russian Arctic at breakneck speed once he became president, calling for the construction of ports, airports, and railways to increase its connectivity.[58]

Since, under UNCLOS, land is the basis of claims to all maritime zones, it comes as no surprise that much of the Arctic Ocean, around which several continents converge, has become the setting for many overlapping claims.[59] Russia was the first country in 2001 to table a submission before the Commission on the Limits of the Continental Shelf (CLCS), requesting an extension of its shelf beyond the 200-nautical-mile zone. However, in 2002, the commission responded that further information was required for a decision to be made.[60]

Russia was sent back to the drawing board, and after years of extensive research, it submitted a new and expanded claim in August 2015, including ownership of the North Pole. According to the *Barents Observer*:[61]

> If approved, the Russian claim will expand the country's territory by 1.2 million square kilometers. Estimates indicate that the area includes 594 oil fields and 159 gas fields, as well as two major nickel fields and more than 350 gold deposits. Initial recoverable fuel resources are estimated to 258 billion tons of fuel equivalent, representing 60 percent of Russia's total hydrocarbon resources.

Norway, which also had requested the extension of its continental shelf in 2006, was the first country to find its claim approved in 2009. Canada followed by making a partial submission in the area, as well as in the Atlantic Ocean in 2013, indicating its desire to table a more complete submission with respect to the Arctic Ocean at a later stage. Denmark, on behalf of Greenland, tabled partial submissions for the extension of the Greenland shelf,[62] with its latest claim made in 2014. This claim also includes ownership of the North Pole, putting it in conflict with Russia.[63] Of course, as discussed in previous chapters, the CLCS is not authorized to resolve overlapping claims.

The United States, a nonparty to UNCLOS, has been left behind, unable to table any formal request. This prompted Secretary of State Hillary Clinton, in her testimony before the Senate Committee on Foreign Relations in May 2012, to urge the Senate to ratify UNCLOS. She set out the economic benefits that would accrue to the United States from doing so in the Arctic. This sentiment was echoed in the U.S. 2013 National Arctic Strategy, which stated that a "continental shelf claim in the Arctic region could extend more than 600 nautical miles from the north coast of Alaska."[64]

The scramble over the Arctic seabed is no different from the scramble in the Antarctic. The big difference, however, is that Arctic powers want the quickest possible decision to be taken on their claims so they can proceed with minerals prospecting and exploitation. In the Antarctic, on the other hand, many have requested that the commission not rule on their claims so as not to prejudice Article IV of the Antarctic Treaty, with mining in the continent being banned in any case.[65] While Russia has tabled the biggest continental shelf claim in the Arctic so far, its claim is overshadowed by Australia's in the Antarctic. Australia's continental shelf claim in Antarctica is the size of the entire Arctic Ocean.[66]

While many decisions are currently pending before the CLCS with respect to the Arctic seabed, it is clear that not all disputes will be resolvable under UNCLOS. As Emmerson states, "UNCLOS does not solve every problem in maritime law. Maritime borders, for example, can still be drawn in different ways, and it is up to adjacent states to determine which borderline they are willing to accept. That applies as much to the seabed as to the sea itself."[67] As will be shown in the next discussion, nowhere is this more visible, in fact, than in the Arctic.

According to the Arctic Human Development Report, a large number of maritime delimitation disputes exist in the Arctic Ocean, some of which have been resolved only after decades of disagreement, and some of which are still pending. The report lists a total of 13 disputes covering different bodies of water ranging from the Beaufort Sea to the Lincoln and the Bering/Chukchi Seas.[68]

Among the most prominent maritime disputes are the conflict between the United States and Russia over an area of the Bering Sea (although both countries signed an agreement to settle the matter in 1990, Russia has not yet ratified it); the competing Norwegian and Russian claims over part of the Barents Sea, which is particularly rich in hydrocarbon resources (this conflict would be amicably settled after four decades); and the competing Danish and Canadian claims over Hans Island, which is located between Greenland and Canada. While the island itself is of unknown mineral potential, its proximity to the Northwest Passage has made it attractive.[69]

That said, the most worrisome dispute at the moment revolves around the control of the Northwest Passage itself. As Dadwal explains, Canada claims the passage as an "inland waterway," arguing that it falls under Canadian sovereignty and should be subject to Ottawa's surveillance, regulation, and control. The United States, the European Union (EU), and many others assert that the passage needs to be treated as "international waters"—a strait connecting two high seas.[70]

To understand the full significance of this dispute, it is important to appreciate what the Northwest Passage represents. It is one of two main sea routes opening up within the Arctic Ocean, the other being the NSR along the Russian and Norwegian coastlines.[71] The passage connects the northern Atlantic and Pacific oceans through the Arctic Ocean, and it was ice free for the first time in 2007. It has the potential to reduce navigation distance significantly relative to travel through the Suez Canal. For example, a trip from London to Yokohama would be around 15,700 kilometers through the Northwest Passage, versus the current 21,200-kilometer trip through the Suez Canal.[72] This dispute would, in fact, be no different than having the ownership of the Suez or Panama Canal being contested.

There are lessons for Antarctica in all of this, with the situation on the seventh continent being far more complex than in the north. In Antarctica, sovereignty disputes are likely to arise over not only water, but also land. All the problems currently being faced in the Arctic are likely to occur in the Antarctic, but on a much greater scale. In Antarctica, sovereignty over land will eventually come to determine sovereignty over water. However, as seen in the Arctic, maritime delimitation disputes can take decades to resolve despite the existence of UNCLOS, and subsequent conflicts could then arise over the management of key waterways. The U.S. absence from UNCLOS will be a complicating factor, whether in the Arctic or in the Antarctic.

Of particular worry at present is the rising militarization of the Arctic. The region is now seeing an increase in military deployment by both Russia[73] and the United States.[74] On February 24, Russian defense minister Sergey Shoigu explained why the Arctic was of paramount importance: "A broad spectrum of potential challenges and threats to our national security is now being formed in the Arctic. Therefore, one of the defense ministry's priorities is to develop military infrastructure in this zone."[75] And, while the Antarctic Treaty provides that the continent shall be used for peaceful purposes only,[76] its greater accessibility could quickly change matters.

Of note is the evolving conflict right at the doorstep of Antarctica. While many think of the 1982 Falklands War as being a past event, the sovereignty of the Falkland Islands continues to be disputed. In fact, in June 2014, the UN Committee on Decolonization approved a new resolution calling on Britain and Argentina to negotiate a solution to their dispute over the islands, essentially supporting Argentina's stance in this long-standing conflict.[77] Meanwhile, with the start of oil drilling operations in the Falklands, an Argentinian judge ordered in June 2015 the seizure of oil company assets.[78] In response, Britain has reinforced its military garrison on the islands.[79] Can Antarctica remain immune to such developments?

## THE ARCTIC ECONOMIC COUNCIL: A MODEL FOR ANTARCTICA

In September 2014, under Canada's chairmanship, the Arctic Council created the Arctic Economic Council (AEC), an independent Arctic-to-Arctic business forum designed to promote business opportunities, trade, and investment.[80] The AEC consists of 42 business representatives appointed by the eight Arctic states and six indigenous organizations. They represent global industries such as mining and shipping, and also more local ones such as reindeer herding. The initiative was born from the opening up of the Arctic and the desire of the Arctic governments to begin to engage economic actors in a systematic way.

In January 2014, the Global Agenda Council on the Arctic of the World Economic Forum issued a publication called *Demystifying the Arctic,* in which it identified the principal challenges facing the region. The first was the need to protect the Arctic's environment—an area in which Antarctica is arguably more advanced through the Madrid Protocol and Conservation of Antarctic Marine Living Resources (CCAMLR). The second was to lure investment to the Arctic, which is in desperate need of infrastructure. It wrote, "A critical deficiency [in the Arctic] and area of great strategic importance is the development of infrastructure projects and logistical hubs. Except for certain areas of Norway and the western Russian Federation, the region remains vastly underserved by transportation, ports, and other critical infrastructure."[81]

The AEC's mandate in many ways responds to the two abovementioned challenges, describing its main goal as sustainable business development. Through the gradual establishment of Working Groups on infrastructure, communications, aviation, energy, mining, tourism, fishing, and human resources, the AEC hopes to begin to lay the foundation for a stronger Arctic economy. It wishes to see greater market connectedness between the Arctic states, rather than the current fragmentation.

To a large extent, this initiative sheds light on what may need to happen in Antarctica for its economic development to take off. Businesses will need to be pulled in to provide ideas and signal to governments the kinds of changes in the regulatory environment that they will need in order to make investments. In Antarctica, the first order of business will be the establishment of property rights, which by definition would require an answer to the sovereignty question. There is no doubt that the development of critical infrastructure would also be among the top priorities for Antarctica, were its economic development to be given a green light.

## ACCESSING THE POLES: WHO BREAKS THE ICE?

It is impossible to speak of the creation or growth of an Arctic or Antarctic economy in the absence of a discussion of icebreaking fleets (i.e., in the absence of information on the ability of countries to literally pierce the ice to access these last frontiers). Sea ice is a defining characteristic of polar oceans; it occupies approximately 7 percent of the world's ocean surface—undergoing significant seasonal change on an annual basis—and constitutes the bulk of the world's reservoir of freshwater.[82] In winter, the Arctic ice pack grows to the size of the entire United States, whereas Antarctica nearly doubles in size. Antarctica's ice volume—continent and ocean included—is so large that it accounts for 90 percent of the world's ice and 70 percent of its freshwater.

Although the ice conditions across the Arctic and the Antarctic vary, experts state that these differences do not necessarily drive or affect icebreaker design.[83] Nuclear-powered icebreakers—which only the Russians build—are far more powerful than their diesel counterparts. They have a much broader range and can operate for several years without refueling. Some argue that this makes nuclear power central to the conquest of the polar regions.[84]

Questions surrounding the ownership of icebreaking fleets, the right size for a country's fleet, and the required investments to maintain the fleet tend to be debated regularly in countries with an interest in the poles. Most countries realize that it is difficult to have a meaningful presence in the polar regions in the complete absence of icebreakers. In fact, there are a broad range of reasons why countries wish to maintain sizeable fleets.

The first is to reconfirm, through their naval presence, their territorial and maritime claims in the poles. The second is to actually conduct research on board these vessels, and to establish the extent of their continental shelf claims. The third is to ensure the security of their citizens in the Arctic (if they have any) and of their research stations, as well as to service these stations, resupplying them with fuel and cargo. The fourth is to protect and pursue their commercial interests, such as defending maritime transit routes in the Arctic, or fishing, oil drilling, or other operations that they may have. Finally, the fifth is to bolster search-and-rescue operations in the poles in the event of emergencies.

A number of incidents of various sorts in the Arctic have reconfirmed the importance of having not only rules to govern frozen waters, but an actual physical presence to monitor developments in these remote regions. Satellite surveillance, of course, is also key. Between 1958 and 1992, for example, Russia dumped 18 nuclear reactors into the Arctic Ocean, with a large number fully loaded with nuclear fuel and requiring cleanup.[85]

Currently, oil and gas companies are becoming the main drivers of innovation in polar shipping technology. Borgenson points to the new types of ships that are emerging, "such as double-acting tankers, which can steam bow first through open water and then turn around and proceed stern first to smash through ice. These new ships can sail unhindered to the Arctic's burgeoning oil and gas fields without the aid of icebreakers." These breakthroughs, he argues, are turning what were once commercially unviable projects into booming businesses, revolutionizing the Arctic.[86] But, for now, icebreakers remain essential.

So who has icebreakers, then, and who (at least technologically speaking) is closest to conquering the poles? The world fleet of icebreakers with greater than 10,000 hp contains 50 ships.[87] Russia has the largest fleet, consisting of 14 government-owned icebreakers, with 3 more under construction as of the end of 2014.[88] Finland, Canada, and Sweden each operate six to seven icebreakers. The United States has only 4 ships, and six other countries, such as Japan and Germany, have 1 to 3.[89] China has one modern icebreaker, as well as a second one that is due to be launched in 2016.[90] South Korea has built an icebreaker to conduct research in both polar regions and to provide logistical support to other nations.[91]

Much has been written about how underresourced the United States is relative to other nations with respect to its icebreaking fleet. In a widely read article published in *Foreign Affairs* in 2008, entitled "Arctic Meltdown," Borgenson argued that the United States was losing competitiveness in the Arctic. Despite funding a navy as large as the next 17 countries in the world combined, of its four icebreakers, only one multipurpose breaker remained seaworthy.[92] In fact, the construction of a new icebreaker can take up to 10 years, with U.S. shipyards not having built an icebreaking vessel in nearly 40 years. According to the U.S. Coast Guard, a new icebreaker can cost up to $1 billion.[93]

Looking at the specifics, the U.S. icebreaking fleet currently consists of two multimission icebreakers, the *Polar Star* and the *Polar Sea,* and two single-mission, scientific research breakers, the *Healy* and the *Palmer.* Of the multimission ships, only the *Polar Star* remains operational, performing missions in both polar areas, including the resupply of the McMurdo Antarctic research station. The *Healy* is primarily used for research in the Arctic, and the *Palmer,* which is operated by the National Science Foundation (NSF), is used for research in Antarctica.[94]

Lack of funds has meant that the United States has had to defer the maintenance of its icebreaking fleet. In fact, both of its multimission breakers had reached the end of their service lives, with Congress agreeing only to the repairs of the *Polar Star.* Following its provision of $57 million for its repair, that ship's service life was extended to 2020. After this date, the United States would no longer have any heavy icebreaking capability.[95] Many in the U.S. Coast Guard, as well as numerous policymakers, have lamented this situation, pointing in particular to the need to service the difficult-to-access McMurdo Station. Studies continue on how to expand McMurdo's fuel reserve, allowing the station to continue operating in the event that annual fuel shipments are missed, and to have itself resupply outgoing ships.[96]

As can be seen from this information, heavy investment is required to physically access the polar regions and to defend a broad range of foreign policy, commercial, and other interests. The United States is falling behind Russia, and overall, few countries have a large icebreaking fleet. Asian countries, especially China, are now beginning to build their own fleets. The continued growth of the Arctic economy, as well as the eventual takeoff of an Antarctic economy, will require sustained and very significant investment in shipping fleets.

## GATEWAYS TO ANTARCTICA: ITS CLOSEST CITIES

Five cities are considered "gateways" to Antarctica: clockwise around the continent, they are Cape Town (South Africa), Christchurch (New Zealand), Hobart (Australia), Punta Arenas (Chile), and Ushuaia (Argentina).[97] Understanding how these port and airport cities operate and the kinds of economic activities that they perform in relation to Antarctica provides a glimpse of the sorts of operations that may gradually shift onto the continent itself, were its economy to pick up. In addition, the main industrial activities in Antarctica at the moment—tourism and fishing—provide insight into at least two areas of potential future growth.

The gateway cities to Antarctica offer, and compete over the provision of, a range of different services. These include services to the growing tourism industry, as well as various support and logistical services to the national research programs operating on the continent itself. While to a large extent, the services they have specialized in have been determined by their relative proximity to the continent, other factors have also been in play, such as the quality of their infrastructure.

The closest gateways to the continent are those of Ushuaia, 1,100 kilometers from the closest point in Antarctica, and Punta Arenas, at 1,371 kilometers. It is no surprise, therefore, that most tourist activity in Antarctica takes one of these two ports as its starting point. Next come Hobart, at 2,609 kilometers; Christchurch, at 2,852 kilometers, and Cape Town, at 3,811 kilometers.[98] With the exception of the latter, all gateway cities belong to countries that have made a territorial claim on the continent.

Competition is fierce among these cities over the servicing of national research programs. For example, with the growth of the Korean polar science program, led by the Korean Polar Research Institute, and Korea's construction of a research base on the continent, several ports have vied to act as its logistical hub. Much of the competition between them will depend on who is able to supply the Korean icebreaker—the *Araon*—with diesel.[99] Recently, China signed a five-year agreement with the Australian government to allow Chinese vessels to be resupplied with food and fuel before heading to Antarctica, and it will gradually be expanded to cover aircraft.[100]

Cape Town is playing an increasingly important role as a gateway city. The growth of the South African National Antarctic Program (SANAP), with its management office in Cape Town, has lent the city credibility as an Antarctic research and logistical hub. In its facilities, SANAP offers research space and communication services to the Russian, Swedish, and Belgian Antarctic research programs. Cape Town's international airport also has recently attracted Antarctic expeditions, with a consortium of 11 national programs operating in the Droning Maud Land forming an air network to facilitate intercontinental travel between Africa and East Antarctica.[101]

Lying at a pretty far distance from the continent, Cape Town is unlikely to become a tourist hub for Antarctic travel, particularly so long as travel relies on shipping, as it currently does. However, it has the potential for developing ship repair and cargo handling and storage facilities, given the strength of its existing shipyards and ports. It may also come to supply consumer goods, such as fresh produce, electrical appliances, and vehicles to meet Antarctica's growing needs.[102]

Christchurch has established itself as an important hub for Antarctic activities through its close collaboration with the U.S. research program. In 2011, 104 flights were scheduled between Christchurch and the McMurdo station. The United States Antarctic Program (USAP) also uses Christchurch as a refueling point for its icebreaker. Another program that benefits from its logistical services is the Italian one—Programma Nazionale di Ricerche in Antartide.

The New Zealand, U.S., and Italian programs are all housed in a working campus located near the Christchurch Airport, which was built in 1991 to provide them with administration, communication, and cargo storage facilities, among other services. A range of educational degrees on the Antarctic is offered by various establishments in Christchurch, where a number of Antarctic museums also operate.[103]

Hobart supports Antarctic-related activities through a deepwater port, storage facilities, and the Tasmanian Polar Network, a business platform supporting the Antarctic economy. It provides goods, such as polar clothing and equipment, and a range of logistical, educational, and research services. Various national research programs operate from Hobart, such the French National Program. Hobart is also being promoted as an ideal meeting and conference location on Antarctic matters. For instance, it hosts an Antarctic Midwinter Festival, which aims to promote the city's Antarctic identity and to attract tourism and trade.[104]

As Antarctica's closest ports, Punta Arenas and Ushuaia have been in direct competition. With the heavy concentration of research stations on the Antarctic Peninsula, it is easy to understand how both cities have attracted interest as logistical hubs. However, due to its superior infrastructure, Punta Arenas has acted as the main hub for Antarctic research programs, with 15 countries choosing it as their base. Many benefit from the air link that is run by the Chilean air force between Arenas and the Chilean Frei Station on King George Island.[105]

In fact, Ushuaia is so unequipped to host research programs that the Argentinian government itself runs its Antarctic program out of other cities. However, 90 percent of cruise tourism to Antarctica is initiated from Ushuaia, since Punta Arenas would involve an extra day of travel.[106] It is unlikely that any other gateway city to Antarctica would be able to match Ushuaia in this regard. For cruise tourism, distance is paramount.

With the Antarctic continent gradually becoming more accessible, the role played by the five gateway cities is certainly likely to grow. In fact, some of the activities currently performed by these cities will be increasingly likely to find a parallel on the continent itself, with tourism, educational, shipping, and other services set to expand. The gateway cities may themselves act as triggers for the birth of an Antarctic economy.

## GATEWAYS TO ANTARCTICA: EXISTING INDUSTRIES

Tourism and fisheries currently constitute the bulk of the Antarctic economy.[107] Tourism, in particular, has undergone exponential growth in recent years, with visitors to the continent coming mostly by ship. In 2013–2014, nearly 28,000 tourists made landings on the continent, 30 percent of whom were American, 13 percent Australian, and 11 percent Chinese. These figures have doubled since 2000, reflecting Asia's growing interest in the continent.[108] Being the most accessible part of the continent, the Antarctic Peninsula has been the main destination.[109]

In general, the tourism market in Antarctica can be divided into five segments: the mass tourism market; the sport, fishing, and hunting market; the ecotourism market; the adventure tourism market; and the cultural heritage market. Visitors to Antarctica with an interest in the latter tend to head to Antarctic Heritage Sites because of their historical significance and ties to pioneering explorers and settlers.[110]

In 1966, Lars Eric Lindblad pioneered an educational cruise travel model to Antarctica, where guests would receive a series of lectures onboard about the environment and history of the continent and would be accompanied by wildlife and other experts throughout their expedition. This model is still in force today and has been replicated in other remote parts of the planet as well.[111]

The growth of the tourism industry has certainly not been without risk. In 2007, for example, a ship called the *Explorer* sank off the Antarctic Peninsula, but all the passengers were rescued. In 2009, a Russian icebreaker was stuck in ice near Antarctica with 101 passengers, 23 staff, and 60 crew on board. Their return to Ushuaia was delayed by eight days.[112] There also has been a fear that the increase in tourism and landings on the continent would come to threaten its pristine environment. Environmentalists have argued that it would also lead to growing pressure for the establishment of wharfs, airstrips, hotels, and other tourism facilities onshore.[113]

Environmental issues were not high on the agenda when the Antarctic Treaty when it was adopted, but today tourism is better regulated. UNEP explains:[114]

> The first tourism-specific regulations were adopted by the Antarctic Treaty as early as 1966 with the emergence of commercial tourism activities. These regulations were setting rules for visits by tourist groups to Antarctic stations maintained by the Parties. In 1975 a Statement of Accepted Practices was adopted. This was modified and expanded several times and resulted in the Tourism Guidelines adopted in 1994 by the Parties to the Antarctic Treaty. Aside from establishing guidelines for tourist expeditions, the Parties to the Antarctic Treaty also require tourist expeditions to submit reports on their visits.

In addition, an industry association known as the International Association of Antarctica Tour Operators (IAATO) formed back in 1991, and it has established a series of guidelines for tourism on the continent. Today, the vast majority of Antarctic tour operators have joined IAATO and abide by its rules.[115] This does not mean, however, that tourism is now risk-free in Antarctica. There is no doubt that the growth of this industry will continue to require careful monitoring and regulation.

In addition, the fishing industry has been growing in Antarctica, with four types of fish being the main object of commercial exploitation: Antarctic krill, mackerel icefish, the Antarctic toothfish, and the Patagonian toothfish. According to Hemmings (2015):[116]

> Around the entire Antarctic and subantarctic, contemporary annual returns from fisheries are now substantial. Within the area managed by the 1980 CCAMLR, we are now talking about a commercial value of approximately USD640 million a year, based upon a doubling of wholesale price per tonne calculated for the commercial catch in 2009/2010.

With many species depending on krill for their survival, and the relative shortness of the Antarctic food chain, much concern has arisen over the commercial exploitation of krill.[117] Growing pharmaceutical interest in the crustacean took place when the omega-3 properties of krill oil were first discovered. However, the harshness of fishing conditions in the Antarctic, together with measures put in place by CCAMLR, has come to slow the exploitation of the species. Krill catch now remains relatively stable, with krill being the world's largest underexploited fish stock.[118] Despite CCAMLAR's existence, however, much concern persists over illegal, underreported, and unregulated (IUU) fishing in the Southern Ocean, and care would need to be exercised if this industry continues to grow.[119]

## FINAL OBSERVATIONS

While the size of the Antarctic economy remains small, developments in the Arctic demonstrate that with melting ice, bigger economies begin to form. Still nascent, the Arctic economy shows how growth is by definition incremental. With a local and an international economy in the Arctic currently developing in separate spheres, growing interest in the Arctic, more active engagement by international community, and sustained business interest can open the way to the kinds of investments and technological advances that will be required to make this an integrated and thriving economic space. In the Arctic, Asian capital is now flowing in, and a contest is clearly under way to settle all unresolved territorial and maritime claims.

Tourism and fishing are two of the main industries that currently exist in Antarctica, with some amount of business revolving the servicing of research stations. However, were countries to open up Antarctica to mining come 2048—the Madrid Protocol's review date for Article 7, which bans mining—this could be the beginning of a new conversation. The sovereignty question will once again become central to who can exploit Antarctica, where, and how. The growing militarization of the Arctic should come as a warning for the Antarctic.

## NOTES

1. Stefansson Arctic Institute, Arctic Human Development Report.
2. See the Global Agenda Council on the Arctic; last accessed March 15, 2016, http://www.weforum.org/content/global-agenda-council-arctic-2014-2016-0.
3. Scott G. Borgenson, "Arctic Meltdown: The Economic and Security Implications of Global Warming," *Foreign Affairs* 87:2 (2008): 66.
4. Stefansson Arctic Institute, Arctic Human Development Report, 260.
5. Olav Schram Stokke, "Asian Stakes and Arctic Governance," in *Arctic: Commerce, Governance, and Policy*, Uttam Kumar Sinha and Jo Inge Bekkevold, eds. (London: Routledge, 2015), 11.
6. Ibid., 12.

7. Ari Phillips, "Antarctica Recorded Its Hottest Temperature on Record This Week," *Climate Progress,* March 28, 2015; http://thinkprogress.org/climate/2015/03/28/3640232/antarctica-breaks-thermostat.

8. Union of Concerned Scientists, "Climate Hot Map: Global Warming Effects Around the World"; last accessed March 15, 2016, http://www.climatehotmap.org/global-warming-solutions/polar-regions.html.

9. National Aeronautics and Space Administration (NASA), "What Is Antarctica?" http://www.nasa.gov/audience/forstudents/5-8/features/nasa-knows/what-is-antarctica-58.html.

10. Borgenson, "Arctic Meltdown," 67.

11. See the discussion on the complexity of exploitation and extraction in Antarctica relative to the Arctic in Caroline E. Foster, "Antarctic Resources and Human Security," in *Antarctic Security in the Twenty-First Century*, 167.

12. David Leary, "From Hydrocarbons to Psychrophiles: The 'Scramble' for Antarctic and Arctic Resources," in *Polar Oceans Governance in an Era of Environmental Change*, 133–34.

13. Ibid.

14. Foster, "Antarctic Resources and Human Security," 167.

15. Leary, "From Hydrocarbons to Psychrophiles," 133–34.

16. See the Arctic Council; http://www.arctic-council.org/index.php/en.

17. Stefansson Arctic Institute, Arctic Human Development Report, 155–156.

18. Ibid.

19. Ibid., 151.

20. Paul R. Josephson, *The Conquest of the Russian Arctic* (Cambridge, MA: Harvard University Press, 2014), 11.

21. Stefansson Arctic Institute, Arctic Human Development Report, 100.

22. Mia M. Bennett, "The Maritime Tiger: Exploring South Korea's Interests and Role in the Arctic," in *Arctic*, 128.

23. "Population Distribution in the Circumpolar Arctic, by Country (Including Indigenous Population)," GRID-Arendal (a center collaborating with UNEP); http://www.grida.no/graphicslib/detail/population-distribution-in-the-circumpolar-arctic-by-country-including-indigenous-population_1282.

24. Stefansson Arctic Institute, Arctic Human Development Report, 157.

25. Ibid.

26. Ibid., 35.

27. Gérard Duhaime and Andrée Caron, "The Economy of the Circumpolar Arctic," *The Economy of the North* 17; http://chaireconditionautochtone.fss.ulaval.ca/documents/pdf/65-Economy-circumpolar-Arctic-2006.pdf.

28. The natural resource sector is usually defined as the capture and processing of resources. The public sector is defined as public administration, defense, utilities, education, health services, and social work.

29. Stefansson Arctic Institute, Arctic Human Development Report, 164.

30. Uttam Kumar Shina and Jo Inge Bekkevold, "Special Editor's Introduction," in *Arctic,* 2.

31. Stefansson Arctic Institute, Arctic Human Development Report, 208.

32. Ibid., 209.

33. See for example, Patrizia Vigni, "The Governance of the Arctic Environment: The EU and US Contribution," European University Institute Working Papers, Academy of European Law, series/report no. EUI AEL; 2013/07; TRANSWORLD Project.

34. Shina and Bekkevold, "Special Editor's Introduction," 1.

35. Stokke, "Asian Stakes and Arctic Governance," 13.

36. Meagan Fitzpatrick, " 'Science and Sovereignty' Key to New Arctic Research Center," *CBC News*, August 23, 2012, http://www.cbc.ca/news/politics/science-and-sovereignty-key-to-new-arctic-research-centre-1.1231516.

37. See discussion on Asian shipbuilding capability in Vijay Sakhuja, "The Polar Code and Arctic Navigation," in *Arctic*, 42.

38. Jo Inge Bekkevold and Kristine Offerdal, "Norway's High North Policy and New Asian Stakeholders," in *Arctic*, 67.

39. Sakhuja, "The Polar Code and Arctic Navigation," 42–43.

40. Ibid.

41. Tom Roseth, "Russia's China Policy in the Arctic," in *Arctic*, 78.

42. Ibid.

43. Ibid., 79–80.

44. Shebonti Ray Dadwal, "Arctic: The Next Great Game in Energy Politics," in *Arctic*, 52.

45. Borgenson, "Arctic Meltdown," 72.

46. Charles Emmerson, *The Future History of the Arctic* (London: Vintage Books, 2011), 98.

47. See the parties to the Antarctic Treaty; http://www.ats.aq/devas/ats_parties.aspx?lang=e.

48. Stokke, "Asian Stakes and Arctic Governance," 8.

49. Dadwal, "Arctic," 55.

50. "Press Statement and Answers to Journalists' Questions Following the APEC Summit," Kremlin; http://en.kremlin.ru/events/president/transcripts/19382.

51. Roseth, "Russia's China Policy in the Arctic," 89.

52. Dadwal, "Arctic," 49–51.

53. Stokke, "Asian Stakes and Arctic Governance," 8.

54. Stefansson Arctic Institute, Arctic Human Development Report, 160.

55. Steven Lee Myers and Clifford Krauss, "Melting Ice Isn't Opening Arctic to Oil Bonanza," *International New York Times*, September 7, 2015.

56. Josephson, *The Conquest of the Russian Arctic,* 331.

57. Emmerson, *The Future History of the Arctic,* 227.

58. Josephson, *The Conquest of the Russian Arctic,* 351.

59. Stefansson Arctic Institute, Arctic Human Development Report, 241.

60. H. P. Rajan, "The Legal Regime of the Arctic and India's Role and Options," in *Arctic,* 141.

61. Atle Staalesen, "Russia Submits Claim for North Pole," *Barents Observer*, August 4, 2015.

62. Rajan, "The Legal Regime of the Arctic," 141.

63. Staalesen, "Russia Submits Claim for North Pole."

64. Stephens, "Polar Continental Shelves," 158.

65. Rajan, "The Legal Regime of the Arctic," 34. The projections indicating that the five Arctic littoral states will gain significantly greater access to their Exclusive Economic Zones (EEZs) by midcentury due to climate change have accelerated the pace of submissions to the CLCS. Greenland's access to its EEZ, for example, is expected to grow by 28 percent, Canada's by 19 percent, and Russia's by 16 percent by midcentury. See Stefansson Arctic Institute, Arctic Human Development Report, 241.

66. Emmerson, *The Future History of the Arctic,* 113.

67. Ibid., 110.

68. Stefansson Arctic Institute, Arctic Human Development Report, 242.

69. For a discussion of these disputes, see Dadwal, "Arctic," 53; and Asim Zia, Ilan Kelman, and Michael H. Glantz, "Artic Melting Tests the United Nations Convention on the Law of the Sea," in *Diplomacy on Ice*, Rebecca Pincus et al., eds. (New Haven, CT: Yale University Press, 2015), 133–136.

70. Dadwal, "Arctic," 53.

71. Sakhuja, "The Polar Code and Arctic Navigation," 38.

72. Frédéric Lasserre and Sébastien Pelletier, "Polar Super Seaways? Maritime Transport in the Arctic: An Analysis of Shipowners' Intentions," *Journal of Transport Geography* 19 (2011): 1465.

73. Shinji Hyodo, "Russia's Strategic Concerns in the Arctic and Its Impact on Japan-Russia Relations," in *Arctic*, 97.

74. Dana Gabriel, "US Arctic Ambitions and the Militarization of the High North," *Global Research Center for Research on Globalization*, July 24, 2013.

75. Jeremy Bender, "Russian Defense Minister Explains Why the Kremlin Is Militarizing the Arctic," *Business Insider UK*, February 26, 2015.

76. It specifically prohibits "any measures of a military nature, such as the establishment of military bases and fortifications, the carrying out of military maneuvers, as well as the testing of any type of weapons." See U.S. Department of State website; http://www.state.gov/t/avc/trty/193967.htm.

77. Rosalba O'Brien, "Argentine Judge Orders Seizure of Falklands Drillers' Assets," *The Telegraph,* June 27, 2015.

78. The companies named in the order were Premier Oil Plc., Falkland Oil and Gas Ltd., Rockhopper Exploration Plc., Noble Energy Inc., and Edison International Spa. Ibid.

79. Kylie MacLellan, "Britain Says to Boost Defense of Disputed Falkland Islands," *Reuters*, March 24, 2015.

80. See the website of Arctic Economic Council; last accessed March 15, 2016, http://www.arctic-council.org/index.php/en/our-work2/8-news-and-events/195-aec-2.

81. Global Agenda Council, *Demystifying the Arctic* (Davos, Switzerland: World Economic Forum, 2014), 5.

82. National Research Council, *Polar Icebreakers in a Changing World: An Assessment of U.S. Needs* (Washington, DC: National Academies Press, 2007), 65.

83. Ibid., 69.

84. Josephson, *The Conquest of the Russian Arctic,* 379.

85. Borgenson, "Arctic Meltdown," 72.

86. Ibid., 71.

87. National Research Council, *Polar Icebreakers in a Changing World,* 58.

88. Shiva Polefka, "Icebreakers: Essential Assets for a Changing Arctic," Center for American Progress, March 5, 2015, https://www.americanprogress.org/issues/green/report/2015/03/05/107328/icebreakers-essential-assets-for-a-changing-arctic.

89. National Research Council, *Polar Icebreakers in a Changing World,* 58.

90. Polefka, "Icebreakers."

91. Gabriela Roldan, "Analyzing the Infrastructure of Antarctic Gateway Cities," University of Canterbury, Gateway Antarctica, Postgraduate Certificate in Antarctic Studies (PCAS) 13 (2010–2011), 10.

92. Borgenson, "Arctic Meltdown," 64.

93. Polefka, "Icebreakers."

94. Ronald O'Rourke, *Coast Guard Polar Icebreaker Modernization: Background and Issues for Congress*, Congressional Research Service Report, September 2, 2015, 2–6.

95. Polefka, "Icebreakers."

96. See discussion in National Research Council, *Polar Icebreakers in a Changing World*, 75–76.

97. See Roldan, "Analyzing the Infrastructure of Antarctic Gateway Cities."

98. Ibid.

99. Ibid., 10.

100. Jane Perlez, "China, Pursuing Strategic Interests, Builds Presence in Antarctica," *International New York Times*, May 3, 2015.

101. The Droning Maud Land Air Network was founded by Belgium, Finland, Germany, India, Japan, the Netherlands, Norway, Russia, South Africa, Sweden, and the United Kingdom. See Roldan, "Analyzing the Infrastructure of Antarctic Gateway Cities," 7.

102. Ibid., 8.

103. Ibid., 9–11.

104. Ibid., 12–15.

105. Ibid., 15–18.

106. Ibid., 19–22.

107. Chapter Seven will discuss the extent of the bioprospecting currently taking place in Antarctica (i.e., the exploitation of its biological resources, such as bacteria and other microorganisms, for industrial purposes). Much of this activity is currently taking place under the radar.

108. See the IAATO website, last accessed March 15, 2016, http://iaato.org/tourism-statistics.

109. C. Michael Hall and Marisk Wouters, "Issues in Antarctic Tourism," in *Polar Tourism: Tourism in the Arctic and Antarctic Regions*, Colin Michael Hall et al., eds. (New York: John Wiley & Sons, 1995), 149.

110. UNEP, *Tourism in the Polar Regions: The Sustainability Challenge*, UNEP's Division of Technology, Industry, and Economics, the Sustainable Consumption and Production Branch, 2007, 14.

111. Ibid., 12.

112. Michael Luck, Patrick T. Maher, and Emma J. Stewart, "Setting the Scene: Polar Cruise Tourism in the 21st Century," in *Cruise Tourism in Polar Regions: Promoting Environmental and Social Sustainability*, Michael Luck et al., eds. (London: Earthscan, 2010), 6.

113. Hall and Wouters, "Issues in Antarctic Tourism," 152.

114. UNEP, Tourism in the Polar Regions, 33.

115. See the IAATO website; last accessed March 15, 2016, http://iaato .org/who-is-iaato.

116. Alan D. Hemmings, “Commercial Harvest in Antarctica,” in Chapter 20 *Exploring the Last Continent: An Introduction to Antarctica*, Daniela Liggett et al., eds. (Cham, Switzerland: Springer International Publishing, 2015).

117. See the website of the Antarctic and Southern Ocean Coalition; last accessed March 15, 2016, http://www.asoc.org/advocacy/antarctic -wildlife-conservation/southern-ocean-fisheries.

118. Stephen Nicol, Jacqueline Foster, and So Kawaguchi, “The Fishery for Antarctic Krill—Recent Developments,” *Fish and Fisheries* 13 (2012), 1.

119. See Rachel Baird, *Aspects of Illegal, Unreported, and Unregulated Fishing in the Southern Ocean* (New York: Springer, 2006).

# SIX

# Lines in the Ice: An Orderly Division of Antarctica

Antarctica is a changing continent. Like the rest of the planet, it is warming, with the climate of the Antarctic Peninsula already making it as habitable as Greenland. The sea ice around the continent is expected to shrink by about one-third by the end of the century, which, together with technological breakthroughs in transportation, is already rendering the continent more accessible.

An Antarctic economy has already started to grow. Today, it mainly revolves around fishing and tourism, with the multiple scientific research stations dotted across the continent generating job opportunities for everything from scientists to architects, builders, electricians, plumbers, cooks, and doctors. Through increased tourism, more travel documentaries, and aggressive marketing by tour operators, more people have now reached Antarctica than anyone could have possibly imagined. Nearly 28,000 tourists visited the continent in 2013–2014 alone,[1] bringing home with them countless photographs and travel stories, which would be enough to arouse anyone's curiosity.

The interest in Antarctica's landmass, its living and nonliving mineral resources, its tourism, and its broader economic potential is continuing to grow. As discussed in Chapter Five, the opening up of the Arctic Ocean, and the growing promise of what is now referred to as the "Arctic economy," are also contributing to taking the multiple possibilities offered by Antarctica out of the hypothetical realm and into reality. The big question now facing the international community is: What will happen to Antarctica when 2048 finally arrives?

## ANTARCTICA IN 2048

Back in the late 1980s, the collapse of the Convention on the Regulation of Antarctic Mineral Resource Activities (CRAMRA) led to the adoption of the Madrid Protocol on Environmental Protection, which succeeded only after a "walkout clause" was built into it by the United States.[2] This is a clause that essentially allows countries to request a review of the ban on minerals exploitation in Antarctica in 2048. The ban can be lifted only if another binding legal regime on Antarctic mineral resource activities is successfully negotiated, with the Madrid Protocol stipulating that this new regime "fully safeguard" the interests of all territorial claimants. If an amendment to the protocol does not enter into force within three years of its adoption, any party has the right to withdraw. The amendment would require no less than unanimity among the parties to the Antarctic Treaty.

Assuming that the world does wait until 2048 to exploit Antarctica, what will happen when this time finally comes? A full spectrum of options could emerge, all the way from turning Antarctica into a World Park (thus taking the Madrid Protocol much further than where it is today)[3] to the other extreme of accepting all seven frozen territorial claims, fighting out the three overlapping claims in the Antarctic Peninsula, and dividing up the unclaimed portion of the continent, which amounts to 15 percent of its surface area.[4] It would have to be split somehow between the United States and Russia, the two countries that publicly reserved their right to make territorial claims but decided to hold their fire when the Antarctic Treaty was signed.

Both extremes are unlikely. If converting Antarctica into a World Park was impossible to achieve in the 1950s, when little was known about its economic potential and the extent to which climate change would make it more accessible and habitable, it will be an even more remote possibility in 2048. Similarly, the notion that the frozen territorial claims would simply be revived and approved is not a serious proposal. The prolonged discussion in the United Nations (UN) on the "Question of Antarctica," which lasted from 1982 to 2005,[5] demonstrates that this could never happen. Also, the gradual expansion of the number of Consultative Parties to the Antarctic Treaty, which now include China, South Korea, and India (all of which have research stations on the continent), demonstrates how unlikely it is that Antarctica will be handed over to the original seven claimants.

Could reviving CRAMRA in 2048—perhaps with some modifications—be the middle ground? That would fall somewhere between the two extremes, allowing mining to go forward, but under CRAMRA's strict environmental safeguards. Therefore, while not assigning sovereignty, it would authorize the principal economic activity that a sovereign nation could be interested in, and while not designating Antarctica as a World Park, it would offer some degree of environmental protection. In fact, when the Madrid Protocol stipulated the negotiation of an alternative minerals regime prior to lifting the ban on mining, a revised CRAMRA was probably what the negotiators had in mind.

This option too is untenable, however, as is explored in greater detail later in this chapter. Interest in Antarctica goes far beyond its minerals. This continent, unlike the Arctic, is a habitable landmass. Reviving CRAMRA would do nothing to resolve the questions that are likely to be increasingly posed about the right to inhabit that land.

To suggest that CRAMRA can somehow be revived is to ignore the lessons that should be learned from a negotiation that took place prior to CRAMRA and which continued until 1994—the United Nations Law of the Sea (UNCLOS) negotiation, and the subsequent renegotiation of its provisions on seabed mining. The concept of the Common Heritage of Mankind, which the developing world sought to embed in this agreement, has been highly problematic, causing the United States—the main country with the technological capacity to mine the seabed—to reject the regime.

The negotiations over other global commons such as the Moon, in the Agreement Governing the Activities of States on the Moon and Other Celestial Bodies (referred to as the Moon Treaty), ran into similar problems. In addition, neither UNCLOS nor the Moon Treaty was intended to cover a potentially habitable landmass (or at least, this question has not presented itself yet with respect to the Moon), and they still struggled. Antarctica is potentially habitable, which is an additional complication. These issues and more will be explored in this chapter.

Furthermore, it is important to remember why the original CRAMRA failed. It collapsed not only because of the strength of the environmental campaign against it, but also because of concerns that it would somehow void the seven territorial claims, because it did not include rules for fair economic competition over mineral resources, and because of the intense opposition of the United Nations.

Since CRAMRA's collapse, the question of territorial sovereignty has certainly not gone away; it has manifested itself in the multiple scientific research stations that have come to cover the continent, the intense competition over the surrounding waters to Antarctica, and the rush to claim various portions of its continental shelf. Sovereignty has also been reinforced in Antarctica through five of the territorial claimants' cross-recognition of each other's claims; they include the United Kingdom, New Zealand, Australia, France, and Norway, leaving out only Chile and Argentina, whose claims overlap with the United Kingdom's.[6]

An equally key point is that CRAMRA was not negotiated under a UN umbrella. CRAMRA was negotiated only by the 16 Antarctic Treaty Consultative Parties, which participated in the final session of the Fourth Special Antarctic Treaty Consultative Meeting—and despite this limited number of players, it still failed. In 1983, the non–Consultative Parties were allowed to attend the negotiations, but only as observers.[7] In fact, it was because of the exclusion of many nations that a 1986 UN General Assembly Resolution issued a warning to the Antarctic Treaty parties, stating that it:[8]

1. Reaffirms that any exploitation of the resources of Antarctica should ensure the maintenance of international peace and security in Antarctica, the protection of its environment, the non-appropriation and conservation of its resources and the international management and *equitable sharing of the benefits of such exploitation*;
2. Calls upon the Antarctic Treaty Consultative Parties to impose a moratorium on the negotiations to establish a minerals regime until such time as *all members of the international community can participate fully in such negotiations* [italics added].

Were CRAMRA to be revived today, it would at the very least have to be renegotiated by today's much larger number of Consultative Parties (29 states), and probably the non–Consultative Parties too (another 24 nations). However, in light of the United Nations's objections back in 1986, as well as subsequent developments under UNCLOS that have empowered the developing world further, it would be highly unlikely that the mineral resources of the continent could be divided up without the full inclusion of the United Nations. This would firmly put Antarctica in an UNCLOS scenario, with that scenario being terribly complicated and still unfolding.

But a completely different solution can be contemplated for Antarctica—one that revolves around the little-known set of islands that lie in the middle of the Arctic Ocean known as Svalbard. Svalbard is the only landmass of any real significance in the Arctic, and its sovereignty was contested until the early 20th century; it is mineral rich and shares many of Antarctica's physical characteristics. Svalbard's unique model of governance deserves attention by the Antarctic policymaking and academic communities because it could act as a basis for an orderly division of the seventh continent.

In fact, the world should be asking not only what "will" happen to Antarctica in 2048, but what "should" happen, because policymakers need to start to include the Antarctic actively in the climate refugee equation. While it is melting (in part because of the climate crisis), Antarctica could gradually come to offering a potentially life-saving solution.

## THE TRAGEDY OF ANTARCTICA: SILENT OCCUPATION OF A DISTANT COMMON

It is often argued that Antarctica belongs, philosophically, conceptually, and perhaps even legally, to a category of areas on Earth called the "global commons." The characteristics normally attributed to a global common are the following: (1) the area must be physically and legally situated beyond the limit of national jurisdiction; (2) it must have no recognized or valid sovereignty claims; (3) it must be geophysically indivisible and not practically enclosable; (4) it must benefit from universal access; (5) the effects of mismanagement or abuse of the area must have universal ramifications; and (6) it requires international management to prevent its environmental deterioration.[9]

This contrasts sharply, of course, with what is referred to as "sovereignty." Judge Max Huber defined the term, in the *Island of Palmas Case,* to essentially mean independence: "independence in regard to a portion of the globe is the right to exercise therein, to the *exclusion* of any other State, the functions of a State" [italics added].[10] Currently, it is clear that sovereignty is not exercised in Antarctica, because "exclusion"—the benchmark set by Judge Huber—does not apply. It would not have been possible, for example, for the Australians to prevent the Chinese from establishing their second Antarctic scientific research station—Zhongshan—in the so-called Australian Antarctic Territory (AAT) in 1989. Similarly, it would have been equally impossible for New Zealand to stop South Korea from opening its Jang Bogo Station in the Ross Dependency in 2014.

While the global commons were once considered finite, including only the portions of our oceans beyond the Exclusive Economic Zones (EEZs), outer space, and Antarctica, technological advances are gradually expanding the realm of what can be conceived of as common or shared, such as the Internet, the electromagnetic spectrum, and even more. In addition, the term can be stretched to include shared resources that require the type of international management regime that global commons typically do, such as our stratospheric ozone layer and our climate system.[11] Antarctica is often compared to these commons and contrasted with the Sahara, the only other significant portion of our planet that is largely closed off to human habitation because of the harshness of its climate (in this case, being too hot). The sovereignty of the Sahara, however, is very much defined and agreed upon.

The legal status of activities taking place in the global commons is often broken down into four principal approaches: (1) *res nullius*, the property of nobody, which will accrue to its first legal taker; (2) *res communis,* the legal opposite of *res nullius,* that the property may be available for use by everyone; (3) Common Heritage of Mankind, a property to which no sovereignty applies, that is used exclusively for peaceful purposes, that enjoys freedom of scientific research, and that shares the economic benefit of any activity taking place within its boundaries with all humanity; and (4) *res publica*, a property that forms part of the public heritage of humanity and essentially becomes governed by a public trust.[12]

The question is: Under which of these approaches has the Antarctic Treaty placed Antarctica? Of course, one could begin by asking whether Antarctica is at all "global." While no one's sovereignty is recognized in Antarctica, the Antarctic Treaty was not negotiated by the United Nations and only has 53 parties. In fact, some have described Antarctica as a "special kind of common" or a "disputed common."[13] However, assuming that Antarctica is to be considered global on the basis that others may join the treaty under its more accessible non–Consultative Party status, then which of the legal regimes discussed here best characterizes the management system under which it was placed?

While the term *Common Heritage of Mankind* does not appear anywhere in the Antarctic Treaty, it appears to be the approach that best captures the nature of the governance regime that was introduced there. None of the seven territorial claims made in Antarctica are internationally recognized, and Antarctica is indeed demilitarized and devoted to science. The issue of benefit sharing lies at the heart of the Common Heritage principle, of course, but that term does not arise in the Antarctic Treaty because the Madrid Protocol closed off the continent to mining.

The concept of the Common Heritage of Mankind is key to understand because it has driven much of the UN discussions on the Question of Antarctica.[14] For many years, it has been the main legal force used from outside the treaty to destabilize it, and there is the risk that it will raise its head again. In fact, the challenge for all international treaties dealing with the global commons, regardless of the particular legal approach in which they are anchored, has been how to provide security of tenure to treaty parties so that the necessary investments can be made in resource exploration and development within the commons.[15] This is the problem that had bedeviled CRAMRA. The Common Heritage principle implicitly incorporated the security of tenure by establishing the condition that the benefits derived from tenure be shared with the rest of humanity.

Before the Antarctic Treaty came into being, a variety of different concepts were used to lay territorial claims on the continent, with discovery among them.[16] However, the United States has historically maintained the position that discovery and exploration have to be followed by effective "occupation" to give adequate title to land. With occupation being difficult to achieve in Antarctica's harsh climate, some scholars have asked whether an exception could or should be made for this special context.[17] Of course, the seven unilateral claims were frozen by the Antarctic Treaty, and the question of occupation is no longer extant today. Having said that, one may legitimately ask whether the numerous scientific research stations that have now come to cover the continent could be recast as "occupation," especially if the Question of Antarctica is once again asked in an international forum (i.e., in 2048).

A major fear with respect to the commons is what is often referred to as the "tragedy of the commons": situations in which individuals acting independently and rationally according to their own self-interest behave contrary to the best interests of the group.[18] Hence, there is a need to craft some sort of an international management regime for common spaces. However, in the case of Antarctica, the problem is not the absence of a regime that would prevent this classic tragedy; rather, it has been the silent occupation of the continent brought about by the very regime created to save it. The Antarctic Treaty has led to the multiplication of scientific research stations across the continent, with other additional forms of occupation going by entirely unnoticed—i.e., the two civilian settlements created on the Antarctic Peninsula that will be discussed next. This occupation is gradually making Antarctica less of a "common."

In previous chapters, this book has discussed the extent of the spread of scientific research stations in Antarctica. Worthy of note is how these stations, which initially started on the more accessible coastal areas of the continent, have crept farther and farther inland.[19] Today, three stations operate year round: the Concordia Station, which is jointly run by Italy and France; the Russian Vostock Station; and the U.S. Amundsen-Scott South Pole Station.[20] In other words, these stations are now, in some ways, permanent settlements on the continent. Also worthy of note is that the United States has research stations or other facilities on the territories of all claimant states,[21] and France has a research station outside its area of claim (Concordia is located in the Australian sector).[22] The Concordia Station represents the only jointly operated station on the continent.[23] All of this demonstrates the importance that countries have attached to flying their national flag in Antarctica.

In 1986, Benedetto Conforti, a magistrate, asked, "If a person is born, gets married, or dies in an Antarctic station belonging to one state but in the territory claimed by another state, where is he (or she) born, married, or dead?"[24] At present, of course, it is not expected that people be born, marry, or die in Antarctica, because the only activity that is officially authorized by the Antarctic Treaty is that of scientific research, and countries tend to carefully screen those they send to the continent. However, even if screened, people could still marry or die there, and so could tourists. In general, people visiting or living in Antarctica fall under the jurisdiction of the country to which they belong, but matters could soon start to get a little murky.

Looking at the map of Antarctica is enough to reveal the full complexity of this situation. Place names have already been ascribed to much of the continent, with most of the Consultative Parties to the Antarctic Treaty having place name commissions, or official bodies, that make the decisions.[25] Based on the statistics of the Standing Committee on Antarctic Geographic Information, which compiles the place names sent to it by national name committees, the United States has named 13,192 places on the continent, the United Kingdom 4,971, Russia 4,008, and Argentina 2,545, with other countries quickly playing catch-up. China has so far named 359 places, South Korea 27, and India 21.[26] Overall, 22 countries alone have assigned a total of 37,510 place names to the continent.[27] This too may factor in in 2048, and has been part of the silent occupation of the Antarctic common—an exercise of soft power on the continent.

Perhaps more curious are the two civilian settlements deliberately established in the Antarctica Peninsula and the South Shetland Islands of Antarctica—the zone of three overlapping territorial claims—by both Argentina and Chile, to which the world seems to have paid little notice. In 1978, some 17 years after the entry into force of the Antarctic Treaty, the Argentinian military junta experimented with the building of the first civilian settlement in Antarctica to reinforce its territorial claim, flying 8–10 families to their Esperanza base on the peninsula.[28] The first person to be born in Antarctica was Emilio Marcos Palma on January 7, 1978, which turned him into an overnight hero back in his home country.[29]

His birth on the contested peninsula prompted Chile to retaliate, orchestrating the counterbirth in 1984 of Juan Pablo Camocho at the Chilean Frei Montalva base in the South Shetland Islands of Antarctica (the civilian settlement in the base was later named Villa Las Estrallas).[30] Since then, many more babies have been born on the continent. Residents of Esperanza and Villa Las Estrallas are now regularly interviewed by local press, fascinated by survival stories in the world's coldest continent.[31] The Chilean experiment has grown over the years, with the *International New York Times* reporter Simon Romero bringing Villa Las Estrellas to the international spotlight in a January 2016 article in which he described its school, cemetery, sports facilities, and other living conditions. He wrote:[32]

> Fewer than 200 people live in this outpost founded in 1984 during the dictatorship of General Augusto Pinochet, when Chile was seeking to bolster its territorial claims in Antarctica. Since then, the tiny hamlet has been at the center of one of Antarctica's most remarkable experiments: exposing entire families to isolation and extreme conditions in an attempt to arrive at a semblance of normal life at the bottom of the planet.

Antarctica is becoming less of a "common" with every passing day. In fact, there is absolutely nothing in the Antarctic Treaty that would prevent the type of human settlements that Chile and Argentina have established. Other than freezing the seven territorial claims and dedicating the continent to peace and science, the treaty is otherwise silent on human settlement. All this begs the question of what would happen if a country or company were to suddenly find a valuable mineral in Antarctica, or if they were to decide to build an even bigger human settlement? With the continent having been conquered by research stations and place names, and with two civilian settlements having potentially opened the door to more, it is time for the international community to turn once again to the issue of ownership of Antarctica.

Interestingly, an increasing number of scholars are now making this point. Several have called for a "decolonization" of Antarctica and an end to the scientific imperialism that the Antarctic Treaty has introduced.[33] Others have argued that the fragile balance struck in Article IV of the treaty may not be the final word, with the Antarctic regime having existed before many of the world's countries came into being, raising ethical and equity issues.[34] Alan Hemmings has questioned whether the Antarctic Treaty, a treaty in which a handful of territorial claimants have a clear upper hand, can survive the 21st century without significant updating.[35] He writes, ". . . having sat through 37 Antarctic Treaty System meetings over 20 years, the role that claimancy plays in the workings of this system is inescapable."[36]

Assessing whether claimants may ever come to abandon their claims, he concludes that it will be highly unlikely for one or two claimants to do so alone because of the signal that this would send to allies within the regime, as well as the implications of such a move on other claimants:[37]

> These concerns may suggest that although it remains possible for a claimant to *unilaterally abandon its claim* to territory in Antarctica, there are political costs to so doing which may make this an unattractive diplomatic option barring very significant political needs. If this is the case, then it may suggest that a change in relation to territorial sovereignty positions may require some form of *collective action by claimants* (or a significant number of them) [italics added].

This can be taken to mean that, come 2048, were sovereignty to be reopened together with the ban on mining, it would most likely be reopened in full, not just in part. How, then, can we prepare ourselves for that time, and is there something to be learned from other global commons?

## THE NEW INTERNATIONAL ECONOMIC ORDER AND THE COMMON HERITAGE OF MANKIND

With the question of sovereignty lurking in the background, and with the collapse of CRAMRA having been driven in no small way by Australia's desire to protect its megaclaim in Antarctica (42 percent of the continent), what kind of revised CRAMRA would work the next time around? CRAMRA's negotiations started in 1982, only to see the agreement collapse in 1988. The negotiations took place against the backdrop of a major international power struggle between the developing and developed countries to establish a new world order—a struggle that continued well past CRAMRA's demise and whose outcome would need to be considered in any attempt for revival.

Hemmings warned that the Antarctic Treaty was created before the birth of most of the world's nations, and this point should indeed prompt some reflection. Between 1945 and 1960, three dozen new states in Asia and Africa achieved autonomy or full independence from their European colonizers.[38] It was the momentum created by the influx of a large number of new states on the international stage that prompted the UN General Assembly to adopt in 1974 a Declaration and Action Program on the Establishment of a New International Economic Order (NIEO).[39]

The aim of the NIEO was to reshape the international system in favor of the disadvantaged, particularly on trade and financial matters, and to allow a more equitable distribution of global resources. The main principles of NIEO include the right of every state to sovereignty over its economy and resources, the right to compensation for past exploitation, the right to technology transfer, the right to preferences in tariffs and trade, and the increased flow of development aid. It also declared that the "seabed and ocean floor and the subsoil, thereof, beyond the limits of national jurisdiction, as well as the resources of the area, are the common heritage of mankind."[40]

Negotiated behind closed doors, CRAMRA was to some extent shielded from the full forces of the NIEO, but even so, it had to contend with UN General Assembly resolutions being fired at it from the outside on the basis of the Common Heritage principle. The NIEO had profound implications for the world's so-called global commons; the seas, the Moon and other celestial bodies, and Antarctica. The principle gradually permeated all negotiations with respect to these entities, turning the management of common spaces into a tug of war between the rich and the poor—a platform for the substantiation of NIEO.

This tug of war is not finished, neither with respect to UNCLOS nor the Moon Treaty, which will be discussed in detail later in this chapter. If and when an attempt to revive CRAMRA is made, it too will be thrust into that conflict. The Antarctic Treaty members may come to regret not having grabbed the original CRAMRA and run with it back in 1988. That very same agreement may become impossible to conclude in today's political climate, as already mentioned. A far more egalitarian regime—with all the challenges that that poses—would need to be created. Given that the United States has refused to ratify what is a far more egalitarian UNCLOS, and that the Moon Treaty has been boycotted by all major players, such a regime probably would never see the light of day.

To the experts on the global commons, the seabed, the Moon, and Antarctica share a set of common characteristics. They all have an inhospitable environment that requires tremendous technological sophistication for exploitation, as well as exorbitant financial investment.[41] Interest in regulating the Moon began in 1966, when the world's two superpowers competed to reach it. That year, the Soviet Union submitted a draft treaty to regulate the Moon, which was considered in the formulation of the Treaty on Principles Governing the Activities of States in the Exploration and Use of Outer Space, Including the Moon and Other Celestial Bodies (nicknamed the "Outer Space Treaty"), adopted in 1967.[42]

That treaty was found to be deficient as early as 1969, when the *Apollo 11* spaceflight, led by U.S. astronauts Neil Armstrong and Buzz Aldrin, succeeded in landing humans on the Moon.[43] A new and separate Moon Treaty had to be concluded in 1979, but it is widely considered to be a failed treaty because it was never ratified by the countries with space exploration programs (e.g., the United States, Russia, and Japan).[44] The problems surrounding the Common Heritage principle could never be overcome.

The circumstances leading to UNCLOS were similar: a competition by major players over offshore resources that prompted increased regulation of the seas. While the oceans have long been subject to the "freedom of the seas" doctrine, a principle set forth in the 17th century that gave nations jurisdiction over a narrow belt surrounding a nation's coastline (12 nautical miles—the length of a cannon shot), the remainder of the seas were considered free to all. By the mid-20th century, however, competition for resources beyond this boundary intensified.

In 1945, for example, responding to pressure from the oil industry, President Harry Truman unilaterally extended U.S. jurisdiction over all natural resources in the U.S. continental shelf—oil, gas, and minerals. Constituting the first major challenge to the freedom of the seas doctrine, other nations followed suit. These skirmishes prompted the convening of the Third UN Conference on the Law of the Sea (UNCLOS III) in 1973 to write a comprehensive treaty for the oceans (with much of the Moon Treaty negotiations running in parallel). UNCLOS was concluded in 1982 (three years after the Moon Treaty).[45] However, because of its problematic inclusion and interpretation of the Common Heritage principle, its seabed mining provisions required renegotiation, taking the world all the way to 1994. Dissatisfied with the outcome, the United States boycotted the regime.

It was in the context of international discussions on the oceans that Malta's ambassador to the United Nations, Arvid Pardo, is said to have detonated a political bomb in 1967. He called for the application of a "Common Heritage of Mankind" principle, and in doing so, sowed the seeds of a concept that would come to trouble not only UNCLOS, but many other negotiations as well, and for many years to come. In an eloquent, three-hour speech before the UN General Assembly, Pardo asked for the oceans to be designated the Common Heritage of Mankind.[46] Speaking of the need to ensure that all of humanity would benefit from the sea's "sunken treasures," he requested the creation of a special agency with adequate powers to administer the oceans and the ocean floor beyond national jurisdiction in the interest of all humanity:[47]

> The sea-bed and the ocean floor are a common heritage of mankind and should be used and exploited for peaceful purposes and for the exclusive benefit of mankind as a whole. The needs of poor countries, representing that part of mankind which is most in need of assistance, should receive preferential consideration in the event of financial benefits being derived from the exploitation of the seabed and ocean floor for commercial purposes.

His speech led to the polarization of negotiating positions. In 1978, a expert from the Jimmy Carter administration argued that the mandate of

UNCLOS III had shifted from the design of a technical regime for seabed mining to the "architectonic construction of the contours of a future international legal, economic, and political order." The negotiation turned into a struggle for control of global institutions and resources, leading to a redefinition of its every concept along ideological lines: centralization versus decentralization, private enterprise versus international institutions, resource policy versus free market economics, equity versus efficiency, and so on.[48] This polarization found its way to other regimes, like the Antarctic. Vidas writes:[49]

> Little did the Maltese representative in the UN General Assembly realize that as early as 1967, when he had argued that the seabed and its resources beyond the limits of national jurisdiction should be proclaimed the common heritage of mankind, that the seed was sown for a later initiative on the 'Question of Antarctica' in the UN. Eventually some fifteen years later, the results of the Third UN Conference on the Law of the Sea (UNCLOS III) served as a stimulus for certain developing countries to link the common heritage of mankind concept to the legal and political status of the Antarctic.

Since Pardo's speech, the concept of the Common Heritage of Mankind has come to be interpreted in a variety of different ways. The main tenets of the concept were set out as indicated in the previous discussion; however, over time, it has also come to be associated with the "non-appropriation" of common resources and their management by some sort of a supranational body in the interest of all humanity, with the final goal being that of universal benefit sharing.[50] In fact, this concept is so polarizing that there is complete disagreement over its status in international law. While Christopher Joyner questions whether it is a legal norm at all,[51] John Noyes argues that it already may have penetrated customary international law.[52]

The Question of Antarctica died in the United Nations when more developing countries joined the Antarctic Treaty, particularly the bigger powers like China, India, and finally Malaysia, whose prime minister had initiated the discussion. It could be argued that the Common Heritage of Mankind lost its force as a counterweight to the Antarctic Treaty with the inclusion of these players. That was perhaps true at the time. Were CRAMRA to be revived today, with an entire UNCLOS having been constructed on the back of the Common Heritage principle, it will no doubt return to haunt the Antarctic Treaty System (ATS) with even greater force.

Gone are the days where a CRAMRA could get away with an agreement exclusively negotiated by 16 Antarctic Treaty Consultative Parties (referred in Article 62 of CRAMRA), the accession to which could take place only by "a State which is a Contracting Party to the Antarctic Treaty" (Article 61), and which could think that it could prohibit any mineral activity from taking place in Antarctica except under its provisions (Article 3). Gone are also the days when simple lip service to encourage joint ventures with the developing world in minerals exploration [Article 41(d)] would satisfy this constituency.[53] The concept of universal benefit sharing would have to be at the heart of any new agreement. The full complexity of the situation that a revised CRAMRA would run into is illustrated next.

## THE MOON AND OTHER CELESTIAL BODIES

The Moon Treaty was negotiated to create a more specific regime for the Moon than the Outer Space Treaty. Article 1 of the latter designates outer space, the Moon, and other celestial bodies as the "province of all mankind," providing that their exploration and use be carried out for the benefit of all humanity, regardless of its degree of economic or scientific development. It stipulates that exploration and use by all states take place "without discrimination of any kind, on a basis of equality."[54] While the Outer Space Treaty made no direct mention of the Common Heritage principle, it clearly sowed the seeds for its subsequent arrival. Negotiated in parallel to UNCLOS, the developing world sought to embed the principle solidly into the Moon Treaty.

Initially, it insisted that a moratorium be placed on the exploitation of the Moon until a legal regime could be negotiated to govern its natural resources, with benefit-sharing arrangements with all of humanity at its core. This negotiating position made little headway, and it was not until a Brazilian compromise proposal succeeded in breaking the deadlock in 1979 that the Moon Treaty was concluded. Under this compromise, the developed world agreed that the Common Heritage principle would be interpreted as requiring the creation of an international regime to govern resource exploitation, in exchange for which the developing world renounced the imposition of an outright moratorium.[55] However, the final wording of the treaty has been fairly ambiguous, leading many to conclude that a moratorium was indeed imposed.[56]

The key provisions of the Moon Treaty[57] currently consist of Article 11(1), which clearly designates the Moon as the Common Heritage of Mankind; Article 11(4), which allows states to exploit and use the Moon without discrimination and on an equal basis; and Article 11(5), which commits states to "establish an international regime . . . to govern the exploitation of the natural resources of the moon *as such exploitation is about to become feasible*" [italics added]. Article 11(7) lists the principles by which this next regime would have to abide, with 11(7d) stipulating that the concept of benefit sharing, especially with the developing world, be given special consideration.

It is Article 11(5) in particular that has created ambiguity with its suggestion that an international regime needs to be negotiated "as exploitation is about to become feasible," leading several industrialized countries to conclude that it had indeed imposed a moratorium, with the Moon negotiations starting to resemble UNCLOS. For them, the treaty had introduced a form of socialism through an absence of property rights, through the requirement that proceeds be shared with the developing world, and through the suggestion that an international organization needed to regulate mining on behalf of humanity. This would paralyze private enterprise and have a chilling effect on further space exploration.[58] A report of the U.S. National Commission on Space concluded at the time that the Moon Treaty had introduced a collectivized international regime, analogous to the seabed mining regime, that would seriously inhibit U.S. enterprise in space.[59]

As of today, the Moon Treaty has been ratified by 16 countries, with only 3 being industrialized countries (Australia, Austria, and Belgium).[60] The main nations involved in space exploration have chosen to abstain. The Outer Space Treaty has been seen as a far more permissive agreement, with no need for countries to voluntarily subject themselves to the moratorium of the Moon regime. In some ways, there are now two competing regimes for the governance of the Moon and other celestial bodies.[61] This demonstrates the kinds of issues that the revival of CRAMRA would be likely to run into, with a prevalent risk of never achieving an effective and universal regime.

## THE "SUNKEN TREASURES" OF THE SEAS

While UNCLOS III was convened to negotiate a comprehensive agreement for the seas and dealt with many more issues than seabed mining, the latter became one of the main conflicts dividing the developed and the developing worlds. At the center of the controversy were potato-sized manganese nodules that can be found at the bottom of the sea that contain important metals and minerals. It was as early as 1874 that British research vessel *HMS Challenger,* on the most important oceanographic expedition of all time, found manganese deposits at a depth of 15,600 feet somewhere between Hawaii and Tahiti. Since then, interest has grown in the resources of the seabed, with the Pacific Ocean being of particular interest.[62]

Negotiations on the creation of seabed mining provisions in UNCLOS were polarized right from the start, with the 1974 NIEO having declared the seabed as the Common Heritage of Mankind. The question became: Who can mine the sea, and under what rules? The developing world called for the establishment of an international authority with exclusive rights to mine the "common heritage area," involving states and private groups only as it saw fit. The developed world, on the other hand, pushed for a much weaker international body that would simply license the states or the private actors wishing to engage in mining. The compromise was the creation of what became known as the "parallel system," where private and collective mining could take place side by side.[63]

Influencing the dynamics of these negotiations was the birth of new nation-states that all wished to assert their presence on the international stage and to reverse the exploitation that they had suffered at the hands of their colonizers. The word *compensation* therefore featured regularly in the negotiation. Of particular concern was another issue also under negotiation in UNCLOS III: the establishment of 200-nautical-mile EEZs. Under the previous freedom of the seas doctrine, territorial jurisdiction would stop at 12 nautical miles. Many landlocked developing countries were up in arms at the suggestion of wider national jurisdiction in the seas, arguing that a new form of colonialism was being introduced.[64]

In a famous letter dated May 5, 1978, Nepal's representative to the conference called for the establishment of a Common Heritage Fund, to be financed through contributions by coastal states and whose benefits were to be used to narrow the gap between the rich and the poor, as well as to assist and transfer technology to the developing world.[65] The letter stated:[66]

> Unfortunately in the more than 10 years since it was enunciated by Mr. Arvid Pardo in the fall of 1967, the common heritage principle has suffered from misinterpretation, from attrition and from neglect. As the concept of the economic zone increased in popularity, some nations insisted, wrongly and tragically, that the economic zone and the common heritage were mutually exclusive. That idea makes a cruel hoax of the concept of the common heritage. For the overwhelming proportion of ocean mineral wealth and of marine species are found within the economic zone.

The Nepalese proposal polarized the negotiations further, with some arguing that it had in fact turned the common heritage principle into a common heritage tax. Doubt was also cast on the number of developing countries supporting Nepal, with many developing coastal states said to have championed the EEZ.[67] Going in a somewhat different direction, some developing countries with land-based mining started to call for production limits on mining from the seabed to protect their economic interests.[68]

Captured in Part XI of UNCLOS, the seabed mining provisions state that the seabed and its resources form part of the Common Heritage of Mankind; an International Seabed Authority would be created and empowered to license mining operations and to itself mine the seabed through a corporation that would be called the "Enterprise"; a 36-member council under the International Seabed Authority would take decisions on applications through a set of elected members; a financing system for the Enterprise would be created so it could mine the seabed on behalf of humanity; mining licensees would transfer mining technology to the Enterprise on a compulsory basis; production controls would be set to ensure that land-based miners would not be disadvantaged by the newly accessed resources from the seabed; equitable benefit sharing would take place; and a compensation fund would be created.[69] The creation of an Enterprise was therefore at the heart of Part XI of UNCLOS. It was to become the commercial arm of the International Seabed Authority, functioning like a mining corporation, to mine the seabed on behalf of all humanity.

Part XI referred throughout to the need to consider the interests not only of the developing world, but also of "developing states and peoples who have not attained full independence or self-governing status." For example, the benefit-sharing provisions of the convention made a specific mention of the need to assist this group (Article 140), and so did the provision dealing with the distribution of any surplus revenue of the convention (Article 162).[70] The United States objected to this on the grounds that that could include the Palestinian Liberation Organization, which it considered a terrorist group, and it came to oppose many of the other provisions too.[71]

Of particular concern to industrialized countries was the concept of the Enterprise—a global UN corporation that would compete against them. The Enterprise would be part of the International Seabed Authority; in other words, it would be attached to the body designed to regulate seabed mining. The industrialized world feared that the international regulator, in forming part of the International Seabed Authority and acting as its operational wing, would compete with the entities that it was required to regulate. This conflict of interest proved unacceptable to many negotiators.[72] Furthermore, the Enterprise came to form part of a complex mining regime that quickly became contentious.

Under a "banking system," any state or private entity making an application to exploit a particular seabed site would have had to submit another commercially equivalent site for approval. The state or private entity would be allowed to exploit only the first site, while the second would constitute a "reserved area" to be mined by the Enterprise (the UN corporation) on behalf of the developing world.[73] The Enterprise, therefore, would benefit from the initial prospecting conducted by the applicant. In addition, the licensee would be required to train and transfer mining technology to the Enterprise. Developed countries objected to having to transfer technology to their competitor, arguing also that the requirement may be impossible to fulfill if the technology is proprietary and not a state's to share.[74]

A whole range of other provisions also proved problematic for the industrialized world, such the cost of establishing and maintaining the International Seabed Authority and its various subsidiary bodies; the decision-making rules contained in the agreement; the production limitation, which they characterized as introducing a socialist approach to resource management; the fee structure foreseen by the agreement (i.e., a fixed application fee, an annual fee upon the entry into force of the contract, and other payments thereafter to the International Seabed Authority); and the compensation fund.[75]

In 1980, Jimmy Carter was defeated by Ronald Reagan in U.S. presidential elections, with the Reagan administration firmly believing in small government and calling for a reappraisal of UNCLOS. As Christopher Garrison characterized it, the administration smelled an "ideological rat."[76] It soon made its conditions known for joining an international seabed mining regime. Principal among them were that it would have to be a regime that encouraged rather than deterred private enterprise; it would have to ensure national access to resources; it would have to provide the United States with a clear decision-making role in the convention; and it would have to refrain from setting undesirable precedents for other international negotiations.[77]

While last-minute attempts were made to appease the United States by recognizing the rights of "Pioneer Investors"—those who had already invested large sums of money in resource prospecting before the convention—the Reagan administration sent a special envoy, Donald Rumsfeld, to lobby governments not to sign it.[78] Despite UNCLOS being a treaty that covers much more than seabed mining, dealing with issues as important and as diverse as the establishment of EEZs and jurisdiction over continental shelves, the United States resented its seabed mining provisions so strongly that it rejected the entire package. This was a more complicated situation than walking away from the Moon Treaty—this had a much higher opportunity cost. James Malone, Assistant Secretary of State for Oceans and International Environment and Scientific Affairs, Special Representative of the President on the Law of the Sea, and chair of the U.S. delegations to UNCLOS III, wrote at the time:[79]

> The "common heritage" is certainly a noble phrase. Unfortunately, the phrase became severely distorted in its meaning during the long Law of the Sea negotiations. It became a guise under which the seabed's wealth was to be governed, regulated, and allocated under the false assumption that every nation has an undivided property interest in the deep seabed, and therefore, each is automatically entitled to its proportionate share of the fruits from those whose efforts produce wealth from what would otherwise be economically valueless. The International Seabed Authority would have been given the unprecedented power to redistribute wealth on an international scale.

UNCLOS was adopted in 1982 but came into force only in 1994, after its provisions on seabed mining were essentially overhauled. In what was characterized as a "masterstroke" of international negotiations, shortly after the adoption of UNCLOS, the United States declared that much of the rights that would have been conferred to it by the regime, such as the EEZ and jurisdiction over the continental shelf, were existing rights under customary international law, so it could already enjoy them.[80] Therefore, joining UNCLOS would only constitute a loss, by limiting its seabed rights. In fact, even before the convention was concluded, in 1980 the United States unilaterally passed the Deep Seabed Hard Mineral Resources Act, authorizing private actors to mine the seabed. Other industrialized countries followed suit, such as France, Germany, and the United Kingdom.[81] Therefore, in 1982, UNCLOS found itself in a similar predicament to the Moon Treaty—an agreement in search of signatories.

Eight years later, two successive UN secretary generals would take it upon themselves to revive UNCLOS. In 1990, Javier Perez de Cuellar convened informal consultations aimed at achieving universal participation in UNCLOS, and Boutros Boutros-Ghali followed suit when he assumed the office in 1992.[82] The latter stressed that the convention would be severely damaged if it entered into force without the participation of industrialized states because they were the major users of the sea, the heaviest polluters, and the most important parties in disputes at sea.[83] In fact, it was well understood by all that in the absence of developed-world participation in UNCLOS, the Enterprise could never come into being because it would not know where to mine without the "banking system," nor would there be any resources to fund it. In fact, the entire funding structure of the convention, dependent on the contributions of the industrialized world, would collapse.

After lengthy negotiations, the international community succeeded in revising Part XI of UNCLOS. So as not to do undo the package that

UNCLOS represented, the revisions stood separately and were called the Agreement Relating to the Implementation of Part XI of the UN Convention on the Law of the Sea (referred to as the 1994 Agreement). The success of the negotiation was aided by the collapse by the Soviet Union, seen as a triumph for a more capitalist approach to a range of international issues, including to seabed mining.[84]

The 1994 Agreement trimmed down the institutions of UNCLOS related to the seabed and their associated cost, revised some of the decision-making rules of its main bodies, slashed the various fees associated with participation in the regime, completely removed the production ceilings and the mandatory technology transfer provisions, and subjected the Enterprise to the same rules as any commercial body, with its operations to begin as joint ventures until it could stand on its own.[85]

At the heart of these revisions, therefore, was a changed role for the Enterprise. While the parallel system, or the banking system, would persist, no country would be obligated to fund the Enterprise. Moreover, it would have to acquire technology on the open market, or else it could ask for contractors or sponsoring states to facilitate acquisition on fair and reasonable terms. As a sign of goodwill toward the developing world, the removal of production limitations was accompanied by a reference to the applicability of the rules of the General Agreement on Tariffs and Trade (GATT) to seabed mining, so that unfair subsidization would not take place relative to land-based mining.[86] Despite these changes, which would subsequently have allowed most of the world to ratify UNCLOS and the law to enter into force, the United States rejected the regime.

## THE COMMON HERITAGE MANKIND IN PRACTICE, AND WHY THE UNITED STATES CONTINUES TO REJECT UNCLOS

Relative to all of these developments, it is clear that CRAMRA was shielded from the full force of NIEO and the Common Heritage principle. However, the UN General Assembly resolutions being directed at it from the outside repeatedly called for the application of this principle and the involvement of the entire United Nations.

It is important to remember that CRAMRA was negotiated behind closed doors. In 1982, CRAMRA's negotiations were attended only by the following developing countries: Argentina, Chile, and South Africa (i.e., the original signatories to the Antarctic Treaty). In 1984, two other developing countries joined the negotiation—Brazil and India—with China and Uruguay participating in 1986 (i.e., shortly before the agreement's collapse).

Therefore, only seven developing countries participated as Consultative Parties in the crafting of CRAMRA, with China having come in at the tail end of the negotiations, when momentum was already building for their collapse. As of 1983, any other developing country participating in the negotiations only did so as an observer, with no decision-making power.[87] This allowed CRAMRA to develop without global participation, and therefore without a global benefit-sharing arrangement.

Experience in the implementation of the 1994 Agreement demonstrates that while the Common Heritage principle was seriously contained, it has had important implications for how the treaty came to function in practice.[88] Despite a decision to postpone the establishment of the Enterprise (reflective of the continued tug of war within UNCLOS on the role of this entity),[89] the developing world has started to reap the benefits of the law anyway by sponsoring the access of companies from industrialized countries to the "reserved areas," and by obtaining training in seabed mining.[90] In addition, the International Seabed Authority is currently defining the concept of "benefit sharing" so it may be applied as soon as exploration turns into actual mining.[91]

In fact, when the concept of reserved areas was being negotiated under UNCLOS, there was concern that the developing world would simply lease its rights to these portions of the seabed to mining companies from the developed world. This was not a concern unique to UNCLOS; the International Telecommunications Union had itself learned several important lessons from regulating the geostationary orbit. When slots were allocated, Ricky Lee writes: "Tonga sought to abuse the first-come first-served process of allocating orbital slots on the geostationary orbit by filing 16 orbital slots between Asia and the Americas . . . It is clear that conferring orbital slots to most developing states would only allow these states to 'lease' their orbital slots to commercial operators for profit." This was referred to as the "paper satellite" problem.[92]

It is interesting to observe that since UNCLOS has come into force, of the handful of exploration licenses granted by the International Seabed Authority, three went to Tonga, Nauru, and Kiribati as sponsoring states.[93] Concerned that it could nevertheless be liable for damage resulting from mining operations as the sponsoring state, Nauru requested an Advisory Opinion from the International Tribunal on the Law of the Sea on its specific obligations. An opinion was delivered in February 2011 in which the tribunal cautioned against the creation of "sponsoring states of convenience." It held that sponsoring states from the developing world had the same responsibilities as those from the developed.[94]

With these types of arrangements in place, one can debate whether the Common Heritage principle has truly served its purpose. Is the developing world being exploited further as a result of the principle? However, regardless of whether developing countries are being involved as true partners in the exploration of the reserved areas or simply leasing their slots (the "paper satellite" problem), they have nevertheless come to form part of the seabed mining process and could stand to gain even more when benefit-sharing arrangements are eventually introduced. In other words, the experience acquired under UNCLOS will be likely to reinforce the developing world's call for a share of Antarctica.

The problem, of course, is that a call for the inclusion of the Common Heritage principle in the revival of CRAMRA could lead to a full UNCLOS rejection scenario, with the United States and possibly other industrialized countries refusing to join the regime. History shows that the United States objected to the direction taken by the UNCLOS negotiation right at the outset. The record of the first congressional hearings held on UNCLOS before the Committee on Merchant Marine and Fisheries demonstrates that members of Congress saw UNCLOS as a totally un-American agreement that was not reflective of U.S. values. In the hearings, congressmen and mining industry representatives asked whether the Enterprise would be "our competitor even while they act as judge, jury, and probably policeman";[95] whether "a bunch of UN politicians and bureaucrats are going to be telling our seabed miners what they can or cannot do in international waters";[96] and whether—given that the United States was the only country with the capacity to mine the seabed—this would be an "international treaty to regulate the activity of one country, the United States."[97]

At a loss for words, a U.S. negotiator being interviewed by Congress at one point responded: "We think that the common heritage of mankind principle, if I can be blunt, is a glittering generality. Therefore, we can leave the phrase alone, provided that the rest of the treaty says the right things."[98] In fact, congressional hearings in relation to UNCLOS would soon turn to the scope of the unilateral action that the United States should take to protect its seabed interests, totally dismissing the ongoing negotiations.[99] Today, the anti-UNCLOS lobby in the United States is continuing to fight any potential ratification of the convention, calling upon Congress to persist with the unilateral route, and to sign mutual recognition agreements for portions of the seabed of mining interest to the United States with countries that possess mining technology.[100]

It is important to recall that the United States was in the driver's seat in the negotiation of the Antarctic Treaty. In fact, it was one of the main architects of the regime, with the treaty being signed in Washington under President Dwight Eisenhower. In addition, the United States is the main scientific power in Antarctica, has the largest station on the continent, and also has stations or facilities in the territories of all claimant states. In addition, the United States is one of the few countries with the technological prowess to conduct mining operations in Antarctica. What are the chances, then, of the United States truly engaging in a revised CRAMRA that would be likely to be broadened to the entire United Nations and anchored in the Common Heritage of Mankind principle? They are probably close to nil. So what next for Antarctica?

## ENTER SVALBARD[101]

It is impossible to imagine Antarctica declared a World Park or, at the other extreme, divided between its claimants, Russia and the United States. With a revival of CRAMRA being equally difficult to envisage, what are the remaining options for 2048? Would an extension of the status quo, by prolonging the ban on mining, actually work? Given the increasing interest in Antarctica's resources, and its changing climate, this possibility would run the risk of delaying the collapse of the ATS. Challenges to the system through a mining operation simply commencing, or through the creation of new or bigger human settlements, could lead to its disorganized breakup, with highly uncertain outcomes. In addition, prompting the orderly division of the continent to make it part of the answer to the climate migration crisis could be in humanity's interest.

There is only one country in the world today that is both an Arctic (through its geography) and an Antarctic nation (through its territorial claim)—Norway. Tucked away in the middle of the Arctic Ocean is Svalbard, a little island archipelago that falls under its jurisdiction, which has a unique governance model that is worthy of consideration for Antarctica. Composed of numerous islands between mainland Norway and the North Pole, and about twice the size of Belgium,[102] Svalbard's status was contested after World War I. Following complicated negotiations, a number of countries agreed on a governance model for Svalbard that essentially turned it into international territory. While the intention of these negotiations was never to "internationalize" Svalbard, this was nevertheless the outcome.

Svalbard's governance model is not problem free. It still causes tensions, with Norway continuing to insist that it has sole sovereignty over the archipelago, and that it is no different than any other part of the country. However, these tensions exist because the jurisdiction of Svalbard was negotiated without internationalization in mind—let alone a precise model for it. The renegotiation of Antarctica's status, with the Svalbard model designated as the broad objective from the very start could preempt and mitigate such tensions.

There is no doubt that extending the Svalbard model to Antarctica will appear unimaginable to some. After all, Svalbard is only a small set of islands. Can the model truly be generalized to an entire continent? In addition, the Norwegian government denies that there is such a thing as a "Svalbard model" to begin with. It claims that Svalbard is Norwegian, full stop. As will be demonstrated next, nothing about the Svalbard model is "small"; in fact, it may be the world's most important and innovative legal experiment in the creation of a territory to which people of different ethnic backgrounds can emigrate and enjoy equal economic rights.

**Figure 6.1 Map of Svalbard**

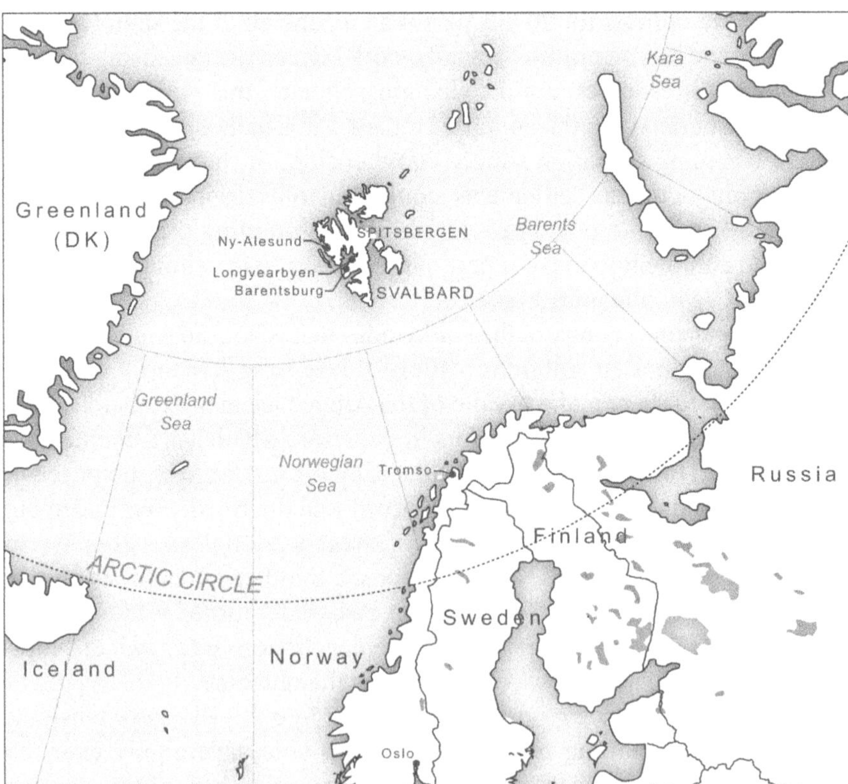


(Joseph Hill)

## THE SVALBARD SOLUTION

At the start of the 20th century, a number of Arctic islands were the subject of territorial disputes: Svalbard, Greenland, Jan Mayen, Franz Josef Land, and Wrangel Island.[103] They were generally considered *terra nullius*, or no man's land. Of these islands, Greenland was the only one to be inhabited by an indigenous population at the time of its discovery. While the sovereignty of all these islands was unambiguously settled, with Greenland handed to Denmark, Jan Mayen to Norway, and Franz Josef Land and Wrangel Island to the Soviet Union, only Svalbard was placed under a special model of governance. Currently, only one set of islands is still in dispute in the Arctic: the tiny Hans Island, which is claimed by both Denmark and Canada.[104] Some would argue that the unambiguous settlement of the sovereignty question over the vast majority of Arctic islands was the outcome of the unsatisfactory arrangement reached in Svalbard, a model that countries later sought to avoid in other parts of the Arctic.[105]

Svalbard was discovered in 1596 by Dutch explorer Willem Barents, after whom the Barents Sea was named. Due to the gulfstream, Svalbard is not as cold as other parts of the Arctic. It has an average temperature of 6°C (42.8°F) in summer and −14°C (6.8°F) in winter.[106] Some 60 percent of the archipelago is nevertheless covered by ice year round, and only 10 percent supports vegetation.[107] Svalbard can be said to have gone through five distinctive economic phases in which different commercial activities prevailed. It gradually went from whaling and sealing to mining, tourism, and research.[108] Currently, Svalbard is the world's northernmost settlement with a permanent civilian population.

At the close of World War I, the Svalbard question was posed due to a widespread sentiment that Norway somehow needed to be compensated for its Allied support and the sacrifice of Norwegian lives and shipping fleets that it made. Norway had only recently gained independence from Sweden, in 1905—another reason for which it wished to clarify the status of the archipelago.[109] At the time, Svalbard's principal inhabitants were Norwegian, and the islands' main employer was the Boston-based Arctic Coal Company.[110] Hence, the frequent reference to Svalbard as a "company town."

Norway's interest in Svalbard was almost immediately met with counterclaims by other countries. The Dutch asserted special rights because of Svalbard's discovery by one of their own, as well as the subsequent semi-permanent Dutch whaling stations established on the islands. This was contested by the Soviet Union, which claimed that its Pomor hunters had reached Svalbard prior to Barents. Also vying for a slice of Svalbard, the

United Kingdom asserted that the Muscovy Company, an English trading company, had claimed Svalbard's biggest island (Spitsbergen) for King James I in 1614.[111]

Particularly challenging in this landscape of claims and counterclaims was Norway's complex relationship with Russia. Russia was an important customer for Norway's fishing industry, and with respect to fishing and hunting rights in the White Sea (the southern inlet of the Barents Sea) and Arctic Ocean, it had the capacity to strangle Norway economically. In addition, there were very close links between Norway's labor movement—one of the most radical in western Europe—and the new postrevolutionary Soviet government.[112]

Even before the close of World War I, Svalbard's sovereignty was discussed in trilateral conferences in 1910 and 1912 between Norway, Sweden, and Russia. Whereas the idea was proposed that Svalbard be run by a "condominium" of all three states, it was abandoned, and all attempts to settle the Svalbard question were shelved until the end of the war.[113] Striking here, of course, is the similarity with Antarctica, for which a condominium approach had also been proposed.[114] As with Antarctica, the approach largely failed because of its exclusion of key players.

It was only in 1919, after the war, that countries once again turned to the Svalbard question at the Paris Peace Conference. The United States, the United Kingdom, France, Italy, and Japan established a Spitsbergen Commission, which quickly produced a Franco-Norwegian treaty proposal. The draft treaty created a mechanism for safeguarding international interests in the archipelago that made the idea of Norwegian sovereignty over Svalbard more palatable. On the basis of this proposal, the Treaty of 9 February 1920 Relating to Spistbergen (Svalbard)—otherwise known as the Svalbard Treaty—was signed in 1920 and entered into force five years later.[115] The Soviet Union held out for a while, using the Svalbard question as a bargaining chip in the recognition of its postrevolutionary government. In 1924, it subscribed to the treaty in exchange for Norway's full recognition of its new government.[116] However, Norwegian-Russian relations with respect to Svalbard have remained tense until this day.

The preamble to the Svalbard Treaty (see the full treaty in Appendix II of this book) starts with the following statement: "Desirous, while recognizing the sovereignty of Norway over the Archipelago of Spitsbergen . . . of seeing these territories *provided with an equitable regime* in order to assure their development and peaceful utilization" [italics added].[117] It emphasizes right from the outset, therefore, the need to ensure equitable access to the island and its resources by all treaty parties. Its main provisions state that:[118]

- Article 1: Norway possesses "full and absolute sovereignty" over the archipelago.
- Articles 2–3: Nationals of all signatory states may undertake commercial activities on a footing of absolute equality, subject to Norwegian legislation. In fact, Article 3 states that the nationals of all parties "shall be admitted under the same conditions of equality to the exercise and practice of *all maritime, industrial, mining, or commercial enterprises both on land and in the territorial waters,* and no monopoly shall be established on any account or for any enterprise whatever" (italics added). It adds that there can be no favorable treatment provided to Norwegian nationals with respect to the export, import, or transit of ships or goods in Svalbard.
- Article 4: Nationals of all parties must have equal access to the services of public wireless telegraphy stations, for communication with ships of all flags.
- Article 6: Parties have a limited period of time to make claims "from taking possession or from occupation of land before the signature of the present Treaty." This has indeed happened. Where valid, title to such land was granted to claimants.
- Article 7: Nationals of all signatory states may acquire, enjoy, and exercise the right of ownership of property, including mineral rights, on terms of "complete equality."
- Article 8: Taxes, dues, and fees levied have to be devoted exclusively to the administration of Svalbard. This was of course designed to ensure that these monies would not flow back to the Norwegian mainland.
- Article 9: Svalbard "may never be used for warlike purposes."

In other words, the Svalbard Treaty created a demilitarized area, to which any national from a signatory country to the treaty could emigrate with an equal right to land and commercial activity. There could be no discrimination in favor of Norwegian citizens. In the Act of 17 July 1925 Relating to Svalbard,[119] the Norwegian government decreed that the Norwegian civil and penal codes, and other Norwegian legislation relating to the administration of justice, would apply to Svalbard. It also provided that a Governor would be appointed to Svalbard by the King of Norway, enjoying "the same authority as a District Governor." In addition, it stipulated that the seizure, sequestration, or any other action relating to property or mining rights shall be performed in Norway by the Court of First Instance. The act was designed to confirm Norway's sovereignty over the archipelago while simultaneously respecting the provisions of the Svalbard Treaty.

As mentioned earlier in this chapter, Judge Huber defined the term *sovereignty* as "independence in regard to a portion of the globe is the right to exercise therein, to the *exclusion* of any other State, the functions of a State" [italics added]. The Svalbard Treaty has led some commentators to conclude that Norway enjoys only limited sovereignty because its rule over the islands does not have an "exclusive character."[120] In fact, as Adam Grydehoj, a Svalbard expert, points out, the strongest opposition to the treaty came from within Norway—namely, its parliament and press, parts of which felt that the treaty and the subsequent Mining Code that was negotiated to complement it restricted Norway's sovereignty.[121] In fact, Svalbard was quick to become one of Norway's main foreign policy headaches, with John Lyng, its foreign affairs minister from 1965 to 1970, saying that his first worry would be Svalbard, were his phone to ring in the middle of the night.[122]

In 1920, the following countries immediately joined the Svalbard Treaty: the United States, the United Kingdom, Denmark, France, Italy, Japan, Norway, the Netherlands, and Sweden. However, membership of the treaty quickly expanded and now includes over 40 countries from places as far from the Arctic as Afghanistan, Egypt, Saudi Arabia, India, and China.[123] Most recently, South Korea joined the treaty.[124]

In Box 6.1, Jonas Store, the former Norwegian minister of foreign affairs, is interviewed on the subject of Svalbard. He affirms in no uncertain terms Norway's sovereignty over the islands, likening the Svalbard Treaty to the requirement for "equal treatment that is accorded to economic operators in the European internal market." What he does not say, of course, is that the Svalbard Treaty is not simply intra-European. However, the Norwegian government has used the European internal market analogy consistently to make the treaty more palatable domestically.

Noteworthy in the interview is how, despite the multiple challenges that the treaty presents for Norway's management of Svalbard, Store considers the governance model to have been a success. At no point does he argue that any particular provision of the treaty needs to be reopened or renegotiated.

The benefits of application of the Svalbard model to Antarctica are clear. The Svalbard model, as presented here, would have one particular and unique advantage for the seventh continent. It can achieve what few other solutions can, in at once reconciling sovereignty with the Common Heritage of Mankind principle. The seven territorial claimants would see their sovereign rights over different sections of the continent confirmed, while at the same time allowing the rest of humanity to benefit economically from Antarctica and to enjoy its living space. In fact, any citizen of the world could freely move to any part of the continent under this model, provided that they abide by the national regulations of the jurisdiction to which they

**Box 6.1 Interview of Jonas Store, Norway's Minister of Foreign Affairs from 2005 to 2012[a]**

1. *Much has been written about the "Svalbard Model" of governance. Has it been a success or failure in your view?*
I would start by saying that the most important thing about Svalbard is that it is an integral part of Norway. Activities as well as regulations on the island have been developed as part of Norwegian legislation, respecting the provisions of the Paris Treaty of 1920 that granted Norway full sovereignty. Norwegian governance has been a success for Svalbard and those who are active in various sectors there, from mining to research and tourism. It has ensured maintenance of peace and stability in the area, and preservation of the area`s distinctive natural habitat and wildlife—well balanced with resident communities and economic activity. This could only be achieved through having clarity regarding sovereignty and governance.

2. *What does Norway's sovereignty over Svalbard amount to in practice?*
It ensures governance and a consistent and firm enforcement of necessary regulations adapted to the Arctic circumstances. Norway has the exclusive right to pass and enforce laws and other regulations for Svalbard under the Svalbard Treaty. In principle, all legislation that applies to the mainland is applicable to Svalbard as well. There is the principle of nondiscrimination of economic actors, in practice not very different from the equal treatment that is accorded to economic operators in the European internal market. Then there are also a few requirements regarding the exercise of authority pertaining to taxes and duties, as well as no military bases.

3. *What rights do other Svalbard Treaty parties enjoy, and can any national of a party to the treaty decide to move to the islands?*
The principal of equal treatment and nondiscrimination is key in the Svalbard Treaty. Again, this principle is well known today in Europe following the internal market of which Norway is also a part, although not a member of the European Union. It ensures equal rights for nationals and companies from signatory nations in many areas, such as hunting and fishing, access and entry, certain types of industrial activities, and property rights—including mineral rights.

4. *Some criticisms have been made of Norway's application of environmental regulations in Svalbard, in particular of its use of these regulations to stop mining activity on the islands. What is your response?*
Even the Treaty of Paris of 1920 has reference to Svalbard's fragile natural environment and the need to preserve it with care. At the same time, Svalbard is one of the places in the world where climate changes are expected to be the greatest—which may have a strong impact on its ecosystem in the future. Therefore, environmental protection must be a key priority. This requires regulation that maintains the fine balance between human activity and preservation of the natural habitat. These regulations are nondiscriminatory.

5. *What about the waters and continental shelf surrounding Svalbard—whose are they?*
The treaty regulating economic activities specifically applies to mainland Svalbard, including the 12-mile territorial sea around the island. Beyond this, we need to follow the principles of the UNCLOS. In accordance with modern international law, in 1977 Norway established a fisheries protection zone of 200 nautical miles around Svalbard. This is a nondiscriminatory zone, which was primarily established to gain control of the fishing in the area in order to protect resources and prevent unregulated fishing. In short, to follow up your question, the waters and continental shelf surrounding Svalbard belong to Norway, as does Svalbard. The content of the Svalbard Treaty, however, only applies to Svalbard and the 12-mile territorial zone around the islands.

6. *There has been some discussion amongst Norwegian politicians last year of housing refugees in Svalbard. How far will this proposal go in your view, and is it a good one?*
I have not seen any serious discussion about such an option.

[a]Interview conducted by the author on January 26, 2016. Jonas Store is currently the leader of the Labor Party.

migrate. Their economic rights would be placed on a par with those of the citizens of the claimant states.

Antarctica, like Svalbard, would continue to be demilitarized. Its infrastructure, such as the "public wireless telegraphy stations" referred to in

Article 4 of the Svalbard Treaty, would be available to all to enjoy on a non-discriminatory basis, and any taxes or charges collected in an Antarctic territory would have to be reinvested in that same jurisdiction. In fact, the model would present a tremendous advantage for the developing world because it would not have to worry, be responsible for, or invest in the exercise of sovereignty over such remote and inhospitable lands, but would simply benefit from those lands and their resources.

For the territorial claimants, on the other hand, this solution is likely to be the closest they could ever get to having their sovereign claims recognized. It is unlikely that any other mechanism (except possibly war) would deliver the same degree of sovereignty. The Svalbard model of sovereignty minus exclusivity is a provocatively feasible potential resolution to the Antarctic problem.

In addition, with all seven territorial claimants, the United States, and Russia being members of the World Trade Organization (WTO),[125] WTO rules would automatically apply to their sections of Antarctica. This would allow fair economic competition for the continent's resources, resolving one of the main concerns that led to the collapse of CRAMRA. At the time, Australia feared that it would find itself competing against heavily subsidized Japanese miners. With Australia's sovereignty confirmed over its Antarctic territory, the automatic extension of WTO rules to Antarctica would address that problem.

The unclaimed portion of Antarctica would probably—and quite logically—be divided between the United States and Russia, the two countries that had reserved their right to table territorial claims at the time of the Antarctic Treaty. As to the dispute over the overlapping territorial claims, it would become much easier to resolve when the sovereignty question is divorced from economic exploitation and occupation of the continent. One can easily imagine the United Kingdom, in particular, dropping its territorial claim in such circumstances, given what it already considered to be its victory in the Falklands. However, this book does not advocate that the Svalbard model be applied to Antarctic without modification. Rather, what is being advocated is the application of the main principle embedded in the model—that of diminished sovereign rule.

In this connection, it is worthwhile to note that a number of Antarctic experts have argued that sovereignty—even in the form of the seven "frozen" territorial claims—has served the continent well over the years. In other words, it may actually be helpful for the continent to have some measure of sovereignty be preserved.

While generally critical of territorial claims in Antarctica, Alan Hemmings, for example, has argued that sovereignty has nevertheless had the benefit of raising the politics of international engagement on the continent to a higher level than might otherwise have been the case. It has "locked in," he says, a level of engagement by the claimant states in the management of Antarctic affairs that might not have materialized under a different legal arrangement.[126] Peter Beck agrees that the frozen claims have acted as a positive force on the continent, fostering what he calls "a kind of dialectical process of conflict and cooperation" that has helped Antarctic Treaty parties turn the regime into a forum for "multilateral collaboration."[127] Francisco Orrega Vicuna has gone even further:[128]

> Sovereignty has become inseparable from cooperation in the Antarctic since it would be difficult to maintain sovereignty unless it were closely linked with the factor of cooperation. Conversely, it can also be said that cooperation on its own, if it does not rely on the support provided by national interest in the Antarctic, including those related to claims in sovereignty, would probably lack the necessary vitality.

What this suggests is that some degree of ownership enhances rather than hinders governance. In fact, this is precisely what has been seen in Svalbard, where the Norwegian government is responsible for protecting the fragile Arctic environment of the archipelago and ensuring the safety of its residents. In Antarctica, sovereignty could prove more vital than in Svalbard because the bulk of the continent (besides the Antarctic Peninsula) is miles from any other territory. Assigning a degree of responsibility to the sovereigns to watch over these distant lands and be held accountable for them could benefit the continent. In fact, in Antarctica, nothing could be worse for the environment than a treaty without enforcement powers (i.e., the current situation), a matter that is explored further in the next chapter.

## THE GRADUAL ECONOMIC DIVERSIFICATION OF SVALBARD AND EXPANSION OF POLITICAL RIGHTS

What is life in Svalbard like, and what would life in Antarctica look like under a Svalbard regime? Whereas the residents of Svalbard enjoy equal economic rights and have done so since the entry into force of the Svalbard Treaty, do they enjoy equal political rights? In other words, if a Svalbard model were applied to Antarctica, would its citizens be governed from afar by the claimant states? Would the citizens of those states enjoy more political rights than foreigners? The Svalbard experience demonstrates that Norwegians, Scandinavians, and foreigners saw their political rights expand along with population growth and the economic diversification of the archipelago.

To understand how Svalbard economy grew, one must turn to the structure of the archipelago. Spitsbergen, the largest island in Svalbard, is home to its main settlement, Longyearbyen. Other important settlements in Spistbergen include the community of Barentsburg, a Russian mining town; and Ny-Alesund, another mining town that progressively became Svalbard's scientific research hub. Prior to the Svalbard Treaty, different towns were essentially ruled by the mining companies themselves.[129]

Svalbard's economic history testifies to gradual economic diversification—an archipelago that gradually transitioned away from whaling and sealing to mining, and subsequently to other economic activities. This did not happen by chance; in the 1970s, it formed part of the Norwegian government's explicit policy of "normalization" in Longyearbyen. It was a policy intended to convert Svalbard's main settlement from a company town to a normal Norwegian municipality.[130]

Mining activities were initiated in Svalbard by American businessman John Munro Longyear (1850–1922) before World War I. He was one of the founders of the Arctic Coal Company, which built the first houses in Longyearbyen.[131] The company went through financial difficulties during the course of the war and was sold to Store Norske, registered in Oslo. In the mid-1970s, it was purchased by the Norwegian state, with most of Longyearbyen's residents continuing to work in mining. Longyearbyen's economy underwent major diversification in the 1990s as a result of the Norwegian government's policy to promote scientific research and tourism in the archipelago. This saw the founding of the first Arctic sciences university and the opening of numerous shops, hotels, bars, and restaurants, making Longyearbyen the first economically diverse settlement in Svalbard.[132]

Ny-Alesund, initially also a mining town, found its fate determined by a mining accident whose seriousness resonated back in the mainland, making continued mining impossible. To help Ny-Alesund diversify, the Norwegian government used a 1964 request by the European Space Research Organization to establish a satellite ground station in Longyearbyen, as the lever, suggesting that the station be opened in Ny-Alesund instead. The town was transformed almost overnight into a scientific research hub. Eager to have science promote informal and peaceful diplomatic activity among the signatories of the Svalbard Treaty, the government turned scientific research into one of the main planks of the Svalbard economy.[133]

Norway's promotion of scientific research in Ny-Alesund was also driven by its desire to prevent the settlement from falling into Russian hands, as Barentsburg had. However, some have questioned whether the government's policy has succeeded in reinforcing Norwegian sovereignty. The strategy is said to have had the unintended consequence of offering many non-Arctic states a "gateway into the Arctic," with more than 10 different countries now running research stations in Ny-Alesund. They include Japan, South Korea, China, and India. China's Huanghe Research Station has been among the most active.[134]

Tourism in Svalbard began to take off even before the government decided to promote it. The opening of an international airport in Longyearbyen in 1975 was a turning point in enhancing access to the archipelago. While much debate had taken place in Norway over whether to expressly promote tourism given Svalbard's fragile environment, the eventual approval from the government turned Longyearbyen into a prime tourist destination. Today, tourists flock into Longyearbyen from all over the world by air and sea, with tourism spreading to all corners of the archipelago.[135]

The gradual economic diversification, or "normalization" of Longyearbyen, was accompanied by population growth, an increasingly complex population structure. From a few hundred residents when Longyearbyen was created, Svalbard currently has 2,667 inhabitants[136] (roughly comparable to the population of the Falkland Islands), from a total of 44 different countries.[137] The bulk of Svalbard's population is concentrated in Longyearbyen (a little over 2,000 inhabitants), and Barentsburg (some 470 residents).[138]

The vast majority of Svalbard's population is transient and of working age. For example, in the Norwegian settlements, only 9 people have lived there since the 1960s, 36 since the 1970s, and 75 since the 1980s.[139] Most of Svalbard's population is between the ages of 20–44 years, with men outnumbering women. In the Norwegian settlements, approximately 18 percent of the population is currently foreign. Roughly half of the foreign population is women, with a third being from Thailand. In the Russian settlements, 80 percent of the residents are Ukrainians.[140]

Despite its transient and special population structure, experts argue that a sense of "local identity" has developed over the years in Longyearbyen, which is reflected in its pubs, restaurants, and sustained tourist activities.[141] However, foreigners living in Svalbard cannot acquire Norwegian citizenship, in large part because of the sizeable Thai population, and the Norwegian government does not provide any social welfare, such as medical services, to Svalbard residents.[142]

Calls for some degree of local democracy grew in Longyearbyen over the years when it became evident that Svalbard's Norwegian governor could no longer run what had become a bigger and more economically complex settlement. During the Cold War, Norway's Svalbard policy had focused on national security, with greater local democracy in the islands seen as a potential hazard.

However, the speed of change in Longyearbyen made the governor's absolute authority untenable, forcing Norway to reassess its position. It agreed in 2002 to establish a Longyearbyen Community Council, with responsibility for "infrastructure in the settlement, social and area planning, economic planning, business work, production of statistics, development and coordination of social services for children, youth and adults, child protection, social work, youth work, child care services. . . ." The governor retained responsibility for these matters in all other parts of the archipelago.[143]

Of note is the fact that the Norwegian government has not sought to promote local democracy in other settlements in Svalbard, especially the Russian settlement of Barentsburg. Local democracy has been seen as less threatening to Norwegian sovereignty in areas where the bulk of the population is Norwegian (i.e., in Longyearbyen). That said, not all of Longyearbyen's residents had supported the call for increased local democracy, in fear of higher taxes and greater bureaucracy. The Svalbard Treaty limits Norway's right to collect taxes to that of financing services in Svalbard, and Longyearbyen has historically enjoyed lower income tax than mainland Norway, as well as no value-added tax.[144]

The implications for Antarctica are clear. Under a Svalbard model of governance, calls for local democracy would be likely to grow. In fact, life in Antarctica will probably evolve along a similar path to Svalbard, through gradual economic diversification and the expansion of human settlements. While claimant states will initially be the ones to govern Antarctica from afar, decentralized governance would impose itself over time. This can be seen as an opportunity for better government, a government closer to the people, that can create a regulatory environment better tailored to the region's needs.

In this regard, it may important to recall the main message that the former prime minister of Greenland, Aleqa Hammond, delivered to the future residents of Antarctica: "Decide your own destiny and your own future—do not let others decide what should happen to you. Everyone talks about conserving and preserving natural resources. You will have your own life, you will live there, and you should be the ones to decide."[145]

The interview in Box 6.2 of Adam Grydehoj, the director of Island Dynamics and a Svalbard expert, provides a sense of what life is like today in Svalbard. Particularly noteworthy is his statement that in Longyearbyen, any Norwegian or Scandinavian citizen can vote in local elections, and that non-Scandinavians can also do so if they have been registered as living in

### Box 6.2 Interview of Adam Grydehoj, Director of Island Dynamics and the Ephinstone Institute, University of Aberdeen, United Kingdom[a]

1. *What is it like to live in Svalbard? Is a Svalbard identity gradually developing?*

Svalbard is a set of islands, little pinpoints on the map. Its main settlements are relatively sparsely populated. Barentsburg and Longyearbyen are close to one another. Ny-Ålesund is quite distant, though. However, to get from Longyearbyen to Barentsbury in winter, you need a snowmobile. To get to Ny-Ålesund, you would need to fly. Boats can make the trip during the summer. Of course, you can't go out without a gun because of polar bears.

Longyearbyen is a strikingly Norwegian town, artificially funded and sustained by the mainland. It has an unusually high number of restaurants, cafes, and bars per capita. It is strikingly nonwild I would say. In terms of identity, I would say most people feel they belong to the specific town they live in. You have to remember that there is a massive population turnover in Svalbard, with most people living there only a few years on contract jobs.

Having said that, a Svalbard identity is forming, but in a way that is specific to circumstances of these islands. From the day a person moves to Svalbard, they are immediately considered "local." This is because all of Svalbard is populated by immigrants who are there on a short-term basis. With everyone being an immigrant, the sense of identity there is different from anywhere else.

2. *Do different nationalities live in different parts of Svalbard, or are they beginning to mix?*

The history of Svalbard is relevant to this question. Whaling was the first commercial activity in Svalbard, followed by mining, and the islands are now principally occupied by Norwegians and Russians. There are no borders, though, between different settlements. The Norwegian state owns most of the land, and the Russian state owns companies that have the right to do things on the land. Initially, it was the coal mining companies that had complete control of Svalbard and that ran everything. However, as Norway gradually liberalized the economy of Longyearbyen, it became more multicultural, multiethnic, and there are now many different nationalities living in Longyearbyen in particular.

People know each other in Longyearbyen; it's a very small town that is predominantly Norwegian. So, yes, people are increasingly mixing.

3. *What language is predominantly spoken in Svalbard?*
Barentsburg is a Russian town, and so it is Russian that is mostly spoken there. In Longyearbyen, with nearly 80 percent of the residents being Norwegian, it is Norwegian that is predominantly spoken, but you do hear an awful lot of English. You have to remember that since most people moving to Svalbard are contract workers, they have no incentive to learn Norwegian, so English continues to be a very important language. In terms of the local authorities, they are all Norwegians and conduct their work in Norwegian. This can create a disconnect between them and the residents.

4. *It is often said of Svalbard that no births or deaths can take place on the islands. Is that true?*
Indeed this is often said; however, this does not come from the Svalbard Treaty, but rather boils down to Norway's plans for Longyearbyen. There are no social services in Longyearbyen, with no social security being provided by the Norwegian government. While there is a hospital in the town, most people are taken south for anything serious. In addition, there are no old-age homes, with the expectation being that people will leave once their contracts are over. So while nothing in the law says that you cannot be born, you cannot get sick, and you cannot die in Svalbard, it is simply understood that the Norwegian state will not be taking care of you. Over the coming decades, it is my sense that pressure will mount to do something about that, in particular as local democracy gets stronger.

5. *What is it like to enter Svalbard? Are there any border controls? Are passports required?*
When you arrive in Svalbard, there are no border controls of any kind, and Svalbard is not part of the Schengen zone. However, the issue isn't how to enter Svalbard, but rather how to get to it. You can only get to Svalbard from the northern Norwegian city of Tromso or from Russia. So Norway has tried to limit the access to Svalbard from Tromso. It can restrict transit visas, for example. Norway's reading of its rights over Svalbard are different to that of almost any other country.

Anyone from any signatory state to the Svalbard Treaty can exercise the same right as Norway on the islands and engage in economic activity on an equal footing to Norwegian citizens, even if Norway has changed the definition of economic activity over time. That said, de facto, Svalbard is open to the citizens of absolutely any country and not just Treaty parties. This simply became the practice over the years.

Part of the trouble for Norway is that when people get to Svalbard, it is impossible to get rid of them. The authorities cannot kick anyone out, except Norwegians perhaps. They could expel a Norwegian criminal, for example, but would have to tread much more carefully were a Russian citizen to be involved.

6. *How much interaction would an average resident of Svalbard have with local authorities?*

It depends which local authority we are talking about. In Longyearbyen, there is a lot of interaction because the local government is elected and the officials live in town. However, there is competition between the local government and the governor. The governor is appointed by Norway, whereas the local government is elected in Svalbard. In fact, some of the residents of Svalbard had been opposed to the formation of a local government because of concerns over higher taxes.

In general, the government in Longyearbyen is not distant from the people. But it is important to remember that Longyearbyen was a company town, so big companies are still very influential. Norway is slowly trying to gain more control of the Russian towns and sometimes does so through the Norwegian courts, which disadvantages the Russians. It has been placing road signs in Norwegians as a symbolic act for example. At some point, the Governor's office restricted the number of helicopters from Longyearbyen to Barentsburg arguing that while tourism is an economic activity, helicopter transport is not and is therefore within its authority to regulate.

7. *To what extent is there local democracy in Longyearbyen and who can take part?*

Longyearbyen residents can vote for the local government if (1) they are Norwegian or other Scandinavian citizens who are at least 16 years old and have been registered as residents of Longyearbyen, or (2) for non-Scandinavian citizens, if they have been registered as residents in Longyearbyen and/or a Norwegian municipality for the previous three years. However, the bigger issue is one of accountability. When the average person lives there for only one election and is not around for the next, to what extent can there be long term planning? Only about 10 percent of the population is stable.

In terms of the types of decisions taken at the local level, I would say that they are like the local decisions in any other Norwegian town. Most of the decisions revolve around how to spend governmental revenue since, under the Svalbard Treaty, it has to be reinvested in Svalbard. However, with the tax rate being very low and the tax base so small, we're not talking about a lot of revenue. On the other hand, in Svalbard, the local government does not have to worry about providing social services.

8. *There has been some discussion amongst Norwegian politicians recently of housing refugees in Svalbard, in particular the Green Party. What are your thoughts on this proposal?*
Refugees, or anyone for that matter, can already go to Svalbard. Norwegian immigration law does not apply to Svalbard. In addition, Norway would have no legal way of getting rid of people. The question isn't whether refugees can be moved or housed in Svalbard, but how refugees would get there. It is about the geographic and not the legal accessibility of Svalbard—the affordability of travel.

[a]Interview conducted by the author on January 26, 2016.

Longyearbyen or a Norwegian municipality over the previous three years. In other words, over time, a Chinese, a Korean, or an Egyptian living and working in the AAT would be likely to see their political rights grow the longer they live in Antarctica. In fact, Grydehoj argues that local democracy will continue to get stronger in Svalbard.

Equally noteworthy is his comment that "Svalbard is open to the citizens of absolutely any country and not just Treaty parties. This simply became the practice over the years." In other words, Svalbard has now internationalized beyond the bounds of the Svalbard Treaty. Antarctica may indeed simply become open to the entire world—to anyone enterprising enough to move to the seventh continent.

## "REALPOLITIK" IN SVALBARD

In 2005, Foreign Affairs Minister Jonas Store presented the Norwegian Government's High North Strategy[146] in Tromso, arguing that a particular kind of politics would be required in the Arctic. Norway would have to implement a "realpolitik" that would reflect the end of the Cold War, the opening up of the Arctic Ocean, and the continued quest for energy resources. He spoke of policymaking in the Arctic in the context of a *handlingsrom* (maneuvering space), in which politicians would confront a *mulighetsrom* (a sphere of possibilities), suggesting that the governance of the Arctic would have to weigh domestic interests against foreign interests in a delicate balancing act.[147]

In fact, the governance of Svalbard has been nothing but a balancing act, particularly in being one of the first experiments of its kind internationally. What is visible in this balancing act is that all sides have had to exercise flexibility in the interest of making the Svalbard Treaty work and adapting it to changing circumstances. Over the years, Norway has tried to assert its sovereignty over the archipelago through measures such as the strengthening of the governor's office, which was gradually expanded and, equipped with helicopters and police officers, an increase of its budget for Svalbard,[148] and the gradual introduction of Norwegian road signs throughout the archipelago, including its Russian settlements.[149] In fact, it has deliberately placed Svalbard policy under the Ministry of Justice and not the Ministry of Foreign Affairs, confirming its position that Svalbard is a domestic matter.[150]

However, it has also known when to exercise restraint, especially in matters of setting the boundaries between environmental protection and development. For example, in his 2002 visit to Norway, Russian president Vladimir Putin is said to have expressed only one concern: Norway's excessive environmental regulation of Svalbard. At the time, Norway had wanted to establish a protected area immediately outside Barentsburg, which Putin believed ran counter to the spirit of the Svalbard Treaty, limiting his country's economic rights. The Norwegian government accepted the concern, dropping its plan.[151]

Thus, while Store stated that Norway applied environmental regulations in Svalbard on a nondiscriminatory basis, clearly some realpolitik has been applied when necessary. On the other hand, there is also evidence of the Russians making an effort to comply with Norwegian regulations. Torbjorn Pedersen writes:[152]

> Even the Russians have increasingly adapted to Norwegian rules. For example, Norwegian traffic signs have appeared in Barentsburg. The

> Russian mining company Trust Arktikugol produces, at least to some extent, area plans and other paperwork to the Governor in accordance with Norwegian regulations. Accidents and crimes in the Russian settlement are now reported to, and investigated by, Norwegian police. In 2006, a trial was held in Barentsburg before a Norwegian court. Russian and Ukrainian miners were tried and convicted of reindeer poaching, arguably a show of Norwegian rule within the Russian society that would have been unthinkable a decade ago.

However, nowhere is friction between the treaty parties as visible as in Svalbard's surrounding waters. This is because the Svalbard Treaty was concluded long before the UNCLOS established a 200-nautical-mile EEZ. As mentioned earlier, while the Svalbard Treaty provided equal economic rights "both on land and in the territorial waters" (Article 3), at the time of the treaty, countries enjoyed jurisdiction over only 12 nautical miles from their shores. Therefore, an argument erupted between Norway and other treaty parties over to how to reconcile the two international agreements. Norway claims that the treaty only covers 12 nautical miles, and that it should be the only one to benefit from the 200-nautical-mile EEZ (a view confirmed by Jonas Store in the interview conducted with him).[153]

Norway's interpretation, of course, has been contested by other fishing nations and countries wanting to exploit the Svalbard continental shelf. While Norwegian officials have taken a firm stance on the matter in public declarations, even going as far as to argue that disagreeing parties can simply go the International Court of Justice (ICJ), they sought to ease the tension by declaring Svalbard's surrounding waters a fisheries protection zone in 1977.[154] This meant that Norwegian environmental protection regulations would apply to these waters on a nondiscriminatory basis, with conservation efforts incumbent upon on all Svalbard Treaty parties. Minister of Foreign Affairs Knut Frydenlund said at the time: "We chose to establish a fisheries zone instead of a full economic zone much due to our aspirations to avoid pushing the controversy amongst Norway and some other states . . . to the extreme."[155] This is another example of realpolitik in Svalbard, although the jury is still out on how long this compromise will hold.

While it is said that Norway's enforcement of environmental regulations in maritime areas adjacent to Svalbard is more lenient than in other territorial waters,[156] there is nevertheless evidence that Norway's "gentle enforcement" has succeeded in achieving compliance. Some have argued that there has been "creeping compliance" with the protection zone as a result of fishers needing the services of the Norwegian coast guard, not simply seeing them as a control body. Their search-and-rescue operations and medical and ice breaking services, for example, are particularly useful to fishers.[157]

Perhaps also in the spirit of realpolitik, Norway's Green Party has recently proposed the creation of a reception center for Syrian refugees in Svalbard, in light of the ongoing civil war in Syria. The proposal appears to have been welcomed by the local government of Longyearbyen, which believes that it has the potential to create jobs and expand the Svalbard economy, in addition to serving a humanitarian goal.[158] While Jonas Store said that he had not heard any serious discussion of this proposal, Adam Grydehoj argued that there are no legal obstacles to housing refugees in Svalbard. Rather, the main impediment is Svalbard's lack of accessibility.

Clearly, the application of the Svalbard model to Antarctica will require even more, rather than less, realpolitik, given the range of countries involved, the size of Antarctica, and of its potential resources. So far, realpolitik has served Svalbard well, making a success of a governance model that few would have thought possible. In Antarctica, this governance model would capitalize on the spirit of cooperation that has already been generated by the existence of the Antarctic Treaty for more than 50 years. A similar approach, in fact, to the territorial waters of Svalbard could be contemplated in Antarctica, with claimants asked to enforce a nondiscriminatory 200-nautical-mile fishery protection zone off their coast. Any exploitation rights within that zone would have to be extended to all countries on a discriminatory basis.

In addition, what now seems like an impossible idea of housing refugees in Svalbard should be seriously considered in Antarctica. While today this proposal may seem inhumane, perhaps even politically incorrect, because of the inhospitable conditions of the Arctic and the Antarctic, turning this idea into a carefully orchestrated plan for the transfer of climate migrants to the seventh continent could provide these people with a new lease on life. In fact, the only political incorrectness would lie in denying helpless climate migrants a chance for a new beginning on what is currently an unowned continent. The harshness of the Antarctic Peninsula's climate would be overcome in the same way that it has been in Greenland—and even so, this harshness is likely to pale in comparison to the alternatives that the refugees will have.

## CONCLUDING REMARKS

The breakup of the Antarctic Treaty is heresy to some. After all, it is widely seen as one of the most successful international agreements negotiated after World War II. The treaty has indeed been successful; it has brought peace and science to the seventh continent. However, as with any good treaty, it should not be set in stone. Good legal regimes require good management by their parties; they also require them to raise the red alert when a fundamental course change to the regime is required. That alert is soon approaching in Antarctica, with treaty parties allowed in 2048 to express themselves on an issue that goes to the very heart of the regime—their ability to mine and economically exploit the seventh continent.

Declaring Antarctica a World Park is even more unlikely today than it was back in 1959, when little was known about the continent's mineral potential and how fast its ice would melt to create living space. In addition, since the conclusion of the Antarctic Treaty, and despite its suspension of territorial claims, countries from all corners of the globe have competed to gain and to strengthen their foothold on the continent. Through a cross-recognition of territorial claims by five of the claimants, the spread of scientific research stations, the assignment of place names, and competition over EEZs and the continental shelf, few will be likely to agree that the continent should be closed to commercial exploitation.

Similarly, awarding Antarctica to its seven territorial claimants is a highly unlikely scenario; the developing world will certainly protest. Even as CRAMRA was being discussed in the 1980s, the United Nations called for a suspension of the negotiations until everyone could participate. In addition, the expansion in the number of parties to the Antarctic Treaty, as well as the rise of research stations from countries as far from the continent as China, South Korea, and India, demonstrates how such a scenario could never materialize.

The revival of CRAMRA, or of a modified CRAMRA, would be equally unlikely to succeed. The Common Heritage of Mankind principle would likely be right at the center of the negotiations, and as seen in the Moon Treaty and UNCLOS, it has the capacity to paralyze treaties and prevent universal participation. Negotiated to build on the Outer Space Treaty, the Moon Treaty committed states to establish an international regime to govern the exploitation of the Moon's natural resources "as such exploitation is about to become feasible," and went into the issue of benefit sharing. Widely interpreted as a moratorium, this resulted in a boycott of the treaty by all countries that currently run space programs.

The UNCLOS negotiations taking place in parallel stumbled over the very same issue—the application of the Common Heritage principle to seabed mining, with the mining provisions negotiated in 1982 requiring another decade to renegotiate. Even then, the United States—the main country with the capacity to mine the seabed—refused to join the 1994 Agreement, leading to a fractured UNCLOS regime. However, the developing world's success in embedding the Common Heritage principle in two of the most important legal regimes dealing with the global commons is likely to strengthen their call for the principle's inclusion in any new and emerging order for Antarctica. This would only extend the tug of war seen between the developed and the developing worlds in the Moon Treaty and UNCLOS to yet another regime, leading to further unsatisfactory outcomes.

The Svalbard model offers an extremely attractive option for Antarctica: sovereignty without exclusivity. It has the capacity to reconcile the Common Heritage principle with the interests of the seven territorial claimants, which also means reconciling the interests of the developed and the developing world. It would put Antarctica within all of humanity's reach, whether in terms of mineral resources or living space, while allowing the exercise of sovereignty. Some degree of sovereignty in a continent as remote as Antarctica could help promote good governance. It would deepen political engagement on the continent, create enforcement capacity for regulatory frameworks, and promote a sense of accountability for whatever transpires in these fa away lands. The Svalbard model would also allow the international community to actively plan and organize the transfer of climate migrants to the seventh continent. It provides a compelling and possibly permanent solution to the Question of Antarctica.

## NOTES

1. See the section "Gateways to Antarctica: Existing Industries," in Chapter 5 of this book, for tourism statistics.
2. See Article 25 of the Madrid Protocol on "Modification or Amendment," in Secretariat of the Antarctic Treaty, *Compilation of Key Documents of the Antarctic Treaty System*.
3. See the section "The Collapse of CRAMRA," in Chapter 4, on the proposal by nongovernmental organizations to designate Antarctica as a World Park.
4. See Figure 3.1 in Chapter 3 for a map of Antarctic territorial claims.
5. See Chapter 3 for a full account of the UN discussions on the Question of Antarctica.
6. Hemmings, "Security Beyond Claims," in *Antarctic Security in the Twenty-First Century*, 87.
7. Wolfrum, *Regulation of Antarctic Mineral Resource Activities,* 2, 6.
8. UN General Assembly, *The Question of Antarctica*, Resolution A/RES/41/88, December 4, 1986.

9. Joyner, *Governing the Frozen Commons*, 45. For a discussion of the Antarctic commons, see also Bernard P. Herber, *Protecting the Antarctic Commons: Problems of Economic Efficiency* (Tucson: Udall Center for Studies in Public Policy, the University of Arizona, 2007).

10. *Island of Palmas Case* (Netherlands, USA), Reports of the International Arbitral Awards, vol. II, April 4, 1928, p. 838.

11. Oran R. Young, "Governing International Spaces: Antarctica and Beyond," in *Science Diplomacy*, 288.

12. Joyner, *Governing the Frozen Commons*, 30–35. These legal concepts are, of course, interlinked; for example, *res communis* preceded today's concepts of public domain and the Common Heritage of Mankind.

13. Ibid., 44.

14. See Chapter 3 for a full account of the UN discussions on the Question of Antarctica.

15. Young, "Governing International Spaces," 291.

16. See the section "Claims of Territorial Sovereignty," in chapter 3, for an overview of the concepts used to lay territorial claims in Antarctica. Also see the section "The United States in the Driver's Seat," for the U.S. views on the distinction between discovery and exploration on the one hand and occupation on the other. For an interesting analysis of the legitimacy of some of the legal concepts used to claim territory in Antarctica, see Benedetto Conforti, "Territorial Claims in Antarctica: A Modern Way to Deal with an Old Problem," *Cornell International Law Journal* 19 (1986): 249–258; and Shirley V. Scott, "Universalism and Title to Territory in Antarctica," *Nordic Journal of International Law* 66 (1997): 33–53.

17. Joyner, Governing the Frozen Commons, 52.

18. Ibid., 29. For a discussion of the tragedy of the commons, see Kathryn Milun, *The Political Uncommons: The Cross-Cultural Logic of the Global Commons* (Burlington: Ashgate Publishing, 2011).

19. See Figure 3.1 in Chapter 3 for a map of Antarctic territorial claims. Of course, the argument is made that inland areas of the continent are just as important to research as the coastal areas and offer entirely different possibilities with respect to astrological research—and I subscribe to that position.

20. Aant Elzinga, "Rallying Around a Flag? On the Persistent Gap in Scientific Internationalism Between Word and Deed," in *The Emerging Politics of Antarctica*, Anne-Marie Brady, ed. (London: Routledge, 2013), 209.

21. Elena Fogarty, "Antarctica: Assessing and Protecting Australia's National Interests," Policy Brief, Lowy Institute for International Policy, August 2011, 5.

22. Elzinga, "Rallying Around a Flag?" 211. France is the only claimant state with a research station outside its claimed territory.

23. Ibid., 210.

24. Conforti, "Territorial Claims in Antarctica," 250.

25. See, for example, the Australian Antarctic Division Place Names Committee; http://www.antarctica.gov.au/law-and-treaty/antarctic-place-names.

26. Australian Antarctic Data Center, *The SCAR Composite Gazetteer;* https://data.aad.gov.au/aadc/gaz/scar/statistics.cfm.

27. Ibid. See also Clancy, Manning, and Brolsma, *Mapping Antarctica*, and Wright, *The Seventh Continent.*

28. "Argentina y Chile hicieron nacer bebés en la Antartida para reclamar el territorio," *Ondas*, December 11, 2008; http://www.radiocable.com/bebes-antartida775.html#sthash.nf5S5Rkz.dpuf [in Spanish].

29. Kara Kovalchik, "First Baby Born in Antarctica," *Not-So-Famous Firsts: Infant Edition,* September 3, 2010; http://mentalfloss.com/article/25676/not-so-famous-firsts-infant-edition.

30. "Argentina y Chile hicieron nacer bebés en la Antartida para reclamar el territorio." See also Camila Bravo, "Chile en la Antártica: A treinta años del primer chileno nacido en el fin del mundo," *24 Oras*, November 25, 2014; http://www.24horas.cl/tendencias/espectaculosycultura/chile-en-la-antartica-a-treinta-anos-del-primer-chileno-nacido-en-el-fin-del-mundo—1506774 [in Spanish].

31. Anahi Aradas, "Antártica: ¿Cómo se vive en una de las zonas más remotas del mundo?," *BBC Mundo*, August 11, 2013; http://www.bbc.com/mundo/noticias/2013/07/130423_sociedad_villa_estrellas_aa [in Spanish].

32. Simon Romero, "Antarctic Life: No Dogs, Few Vegetables, and 'a Little Intense' in the Winter," *International New York Times*, January 6, 2016.

33. See, for example, Sanjay Chaturvedi, "India and Antarctica: Towards Post-Colonial Engagement," in *The Emerging Politics of Antarctica*, 50–74; and Klaus Dodds, "Post-Colonial Antarctica: An Emerging Engagement," *Polar Record* 42 (2006): 59–70.

34. Alan D. Hemmings, "Beyond Claims: Towards a Non-Territorial Antarctic Security Prism for Australia and New Zealand," *New Zealand Yearbook of International Law* 6 (2008): 78, 87.

35. Alan D. Hemmings, "Re-Justifying the Antarctic Treaty System for the 21st Century: Rights, Expectations, and Global Equity," in *Polar Geopolitics? Knowledges, Resources, and Legal Regimes*, Richard C. Powell and Klaus Dodds, eds. (Northampton, UK: Edward Elgar, 2004), 66.

36. Hemmings, "Security Beyond Claims," 84.

37. Ibid., 90.

38. U.S. Department of State, Office of the Historian, "Decolonization of Asia and Africa, 1945–1960"; last accessed March 15, 2016, https://history.state.gov/milestones/1945-1952/asia-and-africa.

39. Christopher Garrison, "Beneath the Surface: The Common Heritage of Mankind," *KEStudies* 1 (2007): 7.

40. Ricky J. Lee, *Law and Regulation of Commercial Mining of Minerals in Outer Space* (New York: Springer, 2012), 225–226.

41. Ibid., 273.

42. Ibid., 109.

43. Ibid., 257–258.

44. Barbara Ellen Heim, "Exploring the Last Frontiers for Mineral Resources: A Comparison of International Law Regarding the Deep Seabed, Outer Space, and Antarctica," *Vanderbilt Journal of Transnational Law* 23 (1990–1991): 833.

45. UN Division for Ocean Affairs and the Law of the Sea, "The UN Law of the Sea: A Historical Perspective"; last accessed March 15, 2016, http://www.un.org/depts/los/convention_agreements/convention_historical_perspective.htm#Historical%20Perspective.

46. Garrison, "Beneath the Surface," 17. For Arvid Pardo's full speech, see the records of the UN General Assembly, Twenty-Second Session, 1515th Meeting, "Agenda Item 92: Examination of the Question of the Reservation Exclusively for Peaceful Purposes of the Seabed and Ocean Floor, and the Subsoil Thereof, Underlying the High Seas Beyond the Limits of Present National Jurisdiction, and the Use of Their Resources in the Interest of Mankind, General Debate," November 1, 1967. http://www.un.org/depts/los/convention_agreements/texts/pardo_ga1967.pdf.

47. UN General Assembly, Twenty-Second Session, 1516th Meeting, "Agenda Item 92: Examination of the Question of the Reservation Exclusively for Peaceful Purposes of the Seabed and Ocean Floor, and the Subsoil Thereof, Underlying the High Seas Beyond the Limits of Present National Jurisdiction, and the Use of Their Resources in the Interest of Mankind, General Debate," November 1, 1967, para. 8 and 13.

48. Garrison, "Beneath the Surface," 24.

49. Davor Vidas, "An Overview," in *Governing the Antarctic: The Effectiveness and Legitimacy of the Antarctic Treaty System* (Cambridge, UK: Cambridge University Press, 1996), 51. The concept of the Common Heritage of Mankind would also extend in 1997 to the UNESCO Universal Declaration on the Human Genome and Human Rights, with Article 1 stating, "The human genome underlies fundamental unity of all members of the

human family, as well as the recognition of their inherent dignity and diversity. In a symbolic sense, it is the heritage of humanity." This is described in Garrison, "Beneath the Surface," 63.

50. Garrison, "Beneath the Surface," 64. See also Joyner, *Governing the Frozen Commons,* 33.

51. Joyner, *Governing the Frozen Commons,* 230, 234.

52. John E. Noyes, "The Common Heritage of Mankind: Past, Present, and Future," *Denver Journal of International Law and Policy* 40 (2011–2012): 455.

53. Wolfrum, *Regulation of Antarctic Mineral Resource Activities,* 104, 130, 141.

54. Garrison, "Beneath the Surface," 58.

55. Lee, *Law and Regulation of Commercial Mining of Minerals in Outer Space,* 259.

56. Ibid., 273–74.

57. Garrison, "Beneath the Surface," 59. The full text of the Moon Treaty (Agreement Governing the Activities of States on the Moon and Other Celestial Bodies) can be found at the UN Office for Outer Space Affairs; last accessed March 15, 2016, http://www.unoosa.org/oosa/SpaceLaw/moon.html.

58. Lee, *Law and Regulation of Commercial Mining of Minerals in Outer Space,* 268–269.

59. Garrison, "Beneath the Surface," 60.

60. For the status of the Moon Treaty, visit the UN Office for Outer Space Affairs; http://www.unoosa.org/oosa/SpaceLaw/moon.html.

61. Lee, *Law and Regulation of Commercial Mining of Minerals in Outer Space*, 274. See also Armel Kerrest, "Outer Space as International Space: Lessons from Antarctica," in *Science Diplomacy*, 133–42.

62. Division for Ocean Affairs and the Law of the Sea, "The UN Law of the Sea."

63. Ibid.

64. Thomas M. Franck, Mohamed El Baradei, and George Aron, "The New Poor: Land-Locked, Shelf-Locked, and Other Geographically Disadvantaged States," *New York University Journal of International Law and Politics* 7 (1974): 33–57.

65. Third UN Conference on the Law of the Sea, "Letter Dated 5 May 1978 from the Representative of Nepal to the President of the Conference," in *Extract from the Official Records of the Third UN Conference on the Law of the Sea,* vol. X. Reports of the Committees and Negotiating Groups, Seventh and Resumed Seventh Session, Document A/CONF. 62/65, 2009, 176.

66. Ibid., 175.

67. William C. Lynch, "The Nepal Proposal for a Common Heritage Fund: Panacea or Pipedream," *California Western International Law Journal* 10 (1980): 42.

68. Garrison, "Beneath the Surface," 42.

69. Lee, *Law and Regulation of Commercial Mining of Minerals in Outer Space,* 246.

70. Full text of UNCLOS, last accessed March 15, 2016, http://www.un.org/depts/los/convention_agreements/texts/unclos/unclos_e.pdf.

71. Lee, *Law and Regulation of Commercial Mining of Minerals in Outer Space,* 254.

72. Ibid., 249.

73. Garrison, "Beneath the Surface," 26.

74. International Seabed Authority, "Legislative History of the Enterprise," Jamaica, 2002, 352.

75. Ibid.

76. Garrison, "Beneath the Surface," 43.

77. Ibid.

78. Ibid., 44.

79. James Malone, "The United States and the Law of the Sea After UNCLOS III," *Journal of Law and Contemporary Problems* 46 (Spring 1983): 32.

80. Garrison, "Beneath the Surface," 44–45.

81. Lee, *Law and Regulation of Commercial Mining of Minerals in Outer Space,* 240.

82. UN General Assembly, "Consultations of the Secretary-General on Outstanding Issues Relating to the Deep Seabed Mining Provisions of the UN Convention on the Law of the Sea," Forty-Eighth Session, A/48/950, June 9, 1994.

83. International Seabed Authority, "Legislative History of the Enterprise," 355.

84. Lee, *Law and Regulation of Commercial Mining of Minerals in Outer Space,* 252.

85. Ibid. Full text of the Agreement Relating to the Implementation of Part XI of the UN Convention on the Law of the Sea, last accessed March 15, 2016, http://www.un.org/depts/los/convention_agreements/texts/unclos/closindxAgree.htm.

86. D. H. Anderson, "Resolution and Agreement Relating to the Implementation of Part XI of the UN Convention on the Law of the Sea: A General Assessment," Max-Planck-Institut für ausländisches öffentliches Recht und Völkerrecht, 1995, 285–287.

87. Wolfrum, *Regulation of Antarctic Mineral Resource Activities,* 95–96.

88. See Michael Wood, "The International Seabed Authority the First Four Years," *Max Planck Yearbook of UN Law,* 3 (1999): 173–241.

89. International Seabed Authority, "Legal and Technical Commission: Issues Relating to the Operation of the Enterprise, in Particular the Legal, Technical, and Financial Implications for the Authority and for States Parties," ISBA/20/LTC/12, Twentieth Session, Kingston, Jamaica, July 14–25, 2014.

90. See how the Pacific Island region describes the benefits derived from the International Seabed Authority in the following presentation: The International Seabed Authority, *SPC-EU EDF10 Deep Sea Minerals (DSM) Project,* Information Brochure 15, 2013; http://gsd.spc.int/dsm/images/pdf_files/dsm_brochures/DSM_Brochure15_ISA.

91. Guest blog post by Dr. James Harrison, "Who Benefits from the Exploitation of Non-Living Resources on the Seabed? Operationalizing the Benefit-Sharing Provisions in the UN Convention on the Law of the Sea," University of Edinburgh, July 1, 2015; http://www.benelexblog.law.ed.ac.uk/2015/07/01/who-benefits-from-the-exploitation-of-non-living-resources-on-the-seabed-operationalizing-the-benefit-sharing-provisions-in-the-un-convention-on-the-law-of-the-sea/.

92. Lee, *Law and Regulation of Commercial Mining of Minerals in Outer Space,* 288–289.

93. International Seabed Authority, "Deep Seabed Mineral Contractors"; last accessed March 15, 2016, https://www.isa.org.jm/deep-seabed-minerals-contractors?page=1&qt-contractors_tabs_alt=0.

94. International Tribunal for the Law of the Sea, "Responsibilities and Obligations of States Sponsoring Persons and Entities with Respect to Activities in the Area," Case No. 17, 2010–2011; https://www.itlos.org/cases/list-of-cases/case-no-17/.

95. Unknown author, Deep Seabed Mining: Hearing Before the Subcommittee on Oceanography of the Committee on Merchant Marine and Fisheries, House of Representatives, Ninety-Fourth Congress (London: Forgotten Books, 2015), 52.

96. Ibid., 40.

97. Ibid., 130.

98. Ibid.

99. Ibid. See some of the later congressional hearings that discuss the need for unilateral action.

100. See, for example, Steven Groves, "The U.S. Can Mine the Deep Seabed Without Joining the UN Convention on the Law of the Sea," The

Heritage Foundation Backgrounder #2746 on UN, December 4, 2012; http://www.heritage.org/research/reports/2012/12/the-us-can-mine-the-deep-seabed-without-joining-the-un-convention-on-the-law-of-the-sea.

101. I wish to thank my friend and colleague Rasmus Gjedssø Bertelsen (inaugural Barents Chair in Politics at the University of Tromsø, the Arctic University of Norway) for suggesting that I explore the Svalbard model in relation to Antarctica.

102. Torbjorn Pedersen, "Norway's Rule on Svalbard: Tightening the Grip on Arctic Islands," *Polar Record* 45 (2009): 148.

103. Patrick Salmon, "Foreign Policy and National Identity: The Norwegian Integrity Treaty 1907–24," *Forsvarsstudier* 1 (1993): 16.

104. Adam Grydehoj, "Informal Diplomacy in Norway's Svalbard Policy: The Intersection of Local Community Development and Arctic International Relations," *Global Change, Peace and Security* 26 (2014): 3; and Adam Grydehoj, "The Globalization of the Arctic: Negotiating Sovereignty and Building Communities in Svalbard, Norway," *Island Studies Journal* 7 (2012): 100.

105. See, for example, Grydehoj, "Informal Diplomacy in Norway's Svalbard Policy," 3.

106. For the history, geography, and climate of Svalbard, see Arvid Viken, "Tourism, Research, and Governance on Svalbard: A Symbiotic Relationship," *Polar Record* 47 (2011): 335–347.

107. Grydehoj, "The Globalization of the Arctic," 100.

108. For an economic history of Svalbard, see Viken, "Tourism, Research, and Governance on Svalbard."

109. Salmon, "Foreign Policy and National Identity," 16–17.

110. Grydehoj, "Informal Diplomacy in Norway's Svalbard Policy," 5.

111. Ibid.

112. Salmon, "Foreign Policy and National Identity," 16–17.

113. Torbjorn Pedersen, "The Svalbard Continental Shelf Controversy: Legal Disputes and Political Rivalries," *Ocean Development and International Law* 37 (2006): 342.

114. See Chapter 3 of this book for details on the condominium proposal for Antarctica.

115. Pedersen, "The Svalbard Continental Shelf Controversy," 342.

116. Salmon, "Foreign Policy and National Identity," 53.

117. The full text of the Treaty of February 9, 1920, relating to Spistbergen (Svalbard) is contained in Appendix II of this book.

118. Ibid.

119. The full Act of July 17, 1925, relating to Svalbard is available at http://app.uio.no/ub/ujur/oversatte-lover/data/lov-19250717-011-eng.pdf.

120. Pedersen, "Norway's Rule on Svalbard," 147.

121. Grydehoj, "The Globalization of the Arctic," 101.

122. Pedersen, "Norway's Rule on Svalbard," 150.

123. For a list of parties, see the Dutch Government Information Service, "Treaty concerning the Archipelago of Spitsbergen, including Bear Island"; https://verdragenbank.overheid.nl/en/Verdrag/Details/004293#Partijen.

124. "North Korea Signs the Svalbard Treaty," *Maritime Executive*, February 6, 2016; http://www.maritime-executive.com/article/north-korea-signs-svalbard-treaty.

125. World Trade Organization (WTO), "Members and Observers of WTO"; last accessed March 15, 2016, https://www.wto.org/english/thewto_e/countries_e/org6_map_e.htm.

126. Hemmings, "Security Beyond Claims," 81.

127. Peter J. Beck, "Who Owns Antarctica? Governing and Managing the Last Continent," *Boundary and Territory Briefing* 1:1 (1994): 18.

128. Ibid.

129. For the main settlements in Svalbard, see Grydehoj, "Informal Diplomacy in Norway's Svalbard Policy," 1–14.

130. On "normalization," see Government of Norway, Ministry of Justice and Public Security, "Development of the Local Community in Longyearbyen"; last accessed March 15, 2016, https://www.regjeringen.no/en/find-document/dep/JD/reports-to-the-storting/19992000/report-no-9-to-the-storting-/14/id456932; and Adam Grydehoj, "Uninherited Heritage: Tradition and Heritage Production in Shetland, Aland, and Svalbard," *International Journal of Heritage Studies* 16 (2010): 83.

131. For the economic history of Svalbard, see Grydehoj, "Uninherited Heritage," 77–89; "Informal Diplomacy in Norway's Svalbard Policy," 1–14; and "The Globalization of the Arctic," 99–118.

132. For the economic history of Svalbard, see the essays by Grydehoj in note 104.

133. Grydehoj, "The Globalization of the Arctic," 105–112.

134. Ibid. For China's active engagement in Svalbard, see Diana Wallis and Steward Arnold, eds., "The Spitsbergen Treaty: Multilateral Governance in the Arctic," *Arctic Papers* 1 (2011): 17.

135. For a history of tourism in Svalbard, see Viken, "Tourism, Research, and Governance on Svalbard," 335–47.

136. Statistik Sentralbyra, Statistics Norway, "Population of Svalbard, 1 July 2015"; https://www.ssb.no/en/befolkning/statistikker/befsvalbard/halvaar/2015-09-24.

137. Statistik Sentralbyra, Statistics Norway, "Population of Svalbard, 1 July 2014"; https://www.ssb.no/en/befolkning/statistikker/befsvalbard/halvaar/2014-09-25.

138. Statistik Sentralbyra, Statistics Norway, "Population of Svalbard, 1 July 2015."

139. Statistik Sentralbyra, Statistics Norway, "Population of Svalbard, 1 July 2012"; https://www.ssb.no/en/befolkning/statistikker/befsvalbard/halvaar/2012-09-20.

140. Statistik Sentralbyra, Statistics Norway, "Population of Svalbard, 1 July 2015."

141. Grydehoj, "Uninherited Heritage," 84.

142. Grydehoj, "The Globalization of the Arctic," 105, 108.

143. Ibid., 106–107.

144. Ibid., 107.

145. See the full interview with the former prime minister of Greenland in Chapter 2 of this book.

146. Norwegian Ministry of Foreign Affairs, "Norwegian Government's High North Strategy," 2005; https://www.regjeringen.no/globalassets/upload/UD/Vedlegg/strategien.pdf.

147. For Store's comments, see Pedersen, "Norway's Rule on Svalbard," 150; and Torbjorn Pedersen, "The Constrained Politics of the Svalbard Offshore Area," *Marine Policy* 32 (2008): 915.

148. Grydehoj, "Informal Diplomacy in Norway's Svalbard Policy," 6; and Pedersen, "Norway's Rule on Svalbard," 148.

149. Grydehoj, "The Globalization of the Arctic," 110.

150. Pedersen, "Norway's Rule on Svalbard," 150.

151. Ibid., 151.

152. Ibid., 148.

153. On the tension between the Svalbard Treaty and UNCLOS, see Torbjorn Pedersen, "International Law and Politics in U.S. Policymaking: The United States and the Svalbard Dispute," *Ocean Development and International Law* 42 (2011): 120–135; Pedersen, "The Svalbard Continental Shelf Controversy," 339–358; Geir Honneland, "Compliance in the Fishery Protection Zone Around Svalbard," *Ocean Development and International Law* 29 (1998): 339–360; and Torbjorn Pedersen and Tore Henriksen, "Svalbard's Maritime Zones: The End of Legal Uncertainty," *International Journal of Marine and Coastal Law* 24 (2009): 141–161.

154. Pedersen, "The Constrained Politics of the Svalbard Offshore Area," 913, 916.

155. Ibid., 916.

156. Ibid.

157. Honneland, "Compliance in the Fishery Protection Zone Around Svalbard," 339–360.

158. See Andrew Griffin, "Norwegian Politicians Propose Putting Refugees on Svalbard—Remote Arctic Islands with More Polar Bears than People," *Independent*, September 13, 2015; and Tess Owen, "Frozen Norwegian Archipelago Might Provide a New Home for Syrian Refugees," *Vice News*, September 14, 2015.

# SEVEN

# Toward Greater Environmental Protection in Antarctica

When the Environment Protocol to the Antarctic Treaty (otherwise known as the Madrid Protocol) entered into force in 1998, it was hailed as a major victory by the international community. Many nongovernmental organizations (NGOs) characterized it as a boon for the environment. They had, after all, contributed to killing the Convention on the Regulation of Antarctic Mineral Resource Activities (CRAMRA), which would have opened Antarctica to mining. Antarctica was now truly dedicated to peace and scientific research. The Madrid Protocol was seen as replacing CRAMRA with stringent environmental regulations, including a ban on mining, and continentwide Environmental Impact Assessments for possible harmful activities. Against this background, it is now common to hear that Antarctica is governed by a strong environmental regime.

Unfortunately, the claim of strong environmental protection in Antarctica is not fully supported by reality. In fact, the Madrid Protocol is based on the assumption that the main threat to the Antarctic environment comes from within Antarctica, hence the need to ban mining, to introduce Environmental Impact Assessments, and to generally reduce the human footprint on the continent. However, as has been well established, the biggest environmental threat to Antarctica comes from outside—i.e., climate change—and little can be done at the moment in Antarctica itself to reverse this problem. While the full threat that climate change poses to the continent may not have been fully known at the time of the Madrid Protocol, once it did become known, parties to the Antarctic Treaty did little to advocate for a strong international agreement on climate change.

Contrary to popular perception, the environmental provisions of the Madrid Protocol are far weaker than the environmental safeguards in the CRAMRA that it displaced.[1] Although CRAMRA would have authorized mining, it would have done so only under extremely stringent environmental provisions, establishing a platform for what could have become a green model for economic growth in Antarctica. Under the Madrid Protocol, all activities that are not explicitly prohibited are allowed. This means that with the exception of mining, the Madrid Protocol provides a de facto authorization of all other human activities on the continent under the auspices of the Antarctica Treaty advocating the pursuit of peace and science. This broad range of activities takes place under far weaker environmental safeguards.

The Madrid Protocol is as weak institutionally as the Antarctic Treaty itself, with the powers of its Committee on Environmental Protection being deliberately contained. In addition, the protocol has failed to designate large enough portions of the continent as protected areas; its Environmental Impact Assessment provisions are weak and sometimes never applied, with their narrow focus no longer responding to what is a fast changing continent; and its liability regime is limited to environmental emergencies, without any obligation for environmental restoration.

Furthermore, two commercial activities are currently proceeding in Antarctica with little to no environmental regulation by the Antarctic Treaty System (ATS), possibly because their magnitude had not been foreseen by negotiators: bioprospecting (the process of discovery and commercialization of new products based on biological resources, such as bacteria) and tourism. Had CRAMRA entered into force, prospecting would have come under stringent regulation and would have been accompanied by an equally serious liability regime. With both commercial activities growing and continuing unchecked, the Madrid Protocol has not shown itself to be responsive to change.

Furthermore, one of the main problems confronting the Madrid Protocol and the Conservation of Antarctic Marine Living Resources (CCAMLR) is that of enforcement because of the jurisdictional vacuum in Antarctica and its surrounding waters. As a result of there being no territorial sovereignty in Antarctica and no recognized Exclusive Economic Zones (EEZs), enforcing the provisions of the protocol and CCAMLR has been challenging, especially vis-à-vis nonparties to the ATS.

In fact, the stringent environmental protection seen in Svalbard could indicate that the world's most fragile environments are best protected under sovereign control. With five of Antarctica's seven claimants being developed countries with a strong record of environmental stewardship (the

United Kingdom, Norway, France, Australia, and New Zealand), and with Argentina and Chile being emerging developing countries that are now joining their ranks, the tragedy of the commons in Antarctica could well be reduced with the application of the Svalbard model.[2]

In addition, whereas the lack of sovereignty in Antarctica could have provided fertile ground for engagement by environmental NGOs—an engagement that would have filled the sovereignty vacuum and strengthened environmental governance on the continent, the opposite has actually occurred. NGO engagement has been uneven, and it has fallen since the demise of CRAMRA and establishment of the Madrid Protocol in the 1990s, especially with the continent being remote and inaccessible. The application of the Svalbard model could incentivize national, not just international, environmental NGOs to step up their involvement in Antarctica, extending their field of operations into what would be offshore territory. With sovereignty will come citizenship, and more sustained participation.

## WHERE IS ANTARCTICA?

In discussing the regulatory regime governing the seventh continent, someone once asked, "Where is Antarctica?" By that, they meant that Antarctica's boundaries have been ill defined by the ATS, with each of its components having a different area of application.[3] With time, this is likely to constitute an important regulatory challenge for the continent.

The Antarctic Treaty applies to "the area south of 60 degrees South Latitude, including all ice shelves."[4] This basically includes Antarctica's landmass and some of its surrounding waters. CCAMLR, on the other hand, has a wider area of application based on the Antarctic Convergence.[5] It applies farther north than the treaty's 60 degrees south designation to include some sub-Antarctic territories.[6] The Antarctic Convergence, which is a natural rather than an artificial boundary, therefore plays an important role in determining CCAMLR's scope. One of the complications facing CCAMLR is that the boundary of the Convergence is shifting with climate change. Evidence suggests that a southward shift of approximately 50 kilometers has already taken place over the past two decades with the rise in temperature.[7]

Meanwhile, the Madrid Protocol applies to "the Antarctic environment and dependent and associated ecosystems," with no coordinates specified.[8] The term *dependent and associated ecosystems* has created ambiguity as to the protocol's area of application relative to the Antarctic landmass. Some have argued that absent further guidance from the parties, it should be read as having identical application to the Antarctic Treaty. However, the use of this specific term in the protocol seems to suggest that the intention was to link it to the biological, rather than the political, boundaries of the Antarctic.[9]

Annex VI of the Madrid Protocol, "Liability Arising from Environmental Emergencies," is limited to "the Antarctic Treaty area."[10] Therefore, it is unclear whether it covers the "dependent and associated ecosystems" of the protocol. For regulators, this situation has created a labyrinth that is sometimes difficult to work through and has translated into multiple and potentially incompatible interpretations in the national legislation of parties.

One of the main unresolved issues that arises from this ambiguity is the scope of application of the ban on mining introduced by the Madrid Protocol. If the protocol is assumed to cover the same area as the Antarctic Treaty, the ban would apply to the Antarctic landmass and its immediately surrounding waters. However, were the term *dependent and associated ecosystems* to be stretched further, it could cover portions of the seabed beyond the limits of national jurisdiction (were such jurisdiction to exist in Antarctica), straying into what would normally be considered the high seas—i.e., the regulatory territory of the United Nations Law of the Sea (UNCLOS).[11]

While the United Kingdom's Antarctic Act of 1994 does not apply the ban on mining to the deep seabed (and therefore implicitly authorizes it), New Zealand explicitly prohibits the mining of the seabed south of 60 degrees. South Africa and many other parties do not address the issue in national legislation, speaking only of the applicability of the ban to the "Antarctic Treaty Area."[12] An additional question is whether the ban applies only to the portion of the Antarctic continental shelf south of 60 degrees, or also to the part above that latitude.[13]

Ambiguity has also stretched to how far the liability regime of Annex VI to the Madrid Protocol extends. Article 1 of the Annex states that it "shall apply to environmental emergencies in the Antarctic Treaty area which relate to scientific research programs, tourism and all other governmental and non-governmental activities . . . including associated logistic support activities."[14] This begs the question of whether pollution caused by one of these activities within the Antarctic Treaty area, but extending beyond it to dependent and associated ecosystems, would be covered by the liability regime.[15] Clearly, the geographic construction of Antarctica from a regulatory point of view has been a challenge that is affecting its governance.[16] Currently, there are many different answers to the question "Where is Antarctica?"

Of course, none of these delimitation issues have been terribly threatening to the Antarctic environment so far. In fact, the feasibility of mining in hostile Antarctic climes has made it unnecessary for anyone to contest the precise boundaries of the mining ban. However, with the level of activity in Antarctica on the rise, the unclear scope of application of various agreements will no doubt come to pose a risk to the environment soon. Tourism, for example, has grown substantially in Antarctica in recent years, and an unclear liability regime could present a serious problem in the event of accidents. It is highly unlikely, however, that parties to the Antarctic Treaty would succeed in resolving these issues, given their implications for territorial claims.

At present, each party has learned to fill the gaps in the ATS using its own ingenuity and sense of environmental justice. This is visible in a separate but related issue: the different sanctions applied by parties to the same offense in the treaty area. For example, under New Zealand law, the illegal taking of a penguin carries a six-month prison sentence, whereas it is punishable by a year in Norway and up to two years in the United Kingdom.[17]

## IS AN ENVIRONMENTAL ACCORD WEAKER THAN A MINING REGIME?

The Madrid Protocol was born out of strong opposition to CRAMRA. In fact, it had its origin in proposals floated by France and Australia at the time of CRAMRA's collapse on the designation of Antarctica as a World Park or a wilderness reserve, proposals originally promulgated by environmental NGOs.[18] The Antarctic Treaty had not itself tackled any environmental issues, having been negotiated primarily from a security prism back in 1959, which also explains the minimalist approach that parties took to its institutional development.[19] Its Secretariat, for example, was only established in 2004.[20]

The only environmental rules that formed part of the ATS, and that had been in force prior to the Madrid Protocol, were the Agreed Measures for the Conservation of Antarctic Fauna and Flora of 1964.[21] These measures had the very broad objective of conserving parts of the Antarctic environment, but had been relatively poorly applied. The implementation of the measures had been left to national authorities, with international monitoring enabled through the exchange of information among states.[22]

The weak enforcement of the Agreed Measures was visible, for example, in the 1981 opening of the Chinese Great Wall Station, where the Chinese brought several doves for release into the Antarctic skies to mark the occasion—a species not indigenous to the continent and therefore prohibited by the measures. Some members of the Chinese delegation also brought in their pet dogs, and others were seen kicking around penguins as footballs for sport. None of these breaches were met with sanctions.[23]

Other violations abounded. In 1985, for instance, the French decided to build an airport at Point Geologie, in Adelie Land, destroying penguin rookeries in the process. These were deliberately blasted away to create the runway. Whereas the evidence at the time pointed to a clear violation of the Agreed Measures, no sanctions were applied. parties to the Antarctic Treaty turned a blind eye, with the undercurrent being that it had occurred on French-claimed territory.[24]

It was during CRAMRA's negotiation that a campaign was actively commenced by NGOs to strengthen environmental protection in Antarctica. This was marked by the Greenpeace voyage to the Southern Ocean to call for the designation of the continent as a World Park.[25] The idea gradually evolved, with environmentalists calling for a World Park anchored in the Common Heritage of Mankind principle, and for an Antarctica managed through a central authority for the benefit of humanity as a whole.[26]

The Antarctic and Southern Ocean Coalition (ASOC) set out the following principles for the park: nonmilitarization, conservation, preservation of the wilderness qualities of the continent, a ban on mining, promotion of scientific research, a ban of nuclear waste, and the establishment of a central management body with the participation of civil society.[27] A far weaker version of these ideas would come to constitute the Madrid Protocol upon the collapse of CRAMRA. Surprisingly, the creation of the Madrid Protocol would turn into a "cut-and-paste" operation from CRAMRA (and an incomplete one at that), CCAMLR, and the Agreed Measures, with little new contribution, and with the spirit of a World Park floating away into oblivion.[28]

In fact, little known about the evolution of Antarctica's governance regime is that environmental issues had occupied center stage in CRAMRA's negotiation. The convention set new standards for environmental protection and had broken entirely new ground in international environmental law. The goal of the convention was not to authorize mining, but rather to establish a regime that would protect Antarctica's fragile environment before, during, and after any mining activity. As Vidas (2000) noted:[29]

> The Consultative Parties' new start after the abandonment of CRAMRA in 1989 can *not* be attributed to CRAMRA's containing insufficient environmental safeguards. These were in fact very stringent. The 'fault'

of CRAMRA may rather have been the failure of the Parties to give it *proper marketing as an environmental protection instrument*, which should have begun already with the choice of title given to that convention. It was a complex combination of economic and political factors that led the Consultative Parties to abandon CRAMRA [italics mine].

Chilean Professor Francisco Orrego Vicuna, a well-known Antarctic scholar, confirmed that "CRAMRA contained a very strong environmental regulatory regime. It set straightforward and demanding environmental standards." In his opinion, it was "amongst the strongest environmental treaties that the world has known, having succeeded in building on the wealth of experience accumulated in international environmental treaty-making."[30]

The convention is said to have broken new legal grounds in covering not only Antarctica's environment, but also its "dependent or associated ecosystems,"[31] going further in this regard for instance than UNCLOS. Its liability regime had also extended to these "dependent or associated ecosystems."[32] In addition, the convention had taken the avant-garde step of making any mining activity contingent on the attainment of the necessary technological capability not only to undertake the mining operation, but also to monitor its impact, and to restore any damage that it may create.[33]

CRAMRA stipulated that decisions on mining activities only be taken after the consideration of "information adequate to enable informed judgments to be made about their possible impacts."[34] Activities could not go forward if found to have a harmful effect on the environment, such as the "degradation of, or substantial risk to, areas of special biological, scientific, historic, aesthetic, or wilderness significance."[35] The regional and global impacts of mining activities would have also had to be assessed.[36]

The convention's environmental regulations would have been enforced through the powers vested in the commission and Regulatory Committees, with one of the main functions of the commission being to protect the environment and to designate areas in which mining activity would be prohibited or restricted. Mining decisions would have required consensus within the commission, meaning that they could have been vetoed by only a single state.[37] In addition, CRAMRA contained an entire chapter devoted to the issue of prospecting;[38] i.e., the exploratory research phase that precedes actual mining, as well as stringent rules on inspection and liability.[39]

When the Madrid Protocol was created, its greatest achievement was cast as its ban on mining, with little attention paid to its remaining provisions which were in fact far weaker than CRAMRA. The protocol established a set of environmental protection principles, largely taken from CRAMRA, also borrowing CRAMRA's scope of application. The 1964

Agreed Measures essentially formed the basis of its Annex II, entitled "Conservation of Antarctic Fauna and Flora."[40]

Consisting of 25 articles, the protocol established a continentwide system for Environmental Impact Assessment, a Committee on Environmental Protection, and a mechanism for inspection, and called for the negotiation of a liability regime. It was complemented with more detailed Annexes on Environmental Impact Assessment (Annex I), the conservation of Antarctic flora and fauna (Annex II), waste disposal and management (Annex III), prevention of marine pollution (Annex IV), protected areas (Annex V), and the liability regime that later followed (Annex VI).[41]

The protocol was seen as establishing a comprehensive, unified, and legally binding instrument for environmental stewardship in Antarctica, covering all human activities on the continent (under the broad aims of the Antarctica Treaty, of course, of a demilitarized continent devoted to peace and science).[42] Article 8 of the protocol clarified that its Environmental Impact Assessment provisions apply to "*any activities undertaken in the Antarctic Treaty* area pursuant to scientific research programs, tourism and all other governmental and non-governmental activities. . . . including associated logistic support activities [italics mine]." With the prohibition on mining in Article 7, the protocol was seen as having covered all possible activities.[43]

Despite how impressive all of this may seem at first glance, the Madrid Protocol's principal weakness lies in the fact that the Environmental Impact Assessment that it introduced for the most harmful kinds of activities (which in such cases is called *Comprehensive Environmental Evaluation*), is to be prepared by the organizer of the activity,[44] with final decisions on that activity essentially taken by the sponsoring party. In other words, nothing in the protocol gives the Antarctic Treaty Consultative Mechanism, or the Committee for Environmental Protection for that matter, veto power over a proposed activity.[45]

This weakness derives from the wording of Article 3(5) of Annex I of the protocol which states that "No final decision shall be taken to proceed with the proposed activity. . . . unless there has been *an opportunity for consideration* of the draft Comprehensive Environmental Evaluation by the ATCM on the advice of the Committee [for Environmental Protection] [italics mine]." It therefore reduces the role of other players to one of giving "consideration" to a proposed activity, rather than any real power to prevent it.

CRAMRA, on the other hand, did not delegate final authorization for a mining activity to the sponsoring party. Rather, a single member of the

CRAMRA Commission could have vetoed a mining operation, preventing it from going forward.[46] Rüdiger Wolfrum stated the following this situation:[47]

> In the negotiation process, however, some of the innovative elements of CRAMRA for the effective protection of the Antarctic environment have been lost. It is difficult to fathom the reason behind such a development. Perhaps Parties to the negotiations felt that the rules designed to protect the Antarctic environment against the potential impact of mineral resource activities were excessive in light of scientific research and tourism. One cannot exclude that the Parties feared that imposing the limitations of CRAMRA *upon all human activities rather than on mineral resource activities* would result in an unacceptable limitation to their claims to national sovereignty in Antarctica [italics mine].

In other words, what Wolfrum argues is that the Madrid Protocol's broader scope of application to all human activities in Antarctica, not just mining, is what resulted in its significant weakening. It is ironic, however, that an Environment Protocol that rose from the ashes of a collapsed mining convention would end up protecting the environment less, rather than more, than the agreement that it displaced. While the protocol completely banned mining, it ended up opening the door to all other human activity under a far less stringent environmental regime. Furthermore, the protocol's inspection regime is worse than CRAMRA's,[48] its liability regime weaker,[49] and its silence on prospecting—an issue to which CRAMRA had devoted an entire chapter, simply inexplicable.[50] All these matters are further explored next.

What this discussion has demonstrated is that negotiators were not willing to generalize CRAMRA's stringent environmental provisions to all spheres of human activity in Antarctica, which Wolfrum attributes to their concern over sovereignty. If this is indeed the case, one must ask what other alternative exists for stronger environmental protection in Antarctica. The Svalbard model, which would assign national sovereignty to all seven territorial claimants, plus probably the United States and Russia for the currently unclaimed portion of Antarctica, is sure to lead to better results.

All these claimants, with the possible exception of Russia, are known for their relatively high environmental standards, and there would be no reason why they would lower these standards in Antarctica. If certain provisions of the Madrid Protocol are considered valuable, they could still be retained under the Svalbard model to provide a continentwide framework within which national environmental regulations could operate.

Particularly puzzling following the adoption of the Madrid Protocol was the reluctance of parties to address the problem of climate change. The climate deniers within the ATS refused to make the issue part of the Antarctic agenda for many years, even in the face of a growing body of evidence of how seriously Antarctica would be affected, and how it, in turn, would affect the rest of the world. Its melting ice could contribute to a global sea level rise that could potentially cause catastrophic damage to low-lying regions across the planet. Alan Hemmings et al. write:[51]

> The historic skepticism of some governments towards the evidence for climate change even kept climate change off the agenda of the Antarctic Treaty Consultative Meeting's environmental advisory body—the Committee for Environmental Protection—until 2008, and sought (unsuccessfully) to discourage the SCAR from undertaking an Antarctic Climate Assessment. Now added as a regular agenda item. . . . discussion of climate change remains essentially confined to climate change effects *in* Antarctica. So, even a high status issue, widely seen as the pre-eminent global security issue of our age, need not easily acquire standing within the Antarctic Treaty System politico-legal fora [italics in original].

In 1999, eight years after the negotiation of the Madrid Protocol and one year after its entry into force, New Zealand's minister of the environment, Simon Upton, said: "Antarctica's greatest defense was isolation, but that isolation has evaporated rapidly."[52] With climate change, his words ring truer every day. The Madrid Protocol is a weak environmental framework that is clearly not responsive to change, especially to the biggest developments likely to face the continent—greater economic activity and climate change. As Antarctica becomes more accessible, the search for a more effective environmental protection regime will need to begin.

## A WEAK COMMITTEE FOR ENVIRONMENTAL PROTECTION

As Antarctica is a continent devoted to science, the framing of scientific questions—including environmental ones—has come to constitute a major source of power on the seventh continent. In fact, in discussing the legacy of the four International Polar Years (IPYs), debate has raged among Antarctic scholars on the relationship between science and power.[53] Whereas mining was banned on the continent, the rendezvous clause of 2048 that is embodied in the Madrid Protocol to review the ban suggests that mining was simply postponed. Science, therefore, plays a key role not only in helping territorial claimants (and potentially other countries too) in establishing a foothold in Antarctica, but also as a discovery and scoping exercise for the resource potential of the continent.

It follows that it has been important for parties to the Antarctic Treaty to not let environmental protection truly get in the way of that scientific discovery. The balance, therefore, between science (almost a code word for unveiling the economic potential of the continent) and environmental protection has been difficult to strike.

Much has been written, and much has already been discussed in this book, on the excessive number of scientific research stations on the continent and the insufficient scientific collaboration among parties. Some have asked why, if Consultative Party status requires a demonstration of a substantial scientific research program on the continent, there is no formal mechanism within the treaty to review whether and how parties continue to fulfill this criterion.[54] In other words, once Consultative Party status is obtained, only a bureaucratic exchange of information on the scope of the science being undertaken actually follows.

So close is the relationship between science and power in Antarctica that when a Committee for Environmental Protection was created by the Madrid Protocol, it was reduced to a mere advisory body to Antarctic Treaty Consultative Meetings (ATCMs).[55] While it can put forward recommendations to these meetings, it is not itself vested with any decision-making power. Its recommendations are discussed within the meetings under the very tight control of territorial claimants. This is evidenced in the fact that the claimants produce approximately 65 percent of all working papers for these Meetings, with a total 10 parties producing 82 percent of all such documents.[56]

It is within this environment that the Madrid Protocol operates. It is a supplement to the Antarctic Treaty that is not designed to compete with it, nor to challenge it, in any way. Wolfrum writes that "the Committee has advisory functions vis-à-vis the ATCM in connection with the *implementation of the Protocol and its Annexes*. All attempts to create an organ with genuine law making or enforcement powers were blocked by those States which regarded the establishment of such an institution as a threat to their sovereign claims to Antarctica [italics mine]."[57]

Celebrating 50 years of the Antarctic Treaty, in 2009 the first three chairs of the Committee for Environmental Protection decided to share their experience with the world.[58] Particularly striking in their account of the challenges that the committee has faced are the following points: (1) how difficult it was for the committee to define its role relative to the Scientific

Committee on Antarctic Research (SCAR) and to the ATCM; (2) the extent of the bureaucracy that the committee has been mired in, preventing it from addressing the real environmental issues facing the continent, and (3) the committee's inability to keep pace with the rapidly changing situation in the Antarctic.

On its relationship with pre-existing institutions, the committee had to address what had become obvious to everyone, i.e., that it was in fact displacing SCAR on environmental issues—the body designed to provide scientific advice to ATCMs. Chapter 4 discussed the deliberate fragmentation of scientific platforms under the ATS, showing that this was part of a deliberate process to prevent the concentration of scientific power.

Gradually, a balance was struck between the role of this new committee and SCAR, with the chairs writing that "the rapid development of environmental management competence within the Committee for Environmental Protection saw SCAR gradually withdraw from its practice of providing environmental advice, and its role has become more focused on scientific advice."[59] However, the existence of two separate bodies has without a doubt reduced their respective roles. The Committee for Environmental Protection also adjusted to the delicate balance between its scientific power and the political power of ATCMs, with the chairs confirming that it is "subservient" to the latter.[60] A subservience that could hardly serve the environment well.

Particularly troubling however, is the extent to which the chairs have felt that the committee was overrun by mundane bureaucratic tasks at the expense of the serious study of the environmental questions confronting the continent. The committee has been generally convened prior to ATCMs, adopting a report for transmission to these meetings. The lengthy negotiations involved in adopting these reports, in addition to the exponential growth in the number of documents exchanged within meetings, has "limited the Committee's ability to make progress on intractable issues or provide adequate time for discussion of high priority issues."[61] This has also affected the committee's ability to keep pace with the changing Antarctic landscape. The chairs expressed their frustration by stating:[62]

> The Antarctic environment has experienced significant change and has been subject to additional pressures. Such pressures are becoming more evident and *arguably more urgent in their need for attention.* Since 1991 ship borne tourists making landings in Antarctica have increased from 6,704 to 32,198. The average annual mean temperature on the Antarctic Peninsula has increased by more than 2.5°C over the past 50 years. There has been a significant loss of ice shelves, and

non-native species have been identified in Antarctica. National program activities have also increased, with eight new bases being established around the continent. Air access to Antarctica has become more prevalent, with approximately 11 new ice runways (permanent and temporary) constructed [italics mine].

## PROTECTED AREAS IN LAND AND WATER: THE WORLD'S LEAST PROTECTED CONTINENT

Annex V of the Madrid Protocol establishes a basis for the designation of an Antarctic Specially Protected Area (ASPA) as any terrestrial or marine area of outstanding environmental, scientific, historic, and aesthetic or wilderness value.[63] Globally, protected areas are a recognized instrument for conserving biodiversity, with the Convention on Biological Diversity (CBD) setting a 17 percent target for the protection of terrestrial land areas under its Strategic Plan for Biodiversity 2011–2020.[64] It is within this framework that the Madrid Protocol has called for ASPAs to be identified "within a systematic environmental-geographical framework" and include "representative examples of major . . . ecosystems."[65]

Any party to the ATS may propose that an area be designated as an ASPA by submitting a proposed Management Plan to the Antarctic Treaty Consultative Mechanism. Entry into such areas has to be made subject to a permit system.[66] The designation of protected areas goes forward once a Management Plan is adopted by the ATCM. Any Consultative Party may block the adoption of a plan.[67] In other words, the Madrid Protocol establishes an extremely high hurdle for the designation of protected areas. A single objection by a Consultative Party is sufficient to stymie the process.

Annex V to the Madrid Protocol was intended to give effect to the main environmental principles embodied in the protocol. Its Article 3, for instance, calls for "The protection of the Antarctic environment and dependent and associated ecosystems and the *intrinsic value* of Antarctica, including its *wilderness and aesthetic values* and its value as an area for the conduct of scientific research [italics mine]." The protocol has been criticized for not containing a single definition of any of these terms, giving rise to disparate interpretations.[68] While some parties see the continent's wilderness as an asset that must be preserved, others see it as unused space that should be utilized.[69]

There has been surprisingly little examination of how effective Antarctica's protected area system has been since its creation. However, recent efforts to examine the issue have found that Antarctica is well below the international benchmark. In 2011, Tina Tin and Alan Hemmings concluded that "it is clear that no substantial [wilderness] area has yet been designated,"[70] and Justin Shaw et al. (2014) expressed the regret that "Antarctica is one of the planet's least protected regions, with only 1.5% of its ice-free area formally designated as specially protected areas."[71]

Shaw and his colleagues went on to remark that this finding was particularly strange in light of the fact that Antarctica had been made a "natural reserve" by the Antarctic Treaty—a provision which could have led Antarctica to be a trendsetter in environmental protection, rather than the reverse.[72] This matter is made all the more surprising by the fact that there are no indigenous people in Antarctica or large-scale commercial activities to impede protected area designations.[73] In light of this, Shaw and his colleagues conclude that "while there is a widespread general perception that Antarctica is well conserved, in practice conservation of terrestrial biodiversity from a continentwide perspective is poorly served by the protected area system."[74]

A particularly disconcerting aspect of the 1.5 percent figure that Shaw and his colleagues present is not only that it falls substantially below the 17 percent target of the CBD, but that it is also well below the percentage of land area protected by some claimant states at home. For example, 18 percent of Australian territory is currently designated as protected.[75] This provides additional evidence in support of granting territorial sovereignty in Antarctica under a Svalbard model. The fact that only one Consultative Party can block the creation of an ASPA is a problem that the assignment of national sovereignty would overcome.

Turning to Antarctic waters, is the situation any better? CCAMLR, of course, is a fairly complex fisheries agreement; like the Madrid Protocol, it as an instrument of the ATS rather than a stand-alone force. CCAMLR has been celebrated by the environmental community as a norm-setter in the management of the high seas because of its incorporation of a precautionary and ecosystem-based management approach. It defines conservation as "rational use," while allowing the scientific and commercial harvesting of living resources to go forward so long as they do not cause changes to the exploited and dependent fish populations or have a significant adverse effect on ecosystems. Its toolbox includes the establishment of Marine Protected Areas.[76]

Like the Antarctic Treaty itself, CCAMLR has had a long history of fending off interest by the United Nations (UN) in the Southern Ocean, especially in keeping the United Nations Food Agriculture Organization (FAO) at bay—an issue finally resolved by making the FAO a cooperating partner of the convention. The Consultative Parties to the Antarctic Treaty wanted to be the main decision-makers on the fate of the Southern Ocean. CCAMLR, which was ratified in 1982, was to some extent the precursor to CRAMRA (with CCAMLR covering marine resources and CRAMRA terrestrial resources).[77]

CCAMLR has had as much of a mixed record in the creation of Marine Protected Areas as has the Madrid Protocol more generally.[78] While the decision-making rule in CCAMLR is slightly different than that of the Madrid Protocol, in practice, it has amounted to the same. Any party to the CCAMLR Commission can object to the establishment of a Marine Protected Area, in which case the conservation measure, in whole or in part, shall not be binding *upon that member*.[79] In other words, other members can still recognize the Marine Protected Area. In practice, though, the objection of one member has been sufficient to kill an entire proposal out of fear that the member remaining outside the arrangement would gain an economic advantage (i.e., would free-ride).

In general, conservation efforts in the Southern Ocean have been hampered by the temptation to exploit its rich living resources. While krill fish continues to be the largest catch, tooth fish brings in approximately 20 times the profit.[80] The changing composition of the CCAMLR Commission over the years has further complicated matters, with fishing states outnumbering nonfishing states by a ratio of around 5:3.[81]

To date, only one protected area has been established by CCAMLR—the South Orkney Islands Southern Shelf Marine Protected Area, created in 2009 at the behest of the United Kingdom (UK). The South Orkney Islands are a group of islands in the Southern Ocean, about 604 kilometers northeast of the tip of the Antarctic Peninsula. While it has been celebrated as the world's first wholly Marine Protected Area on the high seas, covering 94,000 square kilometers in the south Atlantic, evidence suggests that the agreement was reached only when the boundaries of the area were redrawn to exclude valuable fishing grounds—particularly krill fisheries.[82]

Since 2011, repeated attempts have been made to designate the Ross Sea region as a protected area, at the request of New Zealand and the United States.[83] The Ross Sea is considered by many scientists to be the only large intact marine ecosystem remaining on the planet, and it is home to disproportionately high populations of mammals and birds, including one third of all Adélie penguins and one quarter or all emperor penguins. Prior to beginning discussions within the CCAMLR Commission, the United States and New Zealand engaged in extensive negotiation among themselves on the boundaries of the area, with the most valuable fishing areas excised from their proposal. Despite this fact, the Russian delegation defeated the proposal on multiple occasions, most recently in 2015.[84] The main reasons for their opposition appear to be the proposal's continued interference with fishing grounds, now and in the future, its large size, and the duration of the protected area designation. In previous years, the Russians had been supported by an equally skeptical Chinese delegation.

Were sovereignty to be assigned in Antarctica, countries would automatically benefit from a 200-nautical-mile EEZ, and it would be their responsibility to protect these areas. While this would not solve all of CCAMLR's problems, due to its much broader area of application, it would nevertheless cover substantial fishing grounds. For example, had national sovereignty already existed in Antarctica, New Zealand may have very well already succeeded in protecting some portions of the Ross Sea.

Evidence in the Southern Ocean suggests that the far clearer sovereignty situation in sub-Antarctica islands has worked well for conservation, which proves the point made here. In fact, in 2008, ASOC delivered the following verdict to the ATCM:[85]

> Within the greater Antarctic region, but outside the Antarctic Treaty Area, a number of Antarctic Treaty Consultative Parties have designated Marine Protected Areas around their sub-Antarctic islands. Thus Australia has designated the 160,000 km$^2$ Macquarie Island Marine Park and the 65,000 km$^2$ Heard Island and McDonald Islands Marine Reserve; New Zealand has a small marine reserve off the Auckland Islands but is developing marine reserves around all five of its sub-Antarctic island groups; France has designated marine reserves around its Crozet, Kerguelen, Amsterdam and St Paul island groups; and South Africa has proposed an Marine Protected Area around its Prince Edward Islands group. In the case of these states, the actual or proposed areas are substantial. Australia and New Zealand have also achieved World Heritage status for these islands and territorial seas, and South Africa has submitted an application for the Prince Edward Group. Other states with sub-Antarctic islands may also have plans for Marine Protected Areas. *Overall, the picture in the sub-Antarctic is encouraging. It makes a stark contrast with the feeble Marine Protected Area development seen in the Antarctic Treaty Area* [italics mine].

## ENVIRONMENTAL IMPACT ASSESSMENTS—AN INCREASINGLY LIMITED INSTRUMENT FOR A GROWING CONTINENT

All activities in the Antarctic Treaty area, governmental and nongovernmental, and changes to these activities are subject to the Environmental

Impact Assessment provisions of the Madrid Protocol. They include, for instance, expeditions to the continent, the establishment or decommissioning of research stations and facilities, and tourism.[86] Only fishing, sealing, and whaling are excluded from this requirement because they are covered by other treaties,[87] and emergency operations if time does not allow an assessment to be conducted.[88]

The Madrid Protocol breaks down activities based on impact. Activities having "less than a minor or transitory impact" do not require assessment; activities having "a minor or transitory impact" require an Initial Environmental Evaluation: and activities having "more than a minor or transitory impact" require the most stringent assessment—the abovementioned Comprehensive Environmental Evaluation.[89] It is only the latter that needs to be considered by the ATCM. So far, most of the Comprehensive Evaluations that have been undertaken relate to the construction, maintenance, or dismantling of research stations, scientific activities, and the growing transportation infrastructure on the continent (such as surface traverses and airstrips).[90]

While the notion of a continentwide system for environmental impact assessment sounds impressive, and indeed Antarctica is the only continent to which such a requirement applies, the reality is that the instrument suffers from a range of shortcomings. In addition, it is becoming an increasingly limited instrument for this growing continent. One of its main flaws lies in the fact that Comprehensive Environmental Evaluations are to be prepared by the organizer of an activity,[91] with final decisions being taken by the sponsoring party. This matter was previously discussed in this chapter relative to the far stronger provisions of CRAMRA. In negotiating the Madrid Protocol, countries had rejected the proposal of assessments by an independent international authority.[92]

This has resulted in a situation where discussions in ATCMs have been unable to influence the course of events in any meaningful way. In fact, Hemmings et al. write:[93]

> To a considerable extent, the situation in Antarctica is that a proposal by an operating agency to conduct a financially and logistically feasible activity leads inexorably to its realization. Even the high-level Environmental Impact Assessment provided by the Comprehensive Environmental Evaluation *affects the outcome only at the margins* . . . The fact that the proponent of the activity guides the preparation of the Comprehensive Environmental Evaluation document makes it inherently unlikely that serious consideration will ever be given to not proceeding with the activity . . . To expect this is, frankly, to have

> unrealistic expectations about the capacity of the proponent (or the consultant preparing the Comprehensive Environmental Evaluation document) to detach themselves from self-interest [italics mine].

On the positive side, there is evidence that progress has been made in the incorporation of the Environmental Impact Assessment instrument into national legislation, which means that environmental discussions probably take place at the national level in relation to proposed activities.[94] How serious these discussions are, and whether they take place as a mere *fait accompli*, is another matter, of course. However, at the level of the Antarctic Treaty itself, scrutiny is certainly reduced by the weakness of decision-making rules on these assessments. In addition, the current system incentivizes parties to tread softly on the proposals of others, so their own proposals are not subjected to excessive scrutiny when they come up for review.

Environmental impact assessment in Antarctica suffers from a number of other flaws, too. For instance, some activities have been known to go forward prior to their impact assessments being fully considered by the parties. In fact, Roura and Hemmings (2011) argue that the decision to conduct an activity, as well as the specifics of the activity in terms of timing and location, appear to precede the initiation of an impact assessment. The assessments are typically carried out after key decisions have already been made.[95]

The construction of the South Pole traverse is a case in point. The traverse is a 1,650 kilometers overland ice route that connects the U.S. McMurdo Station on Ross Island to its Amundsen-Scott South Pole Station; it is the longest road that currently exists on the continent. It was created to allow fuel and supplies to travel from the U.S. coastal station all the way to its other station at the pole. Construction of the route started in the 2002–2003 summer season, with the Comprehensive Environmental Evaluation only prepared several years later, despite serious environmental concerns.[96]

Particularly complex is the relationship between the impact assessment instrument and scientific activity, which is considered paramount on the continent. For some, it is a code word for resource discovery. In fact, one of the main principles of the Madrid Protocol is that "[a]ctivities shall be planned and conducted in the Antarctic Treaty area *so as to accord priority to scientific research* and to preserve the value of Antarctica as an area for the conduct of such research [italics mine]."[97] This has acted as an unspoken constraint on the application of the impact assessment tool to scientific research or its supporting logistics.[98] So, for instance, when a Comprehensive Environmental Evaluation was finally prepared for the

South Pole traverse, its impact was discussed more in relation to its contribution science and operations than its environmental footprint.[99]

Russia's scientific drilling into subglacial Lake Vostock also revealed the shortcomings of Environmental Impact Assessment provisions. The lake is currently the largest of Antarctica's 400 known subglacial lakes, with Russia's Vostock Station lying at its northern end. Its value for science is expected to be high, with the drilling of the lake being viewed by the Russians as enhancing their scientific prestige on the continent.[100]

The Russians had initiated drilling operations in 1990, prior to the adoption of the Madrid Protocol. However, almost simultaneously with the protocol's entry into force in 1998, they continued to drill to within a very short distance of the lake's surface without an impact assessment. A Comprehensive Environmental Evaluation was presented only in 2003 (i.e., years after the start of operations).[101] The reaction to it was mixed; for instance, the Netherlands stated that it failed to analyze worst-case scenarios. In fact, some of these scenarios later materialized, with Russia reporting in 2007 that its drill was stuck and would require the use of about 200 liters of antifreeze to extract it.[102]

In general, scientific activity in Antarctica has an impact on the environment in two ways: first, in terms of the environmental impact of the research stations themselves, of which there are many (the treatment of their wastewater, garbage, and more); and second, in terms of the impact of the scientific projects they undertake.[103] Hemmings et al. asked, from an environmental point of view, whether there is a need for so many dispersed research stations.[104] In fact, one of the main benefits of assigning sovereignty in Antarctica under the Svalbard model is that it would be likely to lead to a rationalization of scientific activity because the entitlement to the resources of the continent will cease to be linked to establishing research footholds.

Questions have also been asked about the precise scope of environmental impact assessment under the Madrid Protocol, with some wanting to clarify its application to marine resources under CCAMLR to bioprospecting and tourism. With respect to the former, whereas the protocol clearly excludes fishing, some have argued that it nevertheless applies to Antarctica's surrounding waters because the Madrid Protocol covers the entire Antarctic Treaty area and its associated and dependent ecosystems. However, these questions have never been fully resolved.[105]

A growing and far more serious problem, however, is that environmental impact assessments do not adequately address the *cumulative impact* of activities on the continent, or only do so in a perfunctory manner. They assess, for example, the impact of a single tourist activity, not the turning of Antarctic sites into permanent tourist destinations. In addition, they take a myopic, project-oriented approach to the development of the continent, rather than applying a more holistic vision.[106] In fact, they are becoming too narrow an instrument for a continent that is growing fast.

The example of the Fildes Peninsula on King George Island (one of the South Shetland Islands, which lies off the tip of the Antarctic Peninsula and falls within the Antarctic Treaty area) demonstrates the importance of cumulative impact. Currently, it one of the largest ice-free areas in the maritime Antarctic, and it also has become one of the most crowded. According to experts, it "presents a unique example of increasing human pressure due to multiple human uses. Scientific research, station operations, transport, logistics, tourism, nature conservation, and protection of geological and historical values regularly overlap in space and time."[107] What is currently happening on the peninsula testifies to the inadequacy of current impact assessment procedures. In fact, the United Kingdom submitted a working paper to the ATCM in 2014 proposing that assessments consider actual or predicted cumulative impacts such as whether "the proponent consulted other Parries operating within the same geographic area and gained an indication of future plans," and asking that several parties undertake "a joint Environmental Impact Assessment for a geographic area."[108]

Annex V of the Madrid Protocol on Protected Areas had foreseen, in addition to ASPAs, another designation known as Antarctic Specially Managed Areas (ASMAs), for any area where activities are being conducted or may be conducted in the future to assist in the planning and coordination of activities.[109] However, ASMAs were essentially created to allow multiple scientific research stations to coexist in the same sections of the continent. In fact, the suggestion has been made that what the Fildes Peninsula needs is ASMA status.[110]

However, Antarctica is increasingly beginning to need more than what the Madrid Protocol has to offer. What it now seems to need is an urban plan for its most crowded regions—plans that would also pave the way for the future human habitation of the continent, which (as previously described) has already been initiated by Chile and Argentina on a small scale.

The aforementioned UK proposal provides the strongest evidence of this when it speaks of the need to expand impact assessments to the creation of facilities on the continent that are resilient to climate change, and to establish "trigger points" for when fresh impact assessments need to be generated for the same activity.[111] In fact, the extent to which the United Kingdom would like to stretch the impact assessment instrument begs the question of whether we are looking at the right instrument to begin with.

Parties to the Antarctic Treaty need to recognize that there are limits as to what any environmental assessment system can achieve and how far it can be stretched. They need to be more open about the fact that what Antarctica clearly now needs are territorial planning policies. To be effective, such policies require territorial sovereignty. If the current randomness of human activity on the continent continues, as has happened with the Fildes Peninsula, it will be to the continent's environmental detriment. Neither Annex V on protected areas nor Annex I on impact assessment will be able to keep pace.

In fact, ASMAs are the ultimate embodiment of how scientific research reigns supreme in Antarctica and takes precedence over the environment. ASMAs cannot replace urban planning. Some have even questioned if ASMAs are now being used to political ends, with the Chinese requesting that Dome A—the highest and coldest point on the continent—be designated an ASMA when they are the only ones currently operating there. The purpose of ASMAs was to allow coordination among various actors operating in the same geographic location rather than to give one actor a dominant position. Dome A is where China has built its fourth outpost, Antarctic Kunlun Station.[112] In fact, ASMAs are part of a system with which the world needs to progressively dissociate itself in favor of proper territorial planning policies. What Scott writes lends further weight to this argument:[113]

> Antarctic science is . . . prioritized by the Madrid Protocol, which permits exemptions to the rules prohibiting interference with Antarctic native fauna and flora and limiting access to vulnerable areas where such interference or entry is necessary for scientific purposes, as well as providing for the designation of protected areas for the specific purpose of protecting scientific research activities.

In Box 7.1, Kevin Hughes, environmental research and monitoring manager at the British Antarctic Survey, discusses the effectiveness of various aspects of the Madrid Protocol. Hughes is directly responsible for the management of various ASPAs and ASMAs in Antarctica. Particularly interesting here is his comment that while any party can build a research station in Antarctica, so long as it is associated with an impact assessment, any party can block the designation of a protected area. This is because environmental assessments do not specifically require approval by treaty parties, whereas protected areas do. In other words, he sees the balance between science and environmental protection as fundamentally flawed.

### Box 7.1 Interview of Kevin Hughes, Environmental Research and Monitoring Manager at the British Antarctic Survey[a]

1. *To what extent has the Madrid Protocol's regime of Antarctic Specially Protected Areas (ASPAs) and Antarctic Specially Managed Areas (ASMAs) been successful, in your view? Have a sufficient number of ASPAs been designated, and how effective is the enforcement?*
It has been well documented that the current ASPA system is still immature and some way off [from] providing adequate protection for the diverse range of values mentioned in Annex V to the protocol. In general, communication between parties on ASPA visitation to allow coordination to prevent cumulative impacts is poor, despite the requirements of the protocol and the Electronic Information Exchange System.

A more systematic approach to ASPA designation may prove useful, rather than the current system, where parties offer to manage areas that they consider are worth proposing for protection and which are often located near their research stations or other infrastructure. This means more remote sites of value are underrepresented in the protected area system. Also, the quality of values protected in ASPAs near research stations, compared to others sites across the whole continent, is, in some cases, open to question.

ASMAs have recently been a subject of discussion within the ATCM, particularly with regard to the proposed ASMA at Dome A with China as the sole proponent. The use of protected areas to achieve parties' political goals has been a cause for concern in both marine and terrestrial environments.

2. *It is said that the Madrid Protocol's Environmental Impact Assessment (EIA) provisions have never led to a change in any planned activity. It is also said that the outcome of discussions of these assessments by Antarctic Treaty parties is not always taken into account. Do you agree, and how effective is the EIA instrument?*
As the person responsible for much of the EIA and associated permitting work within the British Antarctic Survey (BAS), I feel that these phrases are only half of the story. Each year, we examine around 75 EIAs for all activities undertaken by BAS in the Antarctic. Most of these are preliminary EIAs and have an impact that is less than "minor or transitory." However, we ensure that for all projects,

environment impacts are minimized through appropriate mitigation measures. I cannot comment on the thoroughness of this process when applied by other Treaty parties, but based upon experience during collaborations, some parties have high standards, but this may not always be the case. partly this may be due to different interpretations of the protocol or differences in the enactment of the protocol into domestic legislation. The environmental standards employed within each party's own domestic setting may also influence thinking on what is acceptable in Antarctica and what is not.

The Guidelines on EIA are currently under revision by the Committee for Environmental Protection through an Intersessional Contact Group, led by the United Kingdom and Australia. The United Kingdom has identified substantial flaws in a working paper it submitted to the ATCM entitled *Improvements to the Antarctic Environmental Impact Assessment Process.*

One interesting point is that under the current consensus system, any party can build a research station (with all the associated environmental impacts) anywhere in the treaty area, as long as the associated EIA process has been undertaken. In contrast, any party can block the creation of an Antarctic protected area. This example suggests to me that the balance of priorities concerning 'impact vs. protection' could benefit from reexamination.

3. *Under the provisions of the Antarctic Treaty and the Madrid Protocol, inspections are carried out to ensure compliance. What is your opinion on the regularity of these inspections and their quality?*

The comprehensiveness of inspections may vary and the response by parties that have been inspected and the rate of improvement of infrastructure and environmental performance can be slow. Issues reported during one inspection can sometimes be noted again in subsequent inspections. For example, stations on Fildes Peninsula are inspected regularly compared to some more remote stations, but reports exist to suggest poor environmental performance and clear breaches of the protocol still occur.

Some, usually remote, facilities have never been inspected, and there is no internationally agreed schedule for inspections. This could be an area for improvement. The environmental legacy of past activity is a major issue for some Parties, and finding resources to undertake remediation is a challenge.

4. *If you had to list three main improvements that you would make to the environmental governance of Antarctica, what would they be?*

1. It has been pointed out that a future major challenge to conservation of Antarctic may be the slow pace of decision making within the treaty system. Currently, only 5 or 6 parties undertake the great majority of the work to progress the governance and protection of Antarctica through the treaty system. This puts a substantial pressure on the resources of those parties who actively participate in keeping the treaty system current and fit for purpose. One improvement would be a more dynamic treaty system with more parties actively involved in the work of the Committee for Environmental Protection and ATCM and associated decision-making.
2. Sharing of stations by different parties. Greater use of existing bed space within stations to host scientists from different Treaty parties and other nations interested in Antarctic science.
3. Better monitoring around research stations as described under the protocol.

[a]Interview conducted by the author on February 26, 2016. As of this writing, Kevin Hughes is also deputy chief officer of the SCAR Standing Committee on the Antarctic Treaty System and part of the UK delegation to the Antarctic Treaty Consultative Mechanism, predominantly as an advisor to the Committee for Environmental Protection. He sits as the UK representative on the ASMA No. 4 Deception Island Management Group and is responsible for management of the 12 ASPAs for which the United Kingdom is the proponent.

Hughes also expresses clear concern about the political instrumentalization of both ASMAs and ASPAs. Many of the ASMAs proposed by parties happen to be near their research stations, rather than in areas in real need of protection. He also expresses concern about China's call for an ASPA at Dome A. With respect to Environmental Impact Assessments, he considers due diligence to be done at the national level within the United Kingdom, but he does not respond to the fact that parties are under no obligation to fully address the concerns expressed at the treaty level on their Comprehensive Evaluations, or the fact that these evaluations have come in late, after an activity already started.

Particularly worrying is his comment, "The environmental legacy of past activity is a major issue for some Parties, and finding resources to undertake remediation is a challenge." In other words, while the Madrid Protocol was established as a forward-looking instrument (and was adopted only in 1998), the mistakes of the past are sometimes left unaddressed. For the future, Hughes proposes changes to decision-making procedures under the ATS, which is too slow to respond effectively to the continent's environmental needs, and gives a rationalization of scientific activity with increased monitoring. Overall, this interview paints a picture of a highly incomplete environmental protection framework in Antarctica. His comments on inspection are explored further in the next section.

## JURISDICTION, INSPECTION, AND LIABILITY—THE CHALLENGE DOWN UNDER

Of course, even the strongest environmental protection regime would be bound to fail without proper enforcement. Frequent breaches of the Agreed Measures, for instance, have been described already in this chapter. However, enforcement in Antarctica is particularly difficult to achieve in light of what has been described as its "jurisdictional vacuum."[114] Because none of the territorial claims were recognized by the Antarctic Treaty, an alternative jurisdictional model had to be devised to secure enforcement. Created from the perspective of a continent devoted to peace and science, parties to the treaty considered that a jurisdictional system based on nationality would suffice. It was Britain, in fact, that proposed this solution, with Norway agreeing "that any person in the Antarctic should be subject solely to the penal jurisdiction of the country of which he is a national."[115]

On that basis, Article VIII of the Antarctic Treaty was created to provide that persons authorized to carry out activities in Antarctica "shall be subject *only* to the jurisdiction of the Contracting Parties of which they are nationals in respect of all acts or omissions occurring while they are in Antarctica for the purpose of exercising their functions [italics mine]."[116] While this system served the continent well for over three decades, its limitations started to show with the growth of scientific research, tourism, fishing, and other activities in Antarctica.

Today, it has a number of principal shortcomings. First, there is the inability of a party to enforce the rules of the treaty or the Madrid Protocol over the nationals of another treaty party, even in the face of a violation. The maximum that a party can do is to report the violation to the institutions of the Antarctic Treaty. The second is the inability of parties to enforce these provisions on third-party nationals, vessels, or aircraft.[117] For example, with close to half of all vessels visiting Antarctica on tourist cruise ships flying flags of third parties, there is little that the treaty can do by way of enforcement. Moreover, many of these vessels fly flags of convenience—that is, the flag of the country where the vessel was registered and not that of the company's home country, in order to avoid restrictive financial and other regulations. Currently, parties rely on the International Association of Antarctica Tour Operators (IAATO) to ensure compliance with the Madrid Protocol through tour operators. However, this halfway measure provides no guarantee of enforcement, especially with the constant reflagging of vessels that currently takes place.[118]

The problem of enforcement on the continent has led some scholars to argue that a territorial basis for jurisdiction needs to somehow be reintroduced into the treaty. William Bush, for instance, has suggested that without prejudice to the rules of the Antarctic Treaty, claimant states could be authorized to exercise jurisdiction for enforcement purposes within the areas of their claims. Other parties could accept that this does not confirm their claim but only serves the purposes of an objective treaty regime.[119] Of course, such a system would be highly unlikely to materialize because it would be tantamount to recognizing territorial claims. However, the fact that such options are even being discussed points to the near-impossibility of achieving proper enforcement without territorial sovereignty—another argument in favor of the Svalbard model.

Particularly acute in the jurisdictional problem facing Antarctica has been the plight of the Southern Ocean, one that resulted in the coining of the term "illegal, unreported, and unregulated" that is now widely used to describe the fishing there.[120] The jurisdictional challenge faced in the Southern Ocean has created a fundamental difference in effectiveness between CCAMLR and other regional fisheries agreements. As Alan Brown states, "[A]greements in other parts of the world's oceans aimed at protecting regional fishing resources have usually involved fish stocks *within areas under national jurisdiction*. These agreements could therefore be implemented through the enforcement agencies of the countries involved; however, the situation in the Antarctic is not so clear [italics mine]."[121]

In fact, sovereignty issues are fairly complex under CCAMLR, which applies farther north than the Antarctic Treaty's 60 degrees south to include some sub-Antarctic territories. In fact, when CCAMLR was adopted, the chair of the negotiations was compelled to read a statement essentially confirming the existence of EEZs for all the sub-Antarctic islands that fall under national sovereignty. These EEZs were to be excluded from CCAMLR's coverage.[122]

Currently, five countries exercise recognized sovereignty over various sub-Antarctic islands over whose waters CCAMLR has no automatic regulatory powers (with an additional two countries, the United Kingdom and Argentina, in dispute over a number of islands).[123] Therefore, in terms of overall jurisdiction, there are portions of the Southern Ocean that are regulated by these five powers, and other portions regulated by CCAMLR. However, CCAMLR has only 25 members;[124] which means that the vast majority of the world does not recognize CCAMLR's jurisdiction and considers the Southern Ocean—minus the aforementioned EEZs—to be part of the "high seas" under UNCLOS. This has led some commentators to refer to the Southern Ocean as a "textbook example" of an "unmanaged common."[125]

In fact, the EEZs surrounding the sub-Antarctic islands that fall under national jurisdiction can be said to have been saved from the messy legal situation that prevails in the Southern Ocean. For instance, the Heard and McDonald Islands, which belong to Australia, have come under strong domestic legislation, even if it took Australia a while to pay attention to them. Currently, no foreign fishing vessels can fish in the waters of these islands, and catch limits have been set for various types of fisheries.[126]

Francesco Vicuna has argued that, to close the "jurisdictional loophole in Antarctica," countries need to consider strengthening the responsibility of "flag states," increasing port state control, or both.[127] The former could oblige flag states to ensure that their vessels comply with environmental requirements, while the latter could broaden the powers of port states to inspect and punish vessels departing from or arriving at their ports. Both suggestions are themselves very complex from a legal point of view, requiring examination within UNCLOS.[128]

Vicuna has also been critical of CCAMLR's approach to inspection in the Southern Ocean, arguing that it is based on a narrow interpretation of CCAMLR's jurisdictional powers. CCAMLR, he explains, has adhered to an approach of national inspections with respect to vessels flying the same flag. In other words, it has assigned to vessels inspectors of their own nationality, which has itself weakened enforcement.[129] Therefore, CCAMLR, he argues, is weaker than CRAMRA, which took a more institutionalized approach to the issue. Under CRAMRA, inspectors (called "observers" by the agreement) could be nominated by a member of the commission (i.e., a party) or by the commission itself, and could carry out their inspection operations without prejudice to jurisdictional issues. In other words, an inspector of nationality A could inspect a mining operation of country B.[130] What all this points to is that even under the best of environmental worlds and the strongest of regimes, without territorial sovereignty, proper enforcement will remain an uphill battle in Antarctica.

An examination of the inspections conducted under the Madrid Protocol reinforces this conclusion. Unlike CCAMLR, the Madrid Protocol allows inspectors to be nominated either by its parties or by ATCMs; in other words, it was successfully aligned with its predecessor, CRAMRA. To ensure proper implementation, the protocol allows parties to inspect each others' stations, installations, equipment, ships, and aircraft.[131] However, despite its successful resolution of the jurisdictional issue, its inspection provisions remain weak and are poorly implemented. In fact, when the Madrid Protocol was being negotiated, countries explicitly rejected the concept of an autonomous environmental inspection system, which would have resulted in the creation of an independent organ to ensure compliance.[132]

As Kevin Hughes explained in the interview in Box 7.1, inspections currently suffer from a number of shortcomings, including a concentration of inspections on the more accessible parts of the continent (i.e., the peninsula) at the expense of more remote regions; the failure to establish an internationally agreed schedule of inspections; and a poor response to inspection reports, as evidenced by the Fildes Peninsula. In a paper entitled "How Committed Are We to Monitoring Human Impacts in Antarctica?" he wrote:[133]

> With little formal international scrutiny other than occasional station inspections, a lack of enforcement mechanisms in place to penalize contravention of the provisions of the Antarctic Treaty and its related legal instruments, and a need to maintain good diplomatic relations between Antarctic Treaty Parties, nations are under little pressure to prioritize human impact monitoring.

In fact, in a paper submitted by ASOC to the ATCM in which it reviewed the operation of the Madrid Protocol, it reached similar conclusions. With respect to the inspection of scientific research stations and other facilities, it wrote: "Intriguingly, the findings of recent inspections are in some cases not substantially different from those carried out ten or more years earlier. Aspects of recurrent concern include fuel storage and handling; waste management (particularly waste storage); sewage treatment; management of former sites (abandoned stations); and limited awareness of Environmental Impact Assessments and Environmental Impact Assessment follow-up on the ground."[134] It added that "some research stations appear to be caught in a pre-1991 time warp. For instance, observations made by German scientists at the Fildes Peninsula in King George Island . . . suggested that some things have hardly changed since the late 1980s, or if they improved in the intervening years, they have once again gone back to the old ways."[135] In other words, it was seriously dismayed at how the protocol was being implemented.

With implementation so weak, the issue of liability has grown in importance. How, then, has the Madrid Protocol addressed the question of liability? Annex VI of the protocol, "Liability Arising from Environmental Emergencies," proved to be the most difficult component of the Madrid Protocol to negotiate; it required no less than 13 years to do so, when the protocol itself and its first four annexes had taken only a year.[136] Furthermore, despite its negotiations having been completed in 2005, 11 years later, it has yet to enter into force.

The negotiations had to address a number of complicated issues, the most difficult of which was how a liability regime could operate in Antarctica's jurisdictional vacuum. In fact, customary international law on liability was not transferable to Antarctica in the absence of state jurisdiction.[137] The main jurisdictional questions before the parties therefore were who would be liable, for what, and to whom, requiring a significant amount of legal creativity.[138] Currently, it is clear that even if it were to enter into force, the liability regime could not possibly apply to nonparties of the Antarctic Treaty (i.e., to third parties)—already a major weakness.

Under Annex VI, emergencies are essentially defined as accidents that result in, or threaten to result in, significant and harmful impact on the environment. Operators are required to take "preventative measures" and to prepare "contingency plans" for response to incidents and to take "response action" eventually, when it becomes necessary. The latter is defined as reasonable measures to avoid, minimize, or contain the impact of the environmental emergency, "which to that end may include cleanup in appropriate circumstances."[139] The jurisdictional questions were therefore resolved by designating the "operator" as the party liable to undertake "response action" for "significant and harmful" environmental impact (the "who," "what" and, to some extent, the "to whom" questions).

The negotiating history of the protocol demonstrates that whereas the negotiations had started on a fairly ambitious footing, with ASOC tabling a proposal for a liability regime that would be applicable to "all activities" taking place in Antarctica, the level of ambition was quickly reduced.[140] The United States tabled a paper early in the negotiations to limit the scope of the regime to "environmental emergencies."[141] This view eventually prevailed, with Article 1 of Annex VI, on scope, defining the regime as applying "to environmental emergencies in the Antarctic Treaty area which relate to scientific research programs, tourism and all other governmental and non-governmental activities. . . . It shall apply to all tourist vessels that enter the Antarctic Treaty area. It shall also apply to environmental emergencies in the Antarctic Treaty area which relate to other vessels and activities as may be decided."[142]

Notable here is the omission of the regime's applicability to the "dependent or associated ecosystems" to which the Madrid Protocol extends. In fact, particularly strange about this omission is that the annex does make a reference to these dependent and associated ecosystems in its preamble, as well as in its call for the preparation of contingency plans.[143] However, the fact that they are not specifically mentioned under Article 1 could certainly allow operators to consider themselves relieved of responsibility for these extended portions of the Antarctic environment.

In addition, a long debate appears to have taken place on whether emergencies resulting from scientific activities on the continent should be made immune to liability, with some countries wanting continued precedence given to science. Surprisingly, SCAR, which should have the continent's best interest at heart, itself advocated that view but was eventually overruled.[144] While emergencies arising from scientific activities are now included, the preamble to Annex VI nevertheless reiterates the need to "accord priority to scientific research."[145]

For reasons that will be explained later in this chapter, the current liability regime is fairly weak and falls well below the benchmark set by CRAMRA (even if countries intended to expand CRAMRA's liability regime in the future). In fact, it appears that some countries believed the continent to be at little risk of environmental damage and to not require a liability regime at all, given that the Madrid Protocol had banned mining.[146] The irony, however, about this line of argument is the fact that while the protocol did ban mining, it opened the door to all other human activities. In fact, even with regard to a liability regime limited to emergencies, the United States resisted a stringent instrument for fear that it could have the perverse effect of discouraging parties from forming part of an emergency response, as well as of raising the cost of insurance for operators.[147]

Currently, the principal weaknesses of the regime lie in the fact that it is restricted to environmental emergencies; it does not create state responsibility for environmental emergencies not caused by state actors; it does not cover all types of damage; and it does not provide for environmental restoration. It has also been argued that the regime, as currently worded, could have the unintended effect of discouraging cooperation among parties in responding to emergencies. In fact, some have questioned how an agreement with those characteristics could be called a liability regime at all. It is more of a tool for immediate emergency response than a proper liability regime, even if the preamble to Annex VI refers to the annex as "a step" in the establishment of a broader regime.[148]

In being restricted to environmental emergencies, the annex is currently confined to sudden accidents or catastrophic incidents. As Louise de La

Fayette explains, "It does not cover a cumulative impact from a continuous situation, such as a slow leak of a hazardous substance, or from the repetition of many incidents, such as repeated trampling by hundreds of tourists on a particular piece of ground. Nor would it cover the impact of noise over a long period, causing stress to populations of animals, such as marine mammals, or contamination through the gradual introduction of harmful or alien organisms into pristine ecosystems."[149] CRAMRA's liability regime, on the other hand, had applied to "any Antarctic mineral resource activity," not just mining accidents or emergencies.[150]

It appears that limiting state responsibility in the event of environmental emergencies was one of the principal goals of the negotiators to Annex VI.[151] Article 10 of the Annex on State Liability explicitly states, "A Party *shall not be liable* for the failure of an operator, other than its State operators, to take response action to the extent that the Party took appropriate measures within its competence, including the adoption of laws and regulations, administrative actions and enforcement measures, to ensure compliance [italics mine]."[152] This is well below the standards of CRAMRA, which had established subsidiary or residual liability against the sponsoring state based on the violation of its supervisory obligations.[153]

Furthermore, unlike most other existing liability regimes, the annex does not use the word *damage.* It only refers to "significant and harmful impacts on the Antarctic environment" in its definition of environmental emergencies. Therefore, it fails to cover damage to persons, property, or even scientific research programs, despite the protocol's emphasis on science.[154] CRAMRA, on the other hand, covered "loss of or damage to property of a third party or loss of life or personal injury of a third party."[155]

With respect to response measures, the annex has been criticized for not including firm obligations for cleanup. It defines response measures as reasonable measures that may include cleanup. This hesitancy, coupled with the fact that there is no obligation for any restoration of long-term damage to the environment, are among the most serious weaknesses of the regime.[156] Such an obligation had already existed in CRAMRA, where operators were liable for all costs relating to "prevention, containment, cleanup and removal measures, and action taken to restore the status quo."[157]

Finally, and quite surprisingly, in their desire to make the operator definitively responsible for an accident, negotiators included a number of provisions that appear to discourage other players from intervening immediately to help. This is the case despite the fact that in Antarctica's fragile environment, any party reasonably close to the emergency should be encouraged to act right away.[158] The discouragement comes from Article 5, "Response Action," which establishes a number of hurdles that third parties must cross to take part in a response. It states, for example, that "such other Parties shall not take a response action to an environmental emergency . . . *unless* a threat of significant and harmful impact to the Antarctic environment is imminent . . . or the Party of the operator *has failed*. . . . [italics mine].[159]

Overall, therefore, Annex VI of the Madrid Protocol is very weak, and it never entered into force anyway. Clearly, Antarctica's unique and fragile environment deserves better. In fact, countries should move as fast they possibly can toward the assignment of national sovereignty in Antarctica to avoid this continued tragedy.

## BIOPROSPECTING—MINING IN ANTARCTICA IS WELL UNDER WAY, AS IS TOURISM

A common misconception about the Antarctic Treaty and of the Madrid Protocol is that they have completely closed the seventh continent to commercial activity. In fact, the only two activities that they explicitly prohibit are militarization and mining. To recall, Article I of the Antarctic Treaty states, "Antarctica shall be used for peaceful purposes only. There shall be *prohibited*, inter alia, any *measures of a military nature*"; and Article 7 of the Madrid Protocol adds that "Any activity relating to *mineral resources*, other than scientific research, shall be *prohibited* [italics mine]."[160]

Due to these prohibitions, the illusion of a continent closed to all commercial exploitation and devoted exclusively to peace and science seems to have taken hold, with any commercial enterprise occurring in Antarctica generating a tremendous amount of anxiety among Antarctic policymakers and scholars.[161] However, the fact of the matter is that any human activity, other than the prohibited ones, is currently allowed in Antarctica, provided that the provisions of the ATS are respected, such as that Environmental Impact Assessments be conducted as necessary. In fact, CCAMLR was the first real agreement to allow mining in Antarctica—namely, the mining of living marine resources. This made it the first environmental/commercial exploitation agreement to emerge in the scramble for Antarctic resources.[162]

Other than fishing, however, two commercial activities have been on the rise in Antarctica. The first is bioprospecting.[163] In Antarctica, bioprospecting is well under way and is growing both on land and in the sea. In addition, as previously described in this book, Antarctica now has a booming tourism industry, with tourists flocking to the seventh continent from around the globe.

However, of these two activities, it is bioprospecting that tends to surprise onlookers the most because of the ban on mining. In fact, Article 7 of the Madrid Protocol is silent on whether on its prohibition covers living as well as nonliving mineral resources; it simply bans "any activity related to mineral resources" and does not contain a definition of the term *minerals*. However, because it rose from the ashes of CRAMRA and was designed to undo that specific agreement, the view seems to have prevailed that the type of mining that it bans covers only whatever was previously authorized by CRAMRA. CRAMRA had defined mineral resources as "all *non-living natural non-renewable* resources, including fossil fuels, metallic and non-metallic minerals [italics mine]."[164] In other words, it excluded living resources, so the Madrid Protocol is widely considered to have done so as well, even though nothing in the protocol explicitly says so.

There have been two main reasons behind the tremendous interest in Antarctic bioprospecting. The first is lack of knowledge about Antarctic biota and the desire to discover completely new organisms and expand the boundaries of scientific knowledge. The second is the existence of "extremophiles" in Antarctica, which are novel life forms capable of withstanding extreme cold, aridity, and salinity, and which are some of the most sought-after organisms by industry.[165]

Currently, bioprospecting in Antarctica is shrouded in mystery, with various attempts having been made at collecting information more systematically on its scope. The United Nations University (UNU) was brought in to assist in the process,[166] the Committee on Environmental Protection now includes bioprospecting as an agenda item, and Belgium has proposed the creation of a database on the topic.[167] Despite these efforts, information remains scattered, and some of it is out of date.

Based on publicly available information, the biggest users of Antarctica's genetic resources appear to be the pharmaceutical, food and beverage, biotechnology, chemical, cosmetic, aquaculture and agriculture, and environmental remediation industries. Emerging patterns suggest that over 50 percent of bioprospecting is marine-based, followed by terrestrial (at around 35 percent) and inland water prospecting.[168] In the next discussion, information collected from a Belgian submission to the Antarctic Treaty Consultative Mechanism, the UNU, UNEP, ASOC, and a few other sources is used to provide a basic sense of what bioprospecting in Antarctica has achieved.[169]

In terms of marine bioprospecting, Antarctic krill (*Euphausia superba*) has been the object of the greatest commercial exploitation and the largest number of patents. Patents have covered the use of microorganisms and other derivatives from krill in laundry detergents and other cleaning agents, food processing and food products, dietary supplements (because of the use of krill oil in omega-3 fatty acid supplements), and skin care products.

Krill-derived enzymes and extracts are being explored in the treatment and prevention of acne, infections, inflammations, diabetes, immune reactions, and certain cancers. In fact, it was out of concern for krill fishing that CCAMLR was originally negotiated, with the concern later shifting to tooth fish. Companies in Japan, Spain, Chile, the United States, the United Kingdom, Poland, South Korea, and more, have all been involved in the filing of patents for krill-based inventions.

Sponges and other marine invertebrates have been used in bioprospecting as well. The Antarctic sponge (*Kirkpatrick varialosa*), for instance, is the source of a potential cancer drug called variolin. Cold ocean teleost fish from the Southern Ocean has been a source of antifreeze proteins. The molecule responsible for this phenomenon has been patented and could be used commercially to lengthen the shelf life of frozen foods and to generally improve the survival of biological material, such as in transplant surgery. The cosmetics and personal care industries have used Antarctic marine algae and organisms in their products. For instance, the French cosmetic giant Clarins uses *Durvillea antarctica* algae in its extra-firming skin cream. This product was launched in 1996 and went on to become a best seller.

In Antarctica's terrestrial and inland water environments, extremophiles have also been of great scientific interest. A variety of bacteria have been exploited from these bioma for various industrial applications. For example, *Marinomonas protea* bacteria, sourced from a hypersaline Antarctic lake, contains an antifreeze protein that could keep ice cream soft throughout the thawing and refreezing process. The invention has been patented by Unilever, the owner of two ice cream company giants, Ben & Jerry's and Breyer's.

An Antarctic enzyme derived from *Bacillus* sp., a microorganism found in a volcanic vent in Antarctica, is being used to improve the accuracy of deoxyribonucleic acid (DNA) extraction technology. An Antarctic yeast, *Candida antarctica*, has been used in a wide variety of commercial activities, including by the paper industry. Antarctica's terrestrial algae have also been used by the cosmetics and personal care industries. The green algae *Prasiola crispa*, for instance, has been patented by a German company for use in various skin products, including sunscreen and after-sun creams.

A number of Antarctic scholars have expressed concern about all of these developments. After all, how could the Antarctic Treaty's provisions on the sharing and dissemination of scientific information acquired in Antarctica square with the commercial exploitation of this knowledge? Article III of the Antarctic Treaty states that "information regarding plans for scientific programs in Antarctica *shall be exchanged* to permit maximum economy and efficiency of operations [italics mine],"[170] seeming to suggest that the goal of science in Antarctica is to enhance collective scientific knowledge rather than to exploit that knowledge commercially.[171]

Questions have also been raised as to how the benefits of commercial exploitation should be shared. Would those benefits belong only to those directly involved in the research, or should some of the commercial gain somehow be returned to, and invested in, the seventh continent, given that it is a global common?[172] Some scholars, for example, have warned against science proceeding on two tracks in Antarctica—the purely research-focused track and the commercially oriented one, which could raise issues of ethics and equity among parties to the Antarctic Treaty, and among these parties and the rest of the international community. In fact, the mining of biological resources could generate the same concern among members of the United Nations as has the mining of nonliving resources under CRAMRA. Scholars have claimed that bioprospecting could end up jeopardizing the entire "scientific project of Antarctica."[173] On the other hand, Yves Frenot, the deputy head of the French Polar Institute, has argued that it would be unfortunate if the legal uncertainties surrounding bioprospecting in the Antarctic Treaty were to stymie the search for important breakthroughs like new cancer treatments.[174]

What this discussion demonstrates is that the mining of Antarctica has already started, but this time under the radar, without a mining convention to regulate it. While CCAMLR authorizes the mining of fish, the ATS's silence on bioprospecting has effectively also authorized it for all other living resources. The mining of nonliving resources will surely follow—it is simply a matter of time. Of great concern, however, is that bioprospecting in Antarctica, while covered by the Madrid Protocol, which in theory includes "any activities undertaken in the Antarctic Treaty area pursuant to scientific research programs,"[175] has never been subjected to Environmental Impact Assessments.[176] Parties have simply allowed bioprospecting operations to slip past the protocol, which provides even more evidence of how weak the environmental governance system is when it comes to Antarctica.

Particularly puzzling is that CRAMRA had been very strong on the issue, devoting an entire chapter to the prospecting of nonliving resources. It provided that prospecting could not confer upon any operator the right to Antarctic mineral resources; in other words, the discovery of a valuable resource would not automatically entitle the operator to it. Second, while no authorization by CRAMRA would have been required for prospecting, it stipulated that all such activities be conducted in compliance with the accord. In addition, the CRAMRA Commission would have had to be notified of any prospecting activity nine months prior to its initiation.

Third, sponsoring states would have had to ensure that all operators engaged in prospecting maintain a financial capacity, commensurate with the nature and level of the mining operations and the risks involved. They also would have had to ensure timely response action to these risks under strict liability standards. If more than one operator were to be involved in prospecting, sponsoring states would have to ensure that operators respect each other's rights.[177] It is difficult to fathom, therefore, how the Madrid Protocol, a supposed environmental accord, could have ignored the entire issue when its predecessor, a mining agreement, raised the environmental bar so high. The word *bioprospecting* does not appear in the Madrid Protocol—not even once.

The applicability of the CBD, the treaty that governs biological diversity and bioprospecting at the international level, to Antarctica has been the subject of much academic debate. The CBD was essentially created to ensure that countries to which the biological resources belong share the benefits of their commercial exploitation. In other words, it was designed to prevent companies from exploiting the world's natural resources without sharing some of the profits with the rest of humanity (represented in this case by the sovereign state to which the biological resources belong). Doubt has been expressed, however, as to how the CBD could be applied in Antarctica's jurisdictional vacuum when it is anchored in the principle of national sovereignty.[178]

Furthermore, because it introduced the principle of "benefit sharing" from commercial exploitation, the United States never ratified the CBD, which ended up facing a similar fate as UNCLOS.[179] As demonstrated in Chapter 6, benefit sharing is one of the main strands of the Common Heritage of Mankind principle. Sanjay Chaturvedi writes:[180]

> The growing salience of bioprospecting in Antarctica is likely to raise a number of ethical and equity issues that will not be easy to address and translate into legally binding regulations. An overarching issue of utmost importance here relates to the extent to which biological and genetic resources in the Antarctica are free to be "owned" by anyone or they should be regarded as the "common heritage of mankind."

What this point illustrates is that any attempt to regulate bioprospecting in Antarctica through an international or regional accord that includes benefit sharing is not likely to garner universal support. In addition, as with the mining of nonliving resources, any attempt to negotiate an agreement on the mining of living resources will draw UN attention, with the entire international community asking to be involved. For the same reasons that the Svalbard model was presented as the solution to the revival of mining in Antarctica for nonliving resources, it is also one of the few viable ways forward for the mining of other types of resources (other than armed conflict—not the best solution). In addition, it is the only model that can bring desperately needed environmental protection to the continent.

Tourism is a further case in point. Despite the spectacular rise in the number of tourists visiting the continent and the serious accidents that have taken place in recent years (for instance, the *Bahia Paradiso* was not insured when it sank in 1989 less than a mile off the Antarctic Peninsula),[181] little has been done within the ATS to address this growing commercial activity in a serious way. Some have argued that because tourism is already covered by the Madrid Protocol, nothing further needs to be done—in other words, similar arguments to the ones about bioprospecting. In the end, the regulation of the sector has been left to an industry association, IAATO, which is hardly a satisfactory option. It now sets the Antarctic tourism agenda, with a limited role played by treaty parties.[182]

Once again, Antarctica's jurisdictional vacuum has impeded serious regulation—the regulation of tourism "without a clear recipient state," as Davor Vidas has described it.[183] With many tourist vessels flying the flag of nontreaty parties, the challenge of regulating tourism under the umbrella of the Antarctic Treaty is immense. In fact, some have expressed the concern that to do so could lead to the increased relocation of tourism to nonflag states.[184] This issue also would be far better handled by sovereign states, who could exercise greater control of tourism arrivals and departures from their own territories (including Antarctic territories).

## CIVIL SOCIETY IN ANTARCTICA—WHY THE CONTINENT NEEDS CITIZENS

With the environmental governance of the continent being this problematic, one must ask where civil society is in Antarctica and what role it has played. Relative to other continents, the participation of civil in Antarctica has been limited. The continent is uninhabited, its climate is harsh, and it is fairly inaccessible. Moreover, the Consultative Parties have kept its governance in a tight grip and held civil society at bay. Negotiated from a security prism, essentially for the purposes of disarmament, the Antarctic Treaty was born out of a Cold War mentality, with its deliberations shrouded in secrecy and the participation of civil society not particularly encouraged.[185] The Consultative Parties have considered themselves the main gatekeepers of the seventh continent, and they still do.

NGO participation in Antarctica is said to have gone through three main phases: an initial phase between 1959 and 1970, in which NGO activity was hardly perceptible; a second phase between the 1970s and the birth of the Madrid Protocol, in which engagement rose exponentially to protect Antarctica from mineral exploitation and rising commercial interests; and the current period, one of relative calm, in which the engagement of environmental groups has dropped but the engagement of one business NGO (namely, IAATO) has grown.[186] IAATO's active engagement in Antarctica could herald a new era in the relationship between civil society and the seventh continent.

In the first phase, only one NGO can be said to have been truly active in Antarctica—SCAR.[187] Previously discussed in this book was the birth of SCAR and its evolution over the years. It was established in 1958 and entrusted with coordinating scientific research at the end of the IGY. At the time, it was thrust into the middle of a scientific contest among the world's greatest powers—one that would determine who would stay in Antarctica and who would leave after the IGY was over. Becoming a scientific broker of sorts, SCAR succeeded in gradually establishing itself as an institution capable of initiating and coordinating high-quality, international scientific research on Antarctica and the Southern Ocean, as well as on the role of Antarctica in the global environmental system. SCAR is credited with the adoption of the Agreed Measures and the rise of CCAS and CCAMLR. It was granted observer status under the Antarctic Treaty in 1987.[188]

However, a deliberate fragmentation of science gradually took place in the ATS. SCAR found its role reduced through the creation of the Council of Managers of National Antarctic Programs (COMNAP), and the formation of separate advisory bodies within CCAMLR and the Madrid Protocol. Weakened, but still alive, SCAR was considered by some to be coopted by the Antarctic Treaty from the moment that it obtained observer status, losing much of its independence.[189] In fact, its lobbying to exclude scientific activity from the coverage of the Madrid Protocol's Annex VI on liability has raised eyebrows.

The second phase of NGO engagement was a far more active phase driven by the evidence that was slowly emerging on the continent's mineral potential and the commercial interest that this drew.[190] Particularly active in this phase were two NGOs: ASOC and the International Union for the Conservation of Nature (IUCN). Other NGOs, such as Greenpeace and Friends of the Earth, became engaged only in later years.

Organized along similar lines to SCAR, but with a global rather than a regional focus, the IUCN's membership is drawn from associations with a scientific ethos. As with SCAR, this has given the IUCN legitimacy within the ATS because of the primacy that it accords to science. The IUCN has worked with SCAR on a range of scientific projects, and it has become one of the two conservation organizations permitted to attend ATCMs (ASOC is the other).

NGO lobbying against CRAMRA and against the closed governance regime evolving in the Antarctic defined the second phase. NGOs had to take a position for or against the Question of Antarctica, as posed by the United Nations. While ASOC and Greenpeace were initially attracted to UN discussions, they eventually took the strategic decision to influence the Antarctic Treaty regime from within.[191] While a constituent member of ASOC, Greenpeace entered the Antarctic arena in its own right in 1983 to protest CRAMRA. Being an action-oriented NGO, it established its World Park base in Antarctica to launch its campaign for the designation of the continent as a park. With the Madrid Protocol achieved, and because the base was too costly to maintain, Greenpeace dismantled it in 1992.[192]

With the demise of CRAMRA and the emergence of the Madrid Protocol, NGOs considered their work accomplished, which ushered in the third phase of NGO engagement on the continent.[193] Action-oriented NGOs like Greenpeace gradually withdrew from the scene, leaving ASOC and the IUCN to handle more regular day-to-day contact with the Antarctic Treaty on conservation issues. Around the same time, in 1991, IAATO formed to regulate tourism in Antarctica in the midst of growing concerns by the Antarctic Treaty on the potential impact of tourism on the environment, but also as a reaction to the incapacity of parties to themselves address tourism in a holistic fashion. Today, IAATO is the only active business NGO on the continent.

Overall, the first phase of NGO engagement in Antarctica can be said to have legitimized the regime by strengthening its scientific focus—a role essentially played by SCAR. The second phase can be said to have cast a group of NGOs as agenda-setters within the ATS, having helped bring down the CRAMRA in favor of the Environment Protocol. The current phase is one of redefinition—one in which environmental NGOs have lowered their profile and in which a business NGO, IAATO, has stepped in forcefully.[194] This redefinition itself signals that Antarctica is increasingly becoming open for business.

What this demonstrates is that NGO engagement has gone through phases in an uneven fashion. The poor environmental governance regime established for the continent, in the form of the Madrid Protocol, shows that the engagement was shallow. NGOs simply wanted to see CRAMRA fall, paying little attention to its environmental provisions or to the weak environmental safeguards that would follow in the Madrid Protocol. The current phase is so low profile that hardly anyone outside the continent hears anything about its environment, except when a major chunk of ice collapses and is immediately assumed to be evidence of climate change.

At present, there is no major public outcry by NGOs about the limited environmental protection on the continent, such as the fact that it has so few protected areas or that bioprospecting has by and large bypassed the treaty. NGOs like SCAR and ASOC have been expressing their opinions from within the Antarctic Treaty Consultative Mechanism. Their principal audience are members of the Antarctic Treaty itself, not the outside world. So, then, who will blow the whistle more publicly on the poor environmental governance of the continent—a continent that is likely to be subjected to increased economic pressure and demands for human habitation? How can civil society engagement be stepped up in Antarctica? The assignment of national sovereignty is a sure path toward that goal. Sovereignty will give birth to citizens, who by definition will mobilize greater interest and financial resources for the environmental protection of the continent. Antarctica now needs its own people who think of it as their homeland.

## THE SVALBARD ENVIRONMENTAL PROTECTION MODEL

Were a Svalbard model to be applied to Antarctica, with human settlements expanded, and Antarctica more clearly opened to economic enterprise, including the mining of nonliving resources, would the Antarctic environment suffer? In other words, as inadequate as the Madrid Protocol is in protecting the Antarctic environment, would it nevertheless provide greater environmental security than a Svalbard model? Would such a model actually lead to environmental destruction?

In fact, Svalbard has gone through various phases of environmental protection, starting with fairly weak regulations. Today, however, Svalbard's environment is far better protected than Antarctica's, in spite of human settlement, mining, and the archipelago receiving multiple times the number of tourists. The governor of Svalbard, interviewed in Box 7.2, indicates that protected areas now cover 65 percent of the archipelago's land area and 87 percent of its territorial waters; these figures make Antarctica's mere 1.5 percent of ice-free surface even more disturbing.[195] fact, Svalbard's environment is more protected than Norway's:[196]

> There are seven national parks on Svalbard (out of 44 in Norway as a whole), six nature reserves, 15 bird sanctuaries and one geotope (geological protection area). The first three national parks were created in

> 1973, the next three in 2003 and the last in 2005. In all, they comprise close to 14,500 $km^2$ (24 per cent) of Svalbard's land area. In addition, in excess of 20,000 $km^2$ of marine areas are included in the national parks. In total, 65 per cent of the land area of Svalbard is protected in one way or another in order to conserve its unique nature, landscape and cultural heritage. *In comparison, areas with various types of protection constitute just 17 percent of mainland Norway*, of which 10 per cent are national parks [italics mine].

This is the outcome of the Norwegian government setting a goal in 1995 of making Svalbard one of the best managed wilderness areas in the world.[197] Today, Svalbard's environment is governed by the Svalbard Environmental Protection Act of 2001, which was amended and strengthened in 2012.[198] The act is anchored in a number of environmental principles established right at the outset, such the precautionary principle[199] and the principle of "cumulative environmental effects." It states that "Any activity that is started in Svalbard shall be assessed on the basis of the cumulative effects on the natural environment and cultural heritage that would result."[200] Those are exactly the types of impacts that the Madrid Protocol fails to assess. In addition, these principles include the use of "environmentally sound technology and factor inputs," with the act requiring that any activity in Svalbard be based on the technology that places the least possible pressure on the environment when feasible.[201]

Chapter III of the Act on protected areas provides that these shall be designated by the King of Norway; a far cry from the unanimity required under the Madrid Protocol.[202] Chapter VI stipulates that land-use-planning policies be introduced to improve coordination among various interests—which is arguably exactly what Antarctica now needs. It states that:[203]

The plan shall in so far as is necessary designate:

- Building areas, if necessary with further information about the type of buildings.
- Areas for cabins.
- Cultural heritage areas, areas of natural environment and outdoor recreation areas, separately or in combination.
- Areas for the extraction of raw materials.
- Research areas.
- Roads, airports, harbors, cable-cars, high-voltage power lines and other important elements of the communication system.

- Special areas, including danger areas.
- Areas for specific use or protection of sea and river systems, including areas for traffic, fisheries areas, areas of natural environment and outdoor recreation areas, either singly or in combination.
- Areas that are protected under Chapter III of this Act.

The act introduces an Environmental Impact Assessment requirement for certain designated activities based on the following decision-making rule: "The Governor will circulate the application and the environmental impact statement for comment. If new circumstances that may have a significant bearing on the impact of the activity on the environment are revealed by the environmental impact assessment or in any other way, the Governor may require further assessment of the matter. A permit may not be granted before the Governor has confirmed that the duty to carry out an environmental impact assessment has been fulfilled."[204] In other words, a public consultation process is initiated once an assessment is submitted to the governor, with the governor rather than the operator, as per the Madrid Protocol, being the ultimate decision-maker.

The act then establishes a stringent inspection and enforcement regime, in which the "Governor shall be given unimpeded access to property where activities are being carried out that may have an impact on the environment."[205] With the Svalbard model operating within a far clearer jurisdictional framework than Antarctica's, it is no surprise that the governor is accorded such powers.

Enforcement is achieved in part through a strong liability regime, in which "any person that causes environmental damage" shall incur "the costs of or losses relating to reasonable measures to reduce or mitigate environmental damage or to restore the state of the environment."[206] In other words, the responsibility to restore the environment forms a clear part of Svalbard's liability regime, unlike Annex VI of the Madrid Protocol. The act then establishes the Svalbard Environmental Protection Fund,[207] which is financed from fines collected under the act, in addition to a number of other revenue streams, and is used to protect the environment in the archipelago.

In the 1970s, authorities in Norway were concerned about the impact of tourism on the environment, and they felt ambivalent about whether Svalbard should welcome more tourists. As Arvid Viken writes, tourism had consistently been viewed as a threat to the natural environment, but "largely without qualification or documentation of problems," with the Norwegian Ministry of Industrial Affairs finally changing its position on the subject and embracing tourism in Svalbard in the 1990s.[208] Tourism is now tightly managed, with its negative environmental effects kept under close review. Peter Prokosch writes:[209]

> Today, about 60,000 tourists come to Svalbard by plane or ship and are required to pay a fee of 150 NOK. This fee is placed into the environment fund, which is used every year in a transparent way for education, culture heritage, information, nature conservation and research projects for the management of tourism and protected areas. Therefore, Svalbard is not only one of the best protected wilderness areas, it serves also as [a] leading example of linking tourism to conservation.

A fraction of the size of Antarctica, Svalbard receives approximately 60,000 tourists per year, and it is able to use tourism to reinforce environmental protection on the islands. As discussed in Chapter 5 of this book, the entire Antarctic continent receives only about half this number of visitors.[210] Svalbard has succeeded in linking tourism to conservation through its imposition of a tourism fee—something that Antarctica has not attempted to do. In fact, Margaret Johnston and C. Michael Hall have concluded, "Svalbard is unique in the Arctic world. It has a series of strict and far-reaching regulations which govern travel and environmental protection in the archipelago which are quite distinct from the rest of Norway."[211]

These strong environmental protection regulations were the outcome of sustained campaigning by civil society, such as Friends of the Earth Norway, Birdlife Norway, and various tourism bodies.[212] The engagement shown by local Norwegian NGOs testifies to how sovereignty deepens engagement and reflects what is now needed on the seventh continent.

### Box 7.2 Interview of the Governor of Svalbard[a]

1. *To what extent has the Svalbard Environmental Protection Act, as amended in 2012, succeeded in protecting Svalbard's environment?*
The overall picture is that the Svalbard Environmental Protection Act has been an effective tool in the achievement of the environmental management goals of Svalbard as defined by the Norwegian Parliament. The state of the environment in Svalbard is considered as good. However, climate change is expected to result in rapid changes to ecosystems. The wilderness areas of Svalbard are to a large degree intact, and the populations of earlier over-exploited species like the Walrus and Polar Bears are recovering. The Act provides tools for

the management of human activity, research and tourism, with mining and human settlements strictly regulated.

2. *Has Svalbard's experience in the designation of protected areas been successful? What percentage of Svalbard's surface area is now protected?*
Protected areas cover 65 percent of the land area and 87 percent of territorial waters. Assessments of the protected areas network have been made, concluding that the areas sufficiently represent all main ecosystems in Svalbard.

3. *Is there any evidence that tourism hurts Svalbard's environment?*
Tourism is important for employment and local settlement in Svalbard, especially for Longyearbyen. Tourism and tourist activities are increasing, but are also strictly regulated, due to the Svalbard Environmental Act. In general, tourism is not hurting Svalbard's environment, but it is necessary to monitor tourism and traffic and to implement flanking measures if the situation changes. Certain localized impact of traffic on land, flora, and fauna have been observed.

4. *How is Svalbard's Environmental Protection Fund financed, and how has it contributed to Svalbard's environmental protection?*
The Svalbard Environmental Protection Fund was established pursuant to the Svalbard Environmental Protection Act. The fund's income sources consist of an environmental fee imposed on visitors to the islands, fees for hunting and fishing licenses, penalties for the mishandling of flora and fauna in violation of the Svalbard Environmental Protection Act, and other environmental penalties set by the governor.

The Department of the Environment has set the rates for hunting and fishing licenses, and the environmental fee for visitors was introduced in 2007, and fixed at NOK150 per person per entry (around $18). The fee only applies to trips to Svalbard's territories and inland waters and it is not levied on residents of Svalbard. The fee is directly imposed on the purchase of airline or cruise tickets. For visitors to Svalbard travelling on their own boats, payments is made directly to the Governor's Office, which acts as the Secretariat of the Environmental Protection Fund.

The fund's resources are used to initiate and support projects that contribute to the preservation of Svalbard's distinctive wilderness. Since its establishment in 2007, the fund has allocated NOK 90 million (around $11 million) for environmental protection initiatives in Svalbard, supporting a total of 450 projects. The fund has become an important financial instrument for the financing of environmental initiatives in Svalbard.

[a]Interview conducted by the author on February 23, 2016.

## CONCLUDING REMARKS

The Madrid Protocol creates only an illusion of strong environmental protection in Antarctica. In being attached to the Antarctic Treaty, it operates within its "jurisdictional vacuum," which would prevent even the best environmental accord from achieving proper enforcement. This also reduces CCAMLR's effectiveness. The protocol's Committee for Environmental Protection was made deliberately weak from its inception, with the parties rejecting the creation of a stronger, more independent body. The result has been that even an issue as critical as climate change has not received serious attention, with the exception of studies by SCAR.

The Madrid Protocol is weak on protected areas as well, allowing a single party to block the establishment of such an area, while at the same time authorizing all human activities (except the mining of nonliving resources) to go forward with very limited control. This is due to its weak decision-making rules on Environmental Impact Assessments, which give the final say to the sponsoring state. CCAMLR has not shown itself to be any more forceful on Marine Protected Areas. Furthermore, the Madrid Protocol's inspection regime is weak, with decades-old recommendations meeting with little response, and its rules on liability are inadequate. They are confined to environmental emergencies, and do not cover the cumulative environmental harm to which the continent is subjected. Worse, they carry no firm obligations for cleanup or any obligation for environmental restoration.

Particularly surprising is how much weaker the Madrid Protocol is on the environment relative to its predecessor, CRAMRA, even though CRAMRA is a mining convention. Had CRAMRA been adopted back in the late 1980s, it could have contributed to the creation of a green growth model for Antarctica, allowing economic activity to go forward under stringent environmental safeguards. Today, CRAMRA would be impossible to resurrect in light of the far wider interest by the United Nations that Antarctica would be likely to garner, as well as the demands that would likely be made to incorporate the Common Heritage of Mankind principle. Antarctica's currently environmental protection, therefore, hinges on the inadequate Madrid Protocol.

At present, two commercial activities are proceeding in Antarctica relatively unchecked by the provisions of the protocol; bioprospecting and tourism. By providing an illusion of a comprehensive ban on mining in Antarctica, the protocol has effectively allowed the mining of the continent's living resources to take place under the radar, and in the absence of proper environmental screening. While in theory covered by the Environmental Impact Assessment provisions of the protocol, these have not been applied to bioprospecting, even once.

Tourism is also proceeding relatively unchecked by treaty parties, who do not see the need to elaborate on the rules of the Madrid Protocol and who have left matters in the hands of an industry association, IAATO. Neither bioprospecting nor tourism could be easily addressed within the current framework of the Madrid Protocol. Bioprospecting would be likely to raise the same issues that the revival of CRAMRA would. Tourism would be equally difficult to regulate because most vessels fly the flags of non-treaty parties, and Antarctica's jurisdictional vacuum ties the hands of port states.

In being one of the world's least protected continents, Antarctica is in desperate need of better environmental governance. Continuing to believe that Environmental Impact Assessments, however expanded or redefined, and the protected areas system will address Antarctica's growing problems would be contrary to the evidence. Furthermore, no Environmental Impact Assessment, however broad, will succeed in creating the kinds of territorial planning policies that this expanding continent now clearly needs.

These problems cannot be resolved without stopping the damage that Antarctica is subjected to on a daily basis as a global common. National sovereignty is needed in Antarctica, with the continent's own citizens giving its fragile environment a voice. Without addressing its jurisdictional vacuum and allowing sovereign governments to create and enforce proper environmental regulation in their Antarctic territories, the world would be perpetuating the myth of having a safe and pristine environment on the seventh continent. The Svalbard environmental protection model should serve as an example that can be used in the future to address this issue.

## NOTES

1. Of course, because CRAMRA never entered into force, no one will know whether problems would have arisen in its implementation. However, implementation aside, CRAMRA was stronger as a legal regime than the Madrid Protocol.

2. The tragedy of the commons in Antarctica could be reduced with the application of the Svalbard model. This does not mean, however, that the Svalbard model is perfect. It still faces a number of challenges that were discussed in Chapter 6, which arise from differences in interpretation between parties to the Svalbard Treaty about its degree of application to surrounding waters.

3. Julia Jabour, "Biological Prospecting in the Antarctic: Fair Game?" in *The Emerging Politics of Antarctica*, 244.

4. Article VI of the Antarctic Treaty; see Annex I.

5. The Antarctic Convergence, also known as the Polar Front, is an irregular line encircling Antarctica where the cold, northward-flowing Antarctic waters sink beneath the relatively warmer waters of the sub-Antarctic.

6. Article I.1 of CCAMLR states: "This Convention applies to the Antarctic marine living resources of the area south of 60 degrees South latitude and to the Antarctic marine living resources of the area between the latitude and the Antarctic Convergence which form part of the Antarctic marine ecosystem," in *Compilation of Key Documents of the Antarctic Treaty System*.

7. Ruth Davis, "The Durability of the Antarctic Model and Southern Ocean Governance," in *Polar Oceans Governance in an Era of Environmental Change*, 299.

8. Preamble to the Madrid Protocol, in *Compilation of Key Documents of the Antarctic Treaty System*. All subsequent references to provisions of the Madrid Protocol can be found in this source.

9. Davor Vidas, "The Protocol on Environmental Protection to the Antarctic Treaty: A Ten-Year Review," in *Yearbook of International Cooperation on Environment and Development 2002/03,* Olav Schram Stokke and Øystein B. Thommessen, eds. (London: Earthscan, 2003), 54.

10. Article 1 of Annex VI of the Madrid Protocol.

11. Vidas, "The Protocol on Environmental Protection," 54.

12. Kees Bastmeijer, "Implementing the Environmental Protocol Domestically: An Overview," in *Implementing the Environmental Protection Regime for the Antarctic*, Davor Vidas, ed. (Boston: Kluwer Academic Publishers, 2000), 299.

13. Vidas, "The Protocol on Environmental Protection," 54.

14. Article 1 of Annex VI of the Madrid Protocol.

15. Louise Angelique de la Fayette, "Responding to Environmental Damage in Antarctica," in *Antarctica: Legal and Environmental Challenges for the Future,* Gillian Triggs and Anna Riddell, eds. (London: British Institute of International and Comparative Law, 2007), 134.

16. On the geographic construction of Antarctica, see Dodds, "The Antarctic Peninsula," 99.

17. Bastmeijer, "Implementing the Environmental Protocol Domestically," 300.

18. On how the idea of World Park or a Wilderness Reserve evolved, see Kelly Rigg, "Environmentalists' Perspectives on the Protection of Antarctica," in *The Future of Antarctica: Exploitation versus Preservation,* Grahame Cook, ed. (Manchester: Manchester University Press, 1990), 68–80; and Bernard Stonehouse and Kim Crosbie, "Tourist Impacts and Management in the Antarctic Peninsula Area," in *Polar Tourism*, 230.

19. Joyner, *Governing the Frozen Commons*, 66; see also Francesco Francioni, "Establishment of an Antarctic Treaty Secretariat: Pending Legal Issues," in Vidas, *Implementing the Environmental Protection Regime for the Antarctic*, 125.

20. Dodds, "The Antarctic Peninsula," 110–11.

21. See Annex II to the Madrid Protocol, "Conservation of Antarctic Fauna and Flora," last accessed March 15, 2016, http://www.ats.aq/e/ep_faflo.htm. In addition, the obligation to conduct an Initial Environmental Evaluation and Comprehensive Environmental Evaluation was introduced prior to the 1998 entry into force of the Madrid Protocol through Recommendations XIV-2 and XIV-3 of the 1997 ATCM held in Rio di Janeiro.

22. Joyner, *Governing the Frozen Commons,* 67.

23. Ibid., 109–110.

24. Ibid., 109.

25. Triggs, "Negotiation of a Minerals Regime," 195.

26. Ellen S. Tenenbaum, "A World Park in Antarctica: The Common Heritage of Mankind," *Virginia Environmental Law Journal* 10 (1990–91), 109–36.

27. Rigg, "Environmentalists' Perspectives on the Protection of Antarctica," 73.

28. See Davor Vidas, "Entry into Force of the Environment Protocol and Implementation Issues: An Overview," in Vidas, *Implementing the Environmental Protection Regime for the Antarctic*, 4; Alan Hemmings, "After the Party: The Hollowing of the Antarctic Treaty System and the Governance of Antarctica," Symposium on Antarctic Politics, University of Canterbury, Christchurch, New Zealand, July 8–9, 2010, 2; Patrizia Vigni, *Concorrenza Fra Norme Internazionali: Il Regime Giuridico dell'Antartide*

*Nel Contesto Globale* (Milan: Giuffre Editore, 2005), 207–208; and Vidas, "The Protocol on Environmental Protection," 53.

29. Vidas, "The Protocol on Environmental Protection,"52.

30. Interview conducted by the author on January 20, 2016.

31. See the multiple references to the "Antarctic environment or dependent or associated ecosystems" throughout CRAMRA, such as in its Preamble and Article 3.2(a). The full text of CRAMRA can be found at http://sedac.ciesin.org/entri/texts/acrc/cramra.txt.html. All subsequent references to CRAMRA can be found at this source.

32. Article 8 of CRAMRA.

33. Wolfrum, *Regulation of Antarctic Mineral Resource Activities,* 30–31.

34. Article 4(1) of CRAMRA.

35. Article 4(2)(e) of CRAMRA.

36. Article 4(3) of CRAMRA.

37. On the strength of CRAMRA's environmental provisions, see S. K. N. Blay, "New Trends in the Protection of the Antarctic Environment: The 1991 Madrid Protocol," *American Journal of International Law* 86 (1992), 382–385.

38. See Chapter III of CRAMRA.

39. See Articles 8, 11, and 12 of CRAMRA.

40. See Secretariat of the Antarctic Treaty, *Compilation of Key Documents of the Antarctic Treaty System.*

41. Ibid.

42. Vidas, "Entry into Force of the Environmental Protocol and Implementation Issues," 5. Also see Article 3 of the Madrid Protocol, which makes general references to "activities" on the continent.

43. See Articles 7 and 8 of the Madrid Protocol.

44. The protocol stipulates that the Comprehensive Environmental Evaluation contain "the name and address of the person or organization" that prepared it [Annex I of the protocol, Article 3.2(l)], suggesting that a broad range of institutions—governmental or nongovernmental—could conduct the evaluation. Making matters worse is that, often, the proponent of an activity is the national Antarctic program, which in many jurisdictions is also the agency that has responsibility for reviewing the Environmental Impact Assessment.

45. See Annex I of the Madrid Protocol and the discussion of its weakness in Blay, "New Trends in the Protection of the Antarctic Environment," 392.

46. See Article 22(2) of CRAMRA.

47. Wolfrum, *Regulation of Antarctic Mineral Resource Activities,* 94.

48. See, for example, the discussion in Francesco Orrego Vicuna, “Port State Jurisdiction in Antarctica: A New Approach to Inspection, Control, and Enforcement,” in Vidas, *Implementing the Environmental Protection Regime for the Antarctic*, 53.

49. See, for example, the discussion in Natalie Pilcher, “The Longest Journey in the World: An Investigation into the Liability Annex and its Evolution,” Projects GCAS 10 (2007–8), last accessed March 15, 2016, http://www.anta.canterbury.ac.nz/courses/gcas/reports/Projects%20GCAS%2010.shtml; and Francesco Francioni, “The Madrid Protocol on the Protection of the Antarctic Environment,” *Texas International Law Journal* 28 (1993): 71.

50. See, for example, the discussion in Wolfrum, *Regulation of Antarctic Mineral Resource Activities*, 45–47, 84–94.

51. Alan D. Hemmings, Donald R. Rothwell, and Karen N. Scott, “Antarctic Security in a Global Context,” in *Antarctic Security in the Twenty-First Century*, 331–32. On how climate change has been poorly addressed by parties to the Antarctic Treaty, also see Davis, “The Durability of the Antarctic Model and Southern Ocean Governance,” 297. SCAR has, of course, conducted an important assessment of climate change in Antarctica, which was previously referred to in Chapter 2 of this book. However, the discussion of climate change within the regime remains poor.

52. Bastmeijer, “Implementing the Environmental Protocol Domestically,” 307.

53. The four polar years were the first IPY (1882–1883), the second IPY (1932–33), the third IPY, otherwise known as the International Geophysical Year (IGY; 1957–1958), and the fourth IPY (2007–2008). For instance, see the discussion in Susan Barr and Cornelia Ludecke, eds., *The History of the International Polar Years (IPYs)* (London: Springer, 2010); and Jessica M. Shadian and Monica Tennberg, eds., *Legacies and Challenges in Polar Sciences: Historical, Legal, and Political Reflections on the International Polar Year* (Burlington: Ashgate, 2009).

54. John R. Dudeney and David W. H. Walton, “Leadership in Politics and Science Within the Antarctic Treaty,” *Polar Research* 31 (2012).

55. See Article 2 of the Madrid Protocol on “Functions of the Committee.”

56. Dudeney and Walton, “Leadership in Politics and Science Within the Antarctic Treaty.”

57. Wolfrum, *Regulation of Antarctic Mineral Resource Activities,* 93.

58. Olav Orheim, Anthony Press, and Neil Gilbert, “Managing the Antarctic Environment: The Evolving Role of the Committee for Environmental Protection,” in *Science Diplomacy*, 209–221.

59. Ibid., 213.

60. Ibid., 220.

61. Ibid., 214.

62. Ibid., 217.

63. Article 3 of Annex V to the Madrid Protocol.

64. Justine D. Shaw et al., "Antarctica's Protected Areas Are Inadequate, Unrepresentative, and at Risk," *PLoS Biology* 12(6) (June 17, 2014).

65. Article 3(2) of Annex V to the Madrid Protocol.

66. Article 3(4) of Annex V to the Madrid Protocol.

67. Article 6(1) of Annex V to the Madrid Protocol.

68. Tina Tin and Alan D. Hemmings, "Challenges in Protecting the Wilderness of Antarctica," *USDA Forest Service Proceedings*, RMRS-P-64, 2011, 148.

69. Ibid., 151.

70. Ibid., 149.

71. Shaw et al., "Antarctica's Protected Areas Are Inadequate."

72. Ibid.

73. Tin and Hemmings, "Challenges in Protecting the Wilderness of Antarctica," 147. See also Kevin A. Hughes et al., "Assessing the Effectiveness of Specially Protected Areas for Conservation of Antarctica's Botanical Diversity," *Conservation Biology* 30 (2016): 113–20.

74. Shaw et al., "Antarctica's Protected Areas Are Inadequate."

75. Australian Government, Department of the Environment, "Ownership of Protected Areas," http://www.environment.gov.au/land/nrs/about-nrs/ownership.

76. For descriptions of CCAMLR, see Cassandra M. Brooks, "Competing Values on the Antarctic High Seas: CCAMLR and the Challenge of Marine-Protected Areas," *The Polar Journal* 3 (2013): 280; Marcus Haward and Julia Jabour, "Environmental Change and Governance Challenges in the Southern Ocean," in *Polar Oceans Governance*, 36; Denzil Miller, "Sustainable Management in the Southern Ocean: CCAMLR Science," in *Science Diplomacy*, 103–121; and Marcus Haward, "Marine Resources Management, Security, and the Antarctic Treaty System, An Ongoing Agenda?" in *Antarctic Security in the Twenty-First Century*, 215–37.

77. Olav Schram Stokke, "The Effectiveness of CCAMLR," in Olav Schram Stokke and Davor Vidas, eds., *Governing the Antarctic: The Effectiveness and Legitimacy of the Antarctic Treaty System* (Cambridge, UK: Cambridge University Press, 1996), 124, 131–32.

78. See the history of Southern Ocean Marine Protected Areas in CCAMLR, "Marine Protected Areas"; https://www.ccamlr.org/en/science/marine-protected-areas-mpas.

79. Article IX(6) of CCAMLR.

80. Brooks, "Competing Values on the Antarctic High Seas," 277.

81. Ibid., 295. In 1982, the ratio of fishing to nonfishing states had been 1:2.

82. Ibid., 280; and Rebecca L. Singleton and Callum M. Roberts, "The Contribution of Very Large Marine Protected Areas to Marine Conservation: Giant Leaps or Smoke and Mirrors," *Marine Pollution Bulletin* 87 (2014): 7–10.

83. For the full history of the Ross Sea Marine Protected Area, see Brooks, "Competing Values on the Antarctic High Seas," 283–90.

84. Michael Atkin, "Russia Blocks a Proposal for a Marine Protection Area in Antarctica's Ross Sea," *ABC News*, October 30, 2015.

85. ASOC, "Designation of Marine Protected Areas Within the Antarctic Treaty Area," IP 119, ATCM, Kiev, Ukraine, June 2–13, 2008.

86. Article 8 of the Madrid Protocol and Article VII (5) of the Antarctic Treaty.

87. See the Final Act of the XIth Special ATCM (1991): "With respect to the activities referred to in Article 8, the meeting noted that it was not intended that those activities should include activities undertaken in the Antarctic Treaty area pursuant to the Convention on the Conservation of Antarctic Marine Living Resources or the Convention for the Conservation of Antarctic Seals," at http://www.ats.aq/documents/keydocs/vol_1/vol1_3_AT_Final_Act_Eleventh_SATCM_e.pdf.

88. Article 7 of Annex I to the Madrid Protocol.

89. Article 8 of the Madrid Protocol, and Articles 2 and 3 of Annex I to the protocol.

90. See Alan D. Hemmings and Lorne K. Kriwoken, "High-Level Antarctic EIA Under the Madrid Protocol: State Practice and the Effectiveness of the Comprehensive Environmental Evaluation Process," *International Environmental Agreements: Politics, Law and Economics* 10 (2010): 193; Ricardo M. Roura and Alan D. Hemmings, "Realizing Strategic Environmental Assessment in Antarctica," *Journal of Environmental Assessment Policy and Management* 13 (2011): 487; and Antarctic Treaty, "Environmental Impact Assessment," http://www.ats.aq/e/ep_eia.htm.

91. See note 44.

92. Francioni, "The Madrid Protocol on the Protection of the Antarctic Environment," 63.

93. Ibid., 195.

94. Kees Bastmeijer and Ricardo Roura, "Environmental Impact Assessment in Antarctica," in *Theory and Practice of Transboundary Environmental Impact Assessment,* Kees Bastmeijer and Ricardo Roura, eds. (Boston: Brill/Martinus Nijhoff Publishers, 2008), 13.

95. Roura and Hemmings, "Realizing Strategic Environmental Assessment in Antarctica," 494.

96. Bob Maxwell, "The Controversy Around the Proposal and Formation of the South Pole Traverse: Critical Review," Postgraduate Certificate of Antarctic Studies 14, University of Canterbury, 2011/2012.

97. See Article 3(3) of the Madrid Protocol; also see Francioni, "The Madrid Protocol on the Protection of the Antarctic Environment," 59–61, for the main environmental principles of the Madrid Protocol.

98. See the discussion on the relationship between science and the management of Antarctica in Karen N. Scott, "Scientific Rhetoric and Antarctic Security," in *Antarctic Security in the Twenty-First Century*, 284–306.

99. Herber, *Protecting the Antarctic Commons,* 45.

100. Hemmings and Kriwoken, "High-Level Antarctic EIA Under the Madrid Protocol," 199.

101. Bastmeijer and Roura, "Environmental Impact Assessment in Antarctica," 24–25.

102. Hemmings and Kriwoken, "High-Level Antarctic EIA Under the Madrid Protocol," 201. See also ASOC's appeal to the Russian Duma to stop drilling the lake at ASOC Secretariat, "Appeal to the Duma on Lake Vostock, Antarctica. ASOC Calls on Russia to Re-Consider Penetrating the Largest Known Sub-Glacial Lake," April 14, 2008. http://www.asoc.org/storage/documents/Other_publications/asoc_vostok_statement041408.pdf.

103. See, for example, the discussion by Sanchez and Njaastad, "Future Challenges in Environmental Management of National Antarctic Programs," 287–306; Daniela Liggett et al., "Conclusions: Multiple Dimensions of Human Engagement with the Antarctic Environment," 335–351; and Klein et al., "Long-Term Monitoring of Human Impacts to the Terrestrial Environment at the McMurdo Station," 213–227.

104. Roura and Hemmings, "Realizing Strategic Environmental Assessment in Antarctica," 495.

105. Ibid., 502–503.

106. Ibid., 483, 507–508; also see Hemmings and Kriwoken, "High-Level Antarctic EIA Under the Madrid Protocol," 205.

107. Christina Braun et al., "Environmental Assessment and Management Challenges of the Fildes Peninsula Region," 169–191.

108. United Kingdom, "Improvements to the Antarctic Environmental Impact Assessment Process," Working Paper 24, ATCM, Brasilia, Brazil, 2014.

109. Article 4 of Annex V of the Madrid Protocol.

110. See the proposal made for an ASMA in the Fildes Peninsula in Braun et al., "Environmental Assessment and Management Challenges of the Fildes Peninsula Region," 169–191.

111. United Kingdom, "Improvements to the Antarctic Environmental Impact Assessment Process," Working Paper 24.

112. See China, "Proposal for a New Antarctic Specially Managed Area at Chinese Antarctic Kunlun Station, Dome A," Working Paper 8, ATCM, Brussels, Belgium, 2013; and Jane Qiu, "China Aims High from the Bottom of the World: Country Plans Two World-Class Telescopes for Its Antarctic Observatory," *Nature*, August 29, 2012.

113. Scott, "Scientific Rhetoric and Antarctic Security," 287.

114. Vicuna, "Port State Jurisdiction in Antarctica," 53.

115. Ibid., 45.

116. Article VIII(1) of the Antarctic Treaty.

117. For a comprehensive discussion of the jurisdictional limitations of the Antarctic Treaty, see Vicuna, "Port State Jurisdiction in Antarctica," 45–69.

118. Vidas, "The Protocol on Environmental Protection," 56; and see the more general discussion of the problems associated with tourism in Antarctica in Stonehouse and Crosbie, "Tourist Impacts and Management in the Antarctic Peninsula Area"; and Michael Luck, Patrick T. Maher, and Emma J. Stewart, eds., *Cruise Tourism In Polar Regions: Promoting Environmental and Social Sustainability?* (London: Earthscan, 2010). On the constant reflagging of vessels, see Baird, *Aspects of Illegal, Unreported and Unregulated Fishing in the Southern Ocean*.

119. William Bush, "Means and Methods of Implementation of Antarctic Environmental Regimes and National Environmental Instruments: An Exercise in Comparison," in *Implementing the Environmental Protection Regime for the Antarctic*, 40–41.

120. Stuart Kaye, "IUU Fishing in the Southern Ocean: Challenge and Response," in *Antarctica: Legal and Environmental Challenges for the Future*, 40.

121. Alan Brown, "Some Current Issues Facing the Convention on the Conservation of Antarctic Marine Living Resources (CCAMLR)," in ibid., 87.

122. See the chair's statement in Secretariat of the Antarctic Treaty, *Compilation of Key Documents of the Antarctic Treaty System*, 112–113.

123. Australia possesses the sub-Antarctic islands of Heard and McDonald Islands and Macquarie Island; France possesses Crozet Island and Kerguelen Isles; South Africa possesses Prince Edward Island and Marion Island; Norway possesses Bouvet Island and Peter I Island; New Zealand

possesses Balleny Islands and Scott Islands; and the United Kingdom and Argentina remain in dispute over Shag Rocks and South Georgia Island, South Sandwich Islands, Laurie Island, South Orkney Island, and South Shetland Island. Baird, *Aspects of Illegal, Unreported, and Unregulated Fishing in the Southern Ocean*, 130.

124. CCAMLR, "Members"; https://www.ccamlr.org/en/organisation/members.

125. Baird, *Aspects of Illegal, Unreported and Unregulated Fishing in the Southern Ocean,* 46.

126. Ibid., 191–192.

127. Vicuna, "Port State Jurisdiction in Antarctica," 67.

128. Ibid., 45–69.

129. Ibid., 52.

130. Ibid., 52–53; see also Article 12 of CRAMRA.

131. Article 14 of the Madrid Protocol.

132. Francioni, "The Madrid Protocol on the Protection of the Antarctic Environment," 54. For many years, ASOC had also argued for the creation of an Inspectorate. See discussion of this Inspectorate in David J. Bederman, "The Antarctic and Southern Ocean Coalition's Convention on Antarctic Conservation," *Georgetown International Environmental Law Review* 4 (1991–1992): 47–61.

133. Kevin Hughes, "How Committed Are We to Monitoring Human Impacts in Antarctica?" *Environmental Research Letters* 5 (2010): 3.

134. ASOC, "The Antarctic Environmental Protocol, 1991–2011," ATCM, Buenos Aires, Argentina, June 20–July 1, 2011, 4. See also the more recent joint submission by the UN Environment Program and ASOC, "Review of the Implementation of the Madrid Protocol: Inspections by Parties (Article 14)," ATCM, Hobart, Australia, 2012.

135. ASOC, "The Antarctic Environmental Protocol, 1991–2011," 7.

136. Pilcher, "The Longest Journey in the World."

137. Rene Lefeber, "The Legal Need for an Antarctic Liability Regime," in *Implementing the Environmental Protection Regime for the Antarctic*, 186.

138. Patrizia Vigni, "Liability Regime for Antarctica," *Italian Yearbook of International Law*, vol. XV (Boston: Martinus Nijhoff Publishers, 2005), 218.

139. See Article 2 on "Definitions" of Annex VI to the Madrid Protocol.

140. Pilcher, "The Longest Journey in the World."

141. Fayette, "Responding to Environmental Damage in Antarctica," 130.

142. Article 1 of Annex VI to the Madrid Protocol.

143. Preamble and Article 4 of Annex VI to the Madrid Protocol.

144. Fayette, "Responding to Environmental Damage in Antarctica," 133; and Vigni, "Liability Regime for Antarctica," 222.

145. Preamble to Annex VI of the Madrid Protocol.

146. Lefeber, "The Legal Need for an Antarctic Liability Regime," 196.

147. On the U.S. concerns, see Pilcher, "The Longest Journey in the World."

148. See the doubt expressed on whether this is a liability regime in Fayette, "Responding to Environmental Damage in Antarctica," 131.

149. Ibid., 135.

150. Article 8(1) of CRAMRA.

151. Vigni, "Liability Regime for Antarctica," 230.

152. Article 10 of Annex VI to the Madrid Protocol.

153. Francioni, "The Madrid Protocol on the Protection of the Antarctic Environment," 71.

154. Fayette, "Responding to Environmental Damage in Antarctica," 134–135.

155. Article 8 (c) of CRAMRA.

156. Fayette, "Responding to Environmental Damage in Antarctica," 142.

157. Article 8(d) of CRAMRA.

158. Fayette, "Responding to Environmental Damage in Antarctica," 144.

159. Article 5(b) of Annex VI to the Madrid Protocol.

160. Article I of the Antarctic Treaty and Article 7 of the Madrid Protocol.

161. For ample evidence of this, see the volume of press articles, academic articles, working papers within the ATS, and submissions by NGOs that express surprise at the scope of bioprospecting on the continent. See, for instance, Ian Sample, "Cold Rush Threatens Pristine Antarctic: Scientists Warn That the World's Least Damaged Environment Could Be Changed Forever by the Hunt for Potentially Lucrative Organisms," *The Guardian*, February 2, 2004.

162. Richard A. Herr, "The Changing Role of Non-Governmental Organizations in the Antarctic Treaty System," in *Governing the Antarctic*, 99.

163. See the discussion of the challenges involved in managing bioprospecting in Antarctica in Alan D. Hemmings, "Biological Prospecting in the Antarctic Treaty Area," Intergovernmental Meeting of Experts on Biological Prospecting in the Antarctic Treaty Area, Baarn, the Netherlands, February 3–5, 2009.

164. Article 1(6) of CRAMRA.

165. Sanjay Chaturvedi, "Biological Prospecting in the South Polar Region: Science-Geopolitics Interface," in *Legacies and Challenges in Polar Sciences,* 175.

166. See, for instance, the UNU's attempt to shed light on the scope of bioprospecting taking place in Antarctica in UNU Institute for Advance Studies, *Bioprospecting in Antarctica*, Yokohama, Japan, 2005.

167. Christopher Joyner, "Bioprospecting as a Challenge to the Antarctic Treaty," in *Antarctic Security in the Twenty-First Century,* 202–3. See also Belgium, "An Update on Biological Prospecting in Antarctica, Including the Development of the Antarctic Biological Prospecting Database," Working Paper 11, ATCM, Kiev, Ukraine, June 2–13, 2008.

168. Belgium, "An Update on Biological Prospecting in Antarctica," 4–8.

169. Ibid.; also UNU Institute for Advanced Studies, *Bioprospecting in Antarctica*; UNEP, "Industry Involvement in Antarctic Bioprospecting," Information Paper 106, Committee on Environmental Protection of the Madrid Protocol, Cape Town, 2004; ASOC, "Biological Prospecting"; http://www.asoc.org/advocacy/antarctic-environmental-protection/biological-prospecting; Gary Stix, "Patents on Ice: Antarctica as a Last Frontier for Bioprospectors—and Their Intellectual Property," *Scientific American*, May 1, 2004; National Research Foundation and South African Agency for Science and Technology Advancement, "Biotechnology Research in the Antarctic Region," *Antarctic Biotech Research*, 2009; Belgium, UNU-IAS, "The Role of Ex-Situ Collections in Antarctic Bioprospecting," ATCM, Punta del Este, Uruguay, May 3–14, 2010; and Herber, *Protecting the Antarctic Commons,* 45–46.

170. Article III(1)(a)of the Antarctic Treaty.

171. France and Brazil, for instance, have asked how the provisions of the Antarctic Treaty on science could be reconciled with bioprospecting, even though French companies have been involved in this activity. France, "In Search of a Legal Regime for Bioprospecting in Antarctica," Information Paper 13, ATCM, Edinburgh, United Kingdom, 2006; and Brazil, "Bioprospecting Activities of Brazil in Antarctica: A Short Report," ATCM, Baltimore, USA, April 6–17, 2009.

172. On the issue of sharing the benefits of bioprospecting in Antarctica, see ibid., and UNU Institute for Advanced Studies, *Bioprospecting in Antarctica*, 6; Joyner, "Bioprospecting as a Challenge to the Antarctic Treaty," 207–208; and Patrizia Vigni, "Antarctic Bioprospecting: Is It Compatible with the Value of Antarctica as a Natural Reserve?" in Francesco Francioni and Tulio Scovazzi, eds., *Biotechnology and International Law* (Oxford, UK: Hart Publishing, 2006), 124. The matter has also made its

way to the media; see, for instance, Alister Doyle, "Antarctic Patents Strain Goals of Shared Science," *Reuters*, February 5, 2009.

173. See Joyner, "Bioprospecting as a Challenge to the Antarctic Treaty," 208; and Alan D. Hemmings, "Does Bioprospecting Risk Moral Hazard for Science in the Antarctic Treaty System?" *Ethics in Science and Environmental Politics* 10 (2010): 11.

174. See the views of Yves Frenot, the deputy head of the French Polar Institute, in "Antarctic Patents Strain Goals of Shared Science."

175. Article 8(2) of the Madrid Protocol. Also, on how various experts have confirmed that the provisions of the Madrid Protocol apply to bioprospecting, see Vigni, "Antarctic Bioprospecting," 114; and UNU Institute for Advanced Studies, *Bioprospecting in Antarctica*, 17–18.

176. Chaturvedi, "Biological Prospecting in the South Polar Region," 180.

177. See Chapter III of CRAMRA on Prospecting, and Joyner, "Bioprospecting as a Challenge to the Antarctic Treaty," 211–12.

178. See the discussion on the applicability of the CBD to Antarctica in Morten Walløe Tvedt, "Patent Law and Bioprospecting in Antarctica," *Polar Record* 46 (2010): 6–9; Kevin A. Hughes and Paul D. Bridge, "Potential Impacts of Antarctic Bioprospecting and Associated Commercial Activities upon Antarctica Science and Scientists," *Ethics in Science and Environmental Politics* 10 (2010): 14; Peter Convey, Kevin A. Hughes, and Tina Tin, "Continental Governance and Environmental Management Mechanisms Under the Antarctic Treaty System: Sufficient for the Biodiversity Challenges of this Century," *Biodiversity* 13 (2012): 7; Vigni, "Antarctic Bioprospecting," 134; and UNU Institute for Advanced Studies, *Bioprospecting in Antarctica*, 26.

179. Robert F. Blomquist, "Ratification Resisted: Understanding America's Response to the Convention on Biological Diversity, 1989–2000," *Golden Gate University Law Review* 32 (2002). Also see how the perceived conflict between the concept of benefit sharing and intellectual property rights has played out for decades in various international fora, such as the World Trade Organization (https://www.wto.org/english/tratop_e/trips_e/art27_3b_background_e.htm) and the World Intellectual Property Organization.

180. Chaturvedi, "Biological Prospecting in the South Polar Region," 184. On the Common Heritage challenge that a bioprospecting regime may raise in Antarctica, also see Jabour, "Biological Prospecting in the Antarctic," 249–51.

181. Debra Enzenbacher, "The Regulation of Antarctic Tourism," in *Polar Tourism*, 190.

182. Debra Enzenbacher, "Antarctic Tourism Policy-Making: Current Challenges and Future Prospects," in *Antarctica: Legal and Environmental Challenges for the Future*, 155.

183. Davor Vidas, "The Legitimacy of the Antarctic Tourism Regime," in *Governing the Antarctic*, 295.

184. Mike G. Richardson, "Regulating Tourism in the Antarctic: Issues of Environment and Jurisdiction," in *Implementing the Environmental Protection Regime for the Antarctic*, 85.

185. Harlan Cohen, "Public Participation in Antarctica: The Role of Non-Governmental and Intergovernmental Organizations," in *Science Diplomacy,* 274.

186. For the different phases of NGO engagement in Antarctica, see Herr, "The Changing Role of Non-Governmental Organizations in the Antarctic Treaty System," 91–110. For the broadening participation in Antarctic affairs, see Hemmings, "After the Party," 8–10.

187. Herr, "The Changing Role of Non-Governmental Organizations in the Antarctic Treaty System," 96–98.

188. Peter J. Beck, *The International Politics of Antarctica* (London: Croom Helm, 1986), 41.

189. Herr, "The Changing Role of Non-Governmental Organizations in the Antarctic Treaty System," 107.

190. Ibid., 98–103. On the more active phase of NGO engagement, see Lee Kimball, "The Role of Non-Governmental Organizations in Antarctic Affairs," in Christopher Joyner and Sudhir Chopra, eds., *The Antarctic Legal Regime* (Dordrecht, the Netherlands: Martinus Nijhoff, 1988), 46–47; and Margaret L. Clark, "The Antarctic Environmental Protocol: NGOs in the Protection of Antarctica," in Thomas Princen and Matthias Finger, eds., *Environmental NGOs in World Politics: Linking the Local and the Global* (London: Routledge, 1994), 160–85.

191. Herr, "The Changing Role of Non-Governmental Organizations in the Antarctic Treaty System," 103.

192. John May, The Greenpeace Book of Antarctica: A New View of the Seventh Continent (London: Dorling Kindersley, 1988), 72.

193. Herr, "The Changing Role of Non-Governmental Organizations in the Antarctic Treaty System," 104–106.

194. Ibid., 106–108.

195. Interview conducted by the author on February 23, 2016.

196. Statistics Norway, "This Is Svalbard 2014: What the Figures Say," October 2014; https://www.ssb.no/en/befolkning/artikler-og-publikasjoner/_attachment/213945?_ts=14aca3dea58.

197. Anna Mjöll Guðmundsdóttir and Anna Dóra Sæþórsdóttir, "Tourism Management in Wilderness Areas—Svalbard," *Rannsóknir í félagsvísindum* X (2009): 42.

198. Svalbard Environmental Protection Act; https://www.regjeringen.no/en/dokumenter/svalbard-environmental-protection-act/id173945.

199. Ibid., Chapter II, Section 7.

200. Ibid., Chapter II, Section 8.

201. Ibid., Chapter II, Section 10.

202. Ibid., Chapter III, Section 12.

203. Ibid., Chapter VI, Section 49.

204. Ibid., Chapter VII, Section 59–60.

205. Ibid., Chapter VIII, Section 90.

206. Ibid., Chapter IX, Section 95.

207. Ibid., Chapter IX, Section 98.

208. On the conflicted relationship that the Norwegian government has had about tourism in Svalbard, see the discussion in Viken, "Tourism, Research, and Governance on Svalbard."

209. Peter Prokosch, "Svalbard," *Linking Tourism & Conservation*, March 5, 2015. http://www.ltandc.org/svalbard/.

210. Antarctica is, of course, far more remote and inaccessible than Svalbard, with the tourism industry in the Arctic generally more developed than in the Antarctic.

211. Margaret E. Johnston and C. Michael Hall, "Visitor Management and the Future of Tourism in Polar Regions," in *Polar Tourism,* 307.

212. Ibid.

# EIGHT

# The World Is Not Waiting for 2048

Approximately three decades from now, parties to the Antarctic Treaty will be allowed to express themselves on whether to revisit the ban on mining. This could provide an inroad for what will need to be a much broader conversation on the fate of the seventh continent. In the run-up to 2048, and with the progressive warming of Antarctica, nothing would rule out the continued establishment of human colonies, with some countries tempted to repeat the Argentinian and Chilean experiments of settling pregnant women there as a means of establishing their presence. Pressure to begin mining even before 2048 may mount, given the knowledge acquired through what are now decades of uninterrupted scientific research activity in the continent.

In fact, in August 2011, Australia's Lowy Institute for International Policy published a paper entitled "Antarctica: Assessing and Protecting Australia's National Interests," in which it urged Australia to start readying itself for the 2048 conversation. After all, Australia is the player with the largest stake in Antarctica, claiming 42 percent of the continent. Its continental shelf claim in the Antarctic is the size of the entire Arctic Ocean. The Lowy Institute wrote:[1]

> It seems inevitable that the mineral resource question will be reopened in 2048. This will likely involve a revival of the previous arguments of India and Malaysia that Antarctica should be preserved as the common heritage of mankind, and its resources equitably distributed. It would therefore be prudent for Australia to begin diplomatic consultations with other claimant states to develop a unified position in response

to expected future pressure to relinquish claims. Australia must also be realistic in assessing whether international will to extract hydrocarbons may outweigh interest in maintaining Antarctica as a pristine wilderness in the future. It should therefore develop a diplomatic strategy for pursuing an international commitment that, if extraction were to commence, it would at least be conducted in the most environmentally responsible way.

To prepare for 2048, policymakers need to recognize that Antarctica today is not what it was in 1959, nor is the world around it. The continent has and will continue to become more accessible and inhabitable with the melting of its ice sheets and sea ice. Tourists are now flocking there. The climate on the Antarctic Peninsula is already comparable to the currently inhabited Greenland.

The research stations that have come to dot the continent have demonstrated that Antarctica is inhabitable and that housing and transportation technology are capable of adapting to its special climatic and geological environment. Advances in construction technology have already allowed a near-zero emissions Belgian scientific research station to be established. Employing state-of-the-art technology, it has almost complete energy autonomy. Developments in mining technology in the Arctic are also changing the calculus of what is economically feasible in the continent.

The world around Antarctica has changed as well. Relative to 1959, the developing world is emerging rapidly and will be increasingly likely to ask for its share of the riches of the seventh continent. The Arctic's economic development is now in full swing. Asia has been welcomed into the Arctic for the capital that it is contributing to its infrastructure and mining, and it is becoming increasingly active. The most recent scientific research stations to be opened in Antarctica are Indian, South Korean, and Chinese. Asia is also flexing its shipbuilding prowess. It is developing its own fleets of icebreakers to access the poles, overtaking the United States in the process. With the relentless onslaught of climate change, the international community has given official recognition to the plight of climate migrants. The search for alternative living space may now have to start in earnest.

Preparing for 2048 will require a sober assessment not only of what the Antarctic Treaty System (ATS) has achieved, but also of what it has failed to accomplish. While it has indeed maintained peace on the seventh continent, it also has paved the way for its silent scientific conquest and inadvertently laid the groundwork for its commercial exploitation and human settlement. At present, it is hard to say that Antarctica is virgin territory, unoccupied, as idealists wish to believe it is.

The ATS has in fact opened the door to commercial enterprise under a weak environmental regulatory regime, a so-called Environment Protocol. Its ban on mining has propagated the myth that Antarctica's environment has been left intact, allowing the exploitation of its living resources to blossom under the radar. In fact, the ATS has exposed Antarctica to the full forces of the "tragedy of the commons." Today, the Antarctic environment can no longer be governed through the narrow prisms of instruments such as Environmental Impact Assessments. It needs what every other continent needs—coherent urban and territorial planning policies.

Antarctic policymakers would do well to heed the developments now taking place in the Arctic. With the melting of the Arctic Ocean, it is now open for business, and its infrastructure is expanding. An Antarctic economy would be likely to follow a similar path. As the gateway cities to Antarctica begin to shift more of their polar operations to the continent, it will undergo an even more profound transformation, increasingly resembling other continents. The maritime sovereignty disputes now witnessed in the Arctic, accompanied by growing militarization, will likely pale in comparison to those that will emerge in the Antarctic if it is opened to mining. In Antarctica, the tug of war will extend to land and water.

As 2048 approaches, policymakers must ask what type of governance model will be suited to the Antarctica of the future and the new world order around it. The absence of territorial sovereignty in Antarctica has created a no-man's land and a jurisdictional vacuum that has impeded the creation and enforcement of stringent regulatory frameworks. This problem will be exacerbated when the mineral resources of the continent are more intensively exploited—an inevitable development, and a problem already seen with the gathering pace of bioprospecting. They must ask how an equitable system can be devised for sharing Antarctica's riches with all humanity, while reestablishing a territorial sovereignty to end the tragedy of the commons.

The uniqueness of the Svalbard governance model deserves attention in this context. Establishing sovereignty without exclusive ownership of economic resources, the model has created a novel approach to territorial governance. It could solve many of Antarctica's problems by allowing territorial claimants to realize their claims, while permitting the rest of the international community to share the benefits of the continent's riches. The application of the model would constitute a recognition that only a clear system of territorial rights can stop Antarctica's environmental destruction. It would give birth to Antarctic citizens, who would have a powerful vested interest in the orderly protection and use of the continent.

Some Norwegian politicians have tried to initiate a conversation on the housing refugees in Svalbard, but with little success. The relocation of refugees to Antarctica will inevitably have to be seriously considered. While this idea may seem inhumane, perhaps even politically incorrect, because of the continent's inhospitable conditions, this will change. Turning this idea into a carefully orchestrated plan for the transfer of climate migrants to the continent could provide them with a new lease on life that they could not find elsewhere. The calculus of public opinion will shift rapidly upon realization of the moral bankruptcy of the argument that would deny helpless migrants a new beginning on uninhabited lands. The harshness of the Antarctic Peninsula's climate can be overcome in the same way that it has in Greenland, and even this harshness may not seem that bad in comparison to the alternatives that refugees have, and it is a situation that will improve with time as the climate changes.

The Svalbard model is not perfect. It too had to overcome multiple hurdles. Svalbard's journey toward strong environmental protection did not occur overnight. However, it may become the only governance model feasible when the international community faces, for the first time in human history, the task of dividing a continent and its riches equitably. It may become the only model that successfully addresses the shortfalls experienced by the international community in the governance of other global commons—the seas, outer space, the Moon, and other celestial bodies.

The application of the model could also enhance the scientific project that the international community originally envisioned for Antarctica. It would rationalize scientific activity by disassociating claims to Antarctica's resources from the establishment of research footholds, allowing studies to take place where the scientific questions truly lie. With Norway being the only Arctic and Antarctic power, and the one that actually pioneered the Svalbard model, it is in the best position to study the application of this approach to the seventh continent.

The world is not waiting for 2048.

## NOTE

1. Fogarty, "Antarctica: Assessing and Protecting Australia's National Interests," 10.

# Appendix 1

# The Antarctic Treaty

The Governments of Argentina, Australia, Belgium, Chile, the French Republic, Japan, New Zealand, Norway, the Union of South Africa, the Union of Soviet Socialist Republics, the United Kingdom of Great Britain and Northern Ireland, and the United States of America,

Recognizing that it is in the interest of all mankind that Antarctica shall continue forever to be used exclusively for peaceful purposes and shall not become the scene or object of international discord;

Acknowledging the substantial contributions to scientific knowledge resulting from international cooperation in scientific investigation in Antarctica;

Convinced that the establishment of a firm foundation for the continuation and development of such cooperation on the basis of freedom of scientific investigation in Antarctica as applied during the International Geophysical Year accords with the interests of science and the progress of all mankind;

Convinced also that a treaty ensuring the use of Antarctica for peaceful purposes only and the continuance of international harmony in Antarctica will further the purposes and principles embodied in the Charter of the United Nations;

Have agreed as follows:

## ARTICLE I

1. Antarctica shall be used for peaceful purposes only. There shall be prohibited, *inter alia*, any measures of a military nature, such as the establishment

of military bases and fortifications, the carrying out of military maneuvers, as well as the testing of any type of weapons.

2. The present treaty shall not prevent the use of military personnel or equipment for scientific research or for any other peaceful purposes.

## ARTICLE II

Freedom of scientific investigation in Antarctica and cooperation toward that end, as applied during the International Geophysical Year, shall continue, subject to the provisions of the present treaty.

## ARTICLE III

1. In order to promote international cooperation in scientific investigation in Antarctica, as provided for in Article II of the present treaty, the Contracting Parties agree that, to the greatest extent feasible and practicable:

a. information regarding plans for scientific programs in Antarctica shall be exchanged to permit maximum economy and efficiency of operations;
b. scientific personnel shall be exchanged in Antarctica between expeditions and stations;
c. scientific observations and results from Antarctica shall be exchanged and made freely available.

2. In implementing this Article, every encouragement shall be given to the establishment of cooperative working relations with those Specialized Agencies of the United Nations and other international organizations having a scientific or technical interest in Antarctica.

## ARTICLE IV

1. Nothing contained in the present treaty shall be interpreted as:

a. a renunciation by any Contracting Party of previously asserted rights of or claims to territorial sovereignty in Antarctica;
b. a renunciation or diminution by any Contracting Party of any basis of claim to territorial sovereignty in Antarctica which it may have

whether as a result of its activities or those of its nationals in Antarctica, or otherwise;

c. prejudicing the position of any Contracting Party as regards its recognition or non-recognition of any other States right of or claim or basis of claim to territorial sovereignty in Antarctica.

2. No acts or activities taking place while the present treaty is in force shall constitute a basis for asserting, supporting or denying a claim to territorial sovereignty in Antarctica or create any rights of sovereignty in Antarctica. No new claim, or enlargement of an existing claim, to territorial sovereignty in Antarctica shall be asserted while the present treaty is in force.

## ARTICLE V

1. Any nuclear explosions in Antarctica and the disposal there of radioactive waste material shall be prohibited.

2. In the event of the conclusion of international agreements concerning the use of nuclear energy, including nuclear explosions and the disposal of radioactive waste material, to which all of the Contracting Parties whose representatives are entitled to participate in the meetings provided for under Article IX are parties, the rules established under such agreements shall apply in Antarctica.

## ARTICLE VI

The provisions of the present treaty shall apply to the area south of 60° South Latitude, including all ice shelves, but nothing in the present treaty shall prejudice or in any way affect the rights, or the exercise of the rights, of any State under international law with regard to the high seas within that area.

## ARTICLE VII

1. In order to promote the objectives and ensure the observance of the provisions of the present treaty, each Contracting Party whose representatives are entitled to participate in the meetings referred to in Article IX of the treaty shall have the right to designate observers to carry out any inspection provided for by the present Article. Observers shall be nationals of the

Contracting Parties which designate them. The names of observers shall be communicated to every other Contracting Party having the right to designate observers, and like notice shall be given of the termination of their appointment.

2. Each observer designated in accordance with the provisions of paragraph 1 of this Article shall have complete freedom of access at any time to any or all areas of Antarctica.

3. All areas of Antarctica, including all stations, installations and equipment within those areas, and all ships and aircraft at points of discharging or embarking cargoes or personnel in Antarctica, shall be open at all times to inspection by any observers designated in accordance with paragraph 1 of this Article.

4. Aerial observation may be carried out at any time over any or all areas of Antarctica by any of the Contracting Parties having the right to designate observers.

5. Each Contracting Party shall, at the time when the present treaty enters into force for it, inform the other Contracting Parties, and thereafter shall give them notice in advance, of

a. all expeditions to and within Antarctica, on the part of its ships or nationals, and all expeditions to Antarctica organized in or proceeding from its territory;
b. all stations in Antarctica occupied by its nationals; and
c. any military personnel or equipment intended to be introduced by it into Antarctica subject to the conditions prescribed in paragraph 2 of Article I of the present treaty.

## ARTICLE VIII

1. In order to facilitate the exercise of their functions under the present treaty, and without prejudice to the respective positions of the Contracting Parties relating to jurisdiction over all other persons in Antarctica, observers designated under paragraph 1 of Article VII and scientific personnel exchanged under subparagraph 1(b) of Article III of the treaty, and members of the staffs accompanying any such persons, shall be subject only to the jurisdiction of the Contracting Party of which they are nationals in respect of all acts or omissions occurring while they are in Antarctica for the purpose of exercising their functions.

2. Without prejudice to the provisions of paragraph 1 of this Article, and pending the adoption of measures in pursuance of subparagraph 1(e) of

Article IX, the Contracting Parties concerned in any case of dispute with regard to the exercise of jurisdiction in Antarctica shall immediately consult together with a view to reaching a mutually acceptable solution.

## ARTICLE IX

1. Representatives of the Contracting Parties named in the preamble to the present treaty shall meet at the City of Canberra within two months after the date of entry into force of the treaty, and thereafter at suitable intervals and places, for the purpose of exchanging information, consulting together on matters of common interest pertaining to Antarctica, and formulating and considering, and recommending to their Governments, measures in furtherance of the principles and objectives of the treaty, including measures regarding:

a. use of Antarctica for peaceful purposes only;
b. facilitation of scientific research in Antarctica;
c. facilitation of international scientific cooperation in Antarctica;
d. facilitation of the exercise of the rights of inspection provided for in Article VII of the treaty;
e. questions relating to the exercise of jurisdiction in Antarctica;
f. preservation and conservation of living resources in Antarctica.

2. Each Contracting Party which has become a party to the present treaty by accession under Article XIII shall be entitled to appoint representatives to participate in the meetings referred to in paragraph 1 of the present Article, during such time as that Contracting Party demonstrates its interest in Antarctica by conducting substantial scientific research activity there, such as the establishment of a scientific station or the despatch of a scientific expedition.

3. Reports from the observers referred to in Article VII of the present treaty shall be transmitted to the representatives of the Contracting Parties participating in the meetings referred to in paragraph 1 of the present Article.

4. The measures referred to in paragraph 1 of this Article shall become effective when approved by all the Contracting Parties whose representatives were entitled to participate in the meetings held to consider those measures.

5. Any or all of the rights established in the present treaty may be exercised from the date of entry into force of the treaty whether or not any

measures facilitating the exercise of such rights have been proposed, considered or approved as provided in this Article.

## ARTICLE X

Each of the Contracting Parties undertakes to exert appropriate efforts, consistent with the Charter of the United Nations, to the end that no one engages in any activity in Antarctica contrary to the principles or purposes of the present treaty.

## ARTICLE XI

1. If any dispute arises between two or more of the Contracting Parties concerning the interpretation or application of the present treaty, those Contracting Parties shall consult among themselves with a view to having the dispute resolved by negotiation, inquiry, mediation, conciliation, arbitration, judicial settlement or other peaceful means of their own choice.

2. Any dispute of this character not so resolved shall, with the consent, in each case, of all parties to the dispute, be referred to the International Court of Justice for settlement; but failure to reach agreement on reference to the International Court shall not absolve parties to the dispute from the responsibility of continuing to seek to resolve it by any of the various peaceful means referred to in paragraph 1 of this Article.

## ARTICLE XII

1. (a) The present treaty may be modified or amended at any time by unanimous agreement of the Contracting Parties whose representatives are entitled to participate in the meetings provided for under Article IX. Any such modification or amendment shall enter into force when the depositary Government has received notice from all such Contracting Parties that they have ratified it.

(b) Such modification or amendment shall thereafter enter into force as to any other Contracting Party when notice of ratification by it has been received by the depositary Government. Any such Contracting Party from which no notice of ratification is received within a period of two years from the date of entry into force of the modification or amendment in accordance with the provisions of subparagraph 1(a) of this Article shall be deemed to have withdrawn from the present treaty on the date of the expiration of such period.

2. (a) If after the expiration of thirty years from the date of entry into force of the present treaty, any of the Contracting Parties whose representatives are entitled to participate in the meetings provided for under Article IX so requests by a communication addressed to the depositary Government, a Conference of all the Contracting Parties shall be held as soon as practicable to review the operation of the treaty.

(b) Any modification or amendment to the present treaty which is approved at such a Conference by a majority of the Contracting Parties there represented, including a majority of those whose representatives are entitled to participate in the meetings provided for under Article IX, shall be communicated by the depositary Government to all the Contracting Parties immediately after the termination of the Conference and shall enter into force in accordance with the provisions of paragraph 1 of the present Article.

(c) If any such modification or amendment has not entered into force in accordance with the provisions of subparagraph 1(a) of this Article within a period of two years after the date of its communication to all the Contracting Parties, any Contracting Party may at any time after the expiration of that period give notice to the depositary Government of its withdrawal from the present treaty; and such withdrawal shall take effect two years after the receipt of the notice of the depositary Government.

## ARTICLE XIII

1. The present treaty shall be subject to ratification by the signatory States. It shall be open for accession by any State which is a Member of the United Nations, or by any other State which may be invited to accede to the treaty with the consent of all the Contracting Parties whose representatives are entitled to participate in the meetings provided for under Article IX of the treaty.

2. Ratification of or accession to the present treaty shall be effected by each State in accordance with its constitutional processes.

3. Instruments of ratification and instruments of accession shall be deposited with the Government of the United States of America, hereby designated as the depositary Government.

4. The depositary Government shall inform all signatory and acceding States of the date of each deposit of an instrument of ratification or accession, and the date of entry into force of the treaty and of any modification or amendment thereto.

5. Upon the deposit of instruments of ratification by all the signatory States, the present treaty shall enter into force for those States and for States

which have deposited instruments of accession. Thereafter the treaty shall enter into force for any acceding State upon the deposit of its instrument of accession.

6. The present treaty shall be registered by the depositary Government pursuant to Article 102 of the Charter of the United Nations.

## ARTICLE XIV

The present treaty, done in the English, French, Russian, and Spanish languages, each version being equally authentic, shall be deposited in the archives of the Government of the United States of America, which shall transmit duly certified copies thereof to the Governments of the signatory and acceding States.

**Source**: U.S. Department of State. Available online at http://www.state.gov/t/avc/trty/193967.htm

# Appendix 2

# The Svalbard Treaty[1]

The President of the United States of America; His Majesty the King of Great Britain and Ireland and of the British Dominions beyond the Seas, Emperor of India; His Majesty the King of Denmark; the President of the French Republic; His Majesty the King of Italy; His Majesty the Emperor of Japan; His Majesty the King of Norway; Her Majesty the Queen of the Netherlands; His Majesty the King of Sweden,

Desirous, while recognising the sovereignty of Norway over the Archipelago of Spitsbergen, including Bear Island, of seeing these territories provided with an equitable regime, in order to assure their development and peaceful utilisation,

Have appointed as their respective Plenipotentiaries with a view to concluding a Treaty to this effect:

*[Names of plenipotentiaries not reproduced here.]*

Who, having communicated their full powers, found in good and due form, have agreed as follows:

## ARTICLE 1

The High Contracting Parties undertake to recognize, subject to the stipulations of the present Treaty, the full and absolute sovereignty of Norway over the Archipelago of Spitsbergen, comprising, with Bear Island of Beeren-Eiland, all the islands situated between 10° and 35° longitude East of Greenwich and between 74° and 81° latitude North, especially West Spitsbergen, North-East Land, Barents Island, Edge Island, Wiche Islands,

Hope Island or Hopen-Eiland, and Prince Charles Forland, together with all islands great or small and rocks appertaining thereto.

## ARTICLE 2

Ships and nationals of all the High Contracting Parties shall enjoy equally the rights of fishing and hunting in the territories specified in Article 1 and in their territorial waters.

Norway shall be free to maintain, take or decree suitable measures to ensure the preservation and, if necessary, the re-constitution of the fauna and flora of the said regions, and their territorial waters; it being clearly understood that these measures shall always be applicable equally to the nationals of all the High Contracting Parties without any exemption, privilege or favour whatsoever, direct or indirect to the advantage of any one of them.

Occupiers of land whose rights have been recognized in accordance with the terms of Articles 6 and 7 will enjoy the exclusive right of hunting on their own land: 1) in the neighbourhood of their habitations, houses, stores, factories and installations, constructed for the purpose of developing their property, under conditions laid down by the local police regulations: 2) within a radius of 10 kilometres round the headquarters of their place of business or works; and in both cases, subject always to the observance of regulations made by the Norwegian Government in accordance with the conditions laid down in the present Article.

## ARTICLE 3

The nationals of all the High Contracting Parties shall have equal liberty of access and entry for any reason or object whatever to the waters, fjords, and ports of the territories specified in Article 1; subject to the observance of local laws and regulations, they may carry on there without impediment all maritime, industrial, mining, and commercial operations on a footing of absolute equality.

They shall be admitted under the same conditions of equality to the exercise and practice of all maritime, industrial, mining, or commercial enterprises both on land and in the territorial waters, and no monopoly shall be established on any account or for any enterprise whatever.

Notwithstanding any rules relating to coasting trade which may be in force in Norway, ships of the High Contracting Parties going to or coming from the territories specified in Article 1 shall have the right to put into Norwegian ports on their outward or homeward voyage for the purpose of taking on board or disembarking passengers or cargo going to or coming from the said territories, or for any other purpose.

It is agreed that in every respect, and especially with regard to exports, imports, and transit traffic, the nationals of all the High Contracting Parties, their ships and goods shall not be subject to any charges or restrictions whatever which are not borne by the nationals, ships, or goods which enjoy in Norway the treatment of the most favoured nation; Norwegian nationals, ships or goods being for this purpose assimilated to those of the other High Contracting Parties, and not treated more favourably in any respect.

No charge or restriction shall be imposed on the exportation of any goods to the territories of any of the Contracting Powers other or more onerous than on the exportation of similar goods to the territory of any other Contracting Power (including Norway) or to any other destination.

## ARTICLE 4

All public wireless telegraphy stations established or to be established by, or with the authorization of, the Norwegian Government within the territories referred to in Article 1 shall always be open on a footing of absolute equality to communications from ships of all flags and from nationals of the High Contracting Parties, under the conditions laid down in the Wireless Telegraphy Convention of July 5, 1912, or in the subsequent International Convention which may be concluded to replace it.

Subject to international obligations arising out of a state of war, owners of landed property shall always be at liberty to establish and use for their own purposes wireless telegraphy installations, which shall be free to communicate on private business with fixed or moving wireless stations, including those on board ships and aircraft.

## ARTICLE 5

The High Contracting Parties recognize the utility of establishing an international meteorological station in the territories specifies in Article 1, the organization of which shall form the subject of a subsequent Convention.

Conventions shall also be concluded laying down the conditions under which scientific investigations may be conducted in the said territories.

## ARTICLE 6

Subject to the provisions of the present Article, acquired rights of nationals of the High Contracting Parties shall be recognized.

Claims arising from taking possession or from occupation of land before the signature of the present Treaty shall be dealt with in accordance with the Annex hereto, which will have the same force and effect as the present Treaty.

## ARTICLE 7

With regard to methods of acquisition, enjoyment, and exercise of the right of ownership of property, including mineral rights, in the territories specified in Article 1, Norway undertakes to grant to all nationals of the High Contracting Parties treatment based on complete equality and in conformity with the stipulations of the present Treaty.

Expropriation may be resorted to only on grounds of public utility and on payment of proper compensation.

## ARTICLE 8

Norway undertakes to provide for the territories specified in Article 1 mining regulations which, especially from the point of view of imposts, taxes or charges of any kind, and of general or particular labour conditions, shall exclude all privileges, monopolies or favours for the benefit of the State or of the nationals of any one of the High Contracting Parties, including Norway, and shall guarantee to the paid staff of all categories the remuneration and protection necessary for their physical, moral and intellectual welfare.

Taxes, dues, and duties levied shall be devoted exclusively to the said territories and shall not exceed what is required for the object in view.

So far, particularly, as exportation of minerals is concerned, the Norwegian Government shall have right to levy an export duty which shall not exceed 1 per cent of the maximum value of the minerals exported up to 100 000 tons, and beyond that quantity the duty will be proportionately diminished. The value shall be fixed at the end of the navigation season by calculating the average free on board price obtained.

Three months before the date fixed for their coming into force, the draft mining regulations shall be communicated by the Norwegian Government to the other Contracting Powers. If during this period one or more of the said Powers propose to modify these regulations before they are applied, such proposals shall be communicated by the Norwegian Government to the other Contracting Powers in order that they may be submitted to examination and the decision of a Commission composed of one representative of each of the said Powers. This Commission shall meet at the invitation of the Norwegian Government and shall come to a decision within a period of three months from the date of its first meeting. Its decisions shall be taken by a majority.

## ARTICLE 9

Subject to the rights and duties resulting from the admission of Norway to the League of Nations, Norway undertakes not to create nor to allow the establishment of any naval base in the territories specified in Article 1 and not to construct any fortification in the said territories, which may never be used for warlike purposes.

## ARTICLE 10

Until the recognition by the High Contracting Parties of a Russian Government shall permit Russia to adhere to the present Treaty, Russian nationals and companies shall enjoy the same rights as nationals of the High Contracting Parties.

Claims in the territories specified in Article 1 which they may have to put forward shall be presented under the conditions laid down in the present Treaty (Article 6 and Annex) through the intermediary of the Danish Government, who declare their willingness to lend their good offices for this purpose.

The present Treaty, of which the French and English texts are both authentic, shall be ratified.

Ratifications shall be deposited at Paris as soon as possible.

Powers of which the seat of the Government is outside Europe may confine their action to informing the Government of the French Republic, through their diplomatic representative at Paris, that their ratification has been given, and in this case, they shall transmit the instrument as soon as possible.

The present Treaty will come into force, in so far as the stipulations of Article 8 are concerned, from the date of its ratification by all the signatory Powers; and in all other respects on the same date as the mining regulations provided for in that Article.

Third Powers will be invited by the Government of the French Republic to adhere to the present Treaty duly ratified. This adhesion shall be effected by a communication addressed to the French Government, which will undertake to notify the other Contracting Parties.

In witness whereof the abovenamed Plenipotentiaries have signed the present Treaty.

Done at Paris, the ninth day of February, 1920, in duplicate, one copy to be transmitted to the Government of His Majesty the King of Norway, and one deposited in the archives of the French Republic; authenticated copies will be transmitted to the other Signatory Powers.

**Source**: Office of the Governor of Svalbard. Available online at http://www.sysselmannen.no/Documents/Sysselmannen_dok/English/Legacy/The_Svalbard_Treaty_9ssFy.pdf

## NOTE

1. Also known as the Spitsbergen Treaty.

# Abbreviations

| | |
|---|---|
| AAT | Australian Antarctic Territory |
| ACCE report | Antarctic Climate Change and the Environment report |
| AEC | Arctic Economic Council |
| APEC | Asia Pacific Economic Cooperation summit |
| ASMA | Antarctic Specially Managed Area |
| ASOC | Antarctic Southern Ocean Coalition |
| ASPA | Antarctic Specially Protected Area |
| ATCM | Antarctic Treaty Consultative Meeting |
| ATS | Antarctic Treaty System |
| BAS | British Antarctic Survey |
| BAT | British Antarctic Territory |
| CBD | Convention on Biological Diversity |
| CCAS | Convention for the Conservation of Antarctic Seals |
| CCAMLR | Convention for the Conservation of Antarctic Marine Living Resources |
| CIA | Central Intelligence Agency |
| CLCS | Committee on the Limits of the Continental Shelf |
| COMNAP | Council of Managers of National Antarctic Programs |
| CRAMRA | Convention on the Regulation of Antarctic Mineral Resource Activity |
| EAIS | East Antarctica Ice Sheet |
| EEZ | Exclusive Economic Zone |

| | |
|---|---|
| FAO | Food and Agriculture Organization |
| FTA | Free Trade Agreement |
| GATT | General Agreement on Tariffs and Trade |
| GDP | gross domestic product |
| G-77 | Group of 77 |
| IAATO | International Association of Antarctic Tour Operators |
| ICJ | International Court of Justice |
| IGY | International Geophysical Year |
| IOM | International Organization for Migration |
| IPCC | Intergovernmental Panel on Climate Change |
| IPY | International Polar Year |
| IUCN | International Union for the Conservation of Nature |
| IUU | illegal, underreported, and unregulated (fishing) |
| NASA | National Aeronautics and Space Administration |
| NIEO | New International Economic Order |
| NGO | nongovernmental organization |
| NSF | National Science Foundation |
| NSR | Northern Sea Route |
| OPEC | Organization of Petroleum Exporting Countries |
| R&D | research and development |
| SANAP | South African National Antarctic Program |
| SCAR | Scientific Committee on Antarctic Research |
| UN | United Nations |
| UNCLOS | United Nations Convention on the Law of the Sea |
| UNDP | United Nations Development Programme |
| UNEP | United Nations Environment Program |
| UNESCO | United Nations Educational, Scientific, and Cultural Organization |
| UNFCCC | United Nations Framework Convention on Climate Change |
| UNU | United Nations University |
| US | United States |
| USAP | United States Antarctic Program |
| USGS | United States Geological Survey |
| USSR | Union of Soviet Socialist Republics |
| WAIS | West Antarctica Ice Sheet |
| WTO | World Trade Organization |

# Bibliography

ABB Communications. "Clean Sustainable Energy for Greenland." December 10, 2012, http://www.abb.com/cawp/seitp202/b08ea3b92dc74ac8c1257aaf0047543c.aspx.

Acosta, Donald. "Antarctic Mission for C-17A Globemaster, a Success," *International Business Times*, November 23, 2015, http://www.ibtimes.com.au/antarctic-mission-c-17a-globemaster-success-1485549.

Act of July 17, 1925 Relating to Svalbard, last accessed March 15, 2016, http://app.uio.no/ub/ujur/oversatte-lover/data/lov-19250717-011-eng.pdf.

Advisory Group on Climate Change and Human Mobility. "Human Mobility in the Context of Climate Change, UNFCCC-Paris COP-21." November 2015, http://www.internal-displacement.org/assets/publications/2015/201511-human-mobility-in-the-context-of-climate-change-unfccc-Paris-COP21.pdf.

Agreement Governing the Activities of States on the Moon and Other Celestial Bodies. Last accessed March 15, 2016, http://www.unoosa.org/oosa/SpaceLaw/moon.html.

Anderson, D. H. "Resolution and Agreement Relating to the Implementation of Part XI of the UN Convention on the Law of the Sea: A General Assessment." Max-Planck-Institut für ausländisches öffentliches Recht und Völkerrecht, 1995, 285–87 [in German].

Annex II to the Madrid Protocol, "Conservation of Antarctic Fauna and Flora," last accessed March 15, 2016, http://www.ats.aq/e/ep_faflo.htm. Antarctic and Southern Ocean Coalition. "The Antarctic

Environmental Protocol, 1991–2011." Antarctic Treaty Consultative Meeting, Buenos Aires, Argentina, June 20–July 1, 2011.

Antarctic and Southern Ocean Coalition. "Biological Prospecting." http://www.asoc.org/advocacy/antarctic-environmental-protection/biological-prospecting.

Antarctic and Southern Ocean Coalition. "Designation of Marine Protected Areas Within the Antarctic Treaty Area." ATCM, IP 119. Antarctic Treaty Consultative Meeting, Kiev, Ukraine, June 2–13, 2008.

Antarctic and Southern Ocean Coalition Secretariat. "Appeal to the Duma on Lake Vostock, Antarctica. ASOC Calls on Russia to Reconsider Penetrating the Largest Known Subglacial Lake." April 14, 2008. http://www.asoc.org/storage/documents/Other_publications/asoc_vostok_statement041408.pdf.

Aradas, Anahi. "Antártica: ¿cómo se vive en una de las zonas más remotas del mundo?" *BBC Mundo*, August 11, 2013. http://www.bbc.com/mundo/noticias/2013/07/130423_sociedad_villa_estrellas_aa.

Atkin, Michael. "Russia Blocks a Proposal for a Marine Protection Area in Antarctica's Ross Sea," *ABC News*, October 30, 2015.

Auburn, F. M. "Antarctic Minerals and the Third World," *Journal of Polar Studies* 1:1 (1984): 201–23.

Australian Antarctic Data Center. *The SCAR Composite Gazetteer*, last accessed March 15, 2016, https://data.aad.gov.au/aadc/gaz/scar/statistics.cfm.

Australian Antarctic Division Place Names Committee. Last accessed March 15, 2016, http://www.antarctica.gov.au/law-and-treaty/antarctic-place-names.

Australian Department of the Environment. "Ownership of Protected Areas," last accessed March 15, 2016, http://www.environment.gov.au/land/nrs/about-nrs/ownership.

Baird, Rachel. *Aspects of Illegal, Unreported, and Unregulated Fishing in the Southern Ocean.* New York: Springer, 2006.

Barr, Susan, and Cornelia Ludecke, eds. *The History of the International Polar Years (IPYs)*. London: Springer, 2010.

Bastmeijer, Kees. "Implementing the Environmental Protocol Domestically: An Overview." In Vidas, *Implementing the Environmental Protection Regime for the Antarctic*, 287–307.

Bastmeijer, Kees, and Ricardo Roura. "Environmental Impact Assessment in Antarctica." *Theory and Practice of Transboundary Environmental Impact Assessment*, 175–219.

Bastmeijer, Kees, and Ricardo Roura, eds. *Theory and Practice of Transboundary Environmental Impact Assessment*. Boston: Brill/Martinus Nijhoff Publishers, 2008.

Beck, Peter J. "Convention on the Regulation of Antarctic Mineral Resource Activities: A Major Addition to the Antarctic Treaty System," *Polar Record* 25(152): 19–32.

Beck, Peter J. *The International Politics of Antarctica*. London: Croom Helm, 1986.

Beck, Peter J. "Twenty Years On: The UN and the 'Question of Antarctica' 1983–2003," *Polar Record* 40 (2004): 205–212.

Beck, Peter J. "Who Owns Antarctica? Governing and Managing the Last Continent," *Boundary and Territory Briefing* 1:1 (1994): 1–59.

Bederman, David J. "The Antarctic and Southern Ocean Coalition's Convention on Antarctic Conservation," *Georgetown International Environmental Law Review* 4 (1991–1992): 47–61.

Bekkevold, Jo Inge, and Kristine Offerdal. "Norway's High North Policy and New Asian Stakeholders." In Sinha, *Arctic*, 59–74.

Belgium. "An Update on Biological Prospecting in Antarctica, Including the Development of the Antarctic Biological Prospecting Database." Working Paper 11. Antarctic Treaty Consultative Meeting, Kyiv, Ukraine, June 2–13, 2008.

Belgium. UNU-IAS. "The Role of Ex-Situ Collections in Antarctic Bioprospecting." Antarctic Treaty Consultative Meeting, Punta del Este, Uruguay, May 3–14, 2010.

Bender, Jeremy. "Russian Defense Minister Explains Why the Kremlin Is Militarizing the Arctic," *Business Insider UK*, February 26, 2015.

Bennett, Mia M. "The Maritime Tiger: Exploring South Korea's Interests and Role in the Arctic," In Sinha, *Arctic*, 120–37.

Bergin, Anthony. "The Politics of Antarctic Minerals: The Greening of White Australia," *Australian Journal of Political Science* 26 (1991): 216–39.

Bergin, Anthony, and Anthony Press. "Funding Cuts Threaten Our Standing in Antarctica," *The Australian*, June 8, 2013, http://www.theaustralian.com.au/news/inquirer/funding-cuts-threaten-our-standing-in-antarctica/story-e6frg6z6-1226659489671.

Bergsager, Egil. "Basic Conditions for the Exploration and Exploitation of Mineral Resources in Antarctica: Options and Precedents." In Vicuna, *Antarctic Resources Policy*, 167–83.

Berkman, Paul A. "President Eisenhower, the Antarctic Treaty, and the Origin of International Space." In Berkman, *Science Diplomacy*, 17–27.

Berkman, Paul A., Michael A Lang, David W. H. Walton, Oran R. Young, eds. *Science Diplomacy: Antarctica, Science, and the Governance of International Spaces*. Washington, DC: Smithsonian Institution Scholarly Press, 2011.

Bertrand, Kenneth J. *Americans in Antarctica, 1775–1948*. New York: American Geographical Society, 1971.

Blay, S. K. N. "New Trends in the Protection of the Antarctic Environment: The 1991 Madrid Protocol," *American Journal of International Law* 86 (1992): 382–85.

Blay, S. K. N., and B. M. Tsamenyi. "Australia and the Convention for the Regulation of Antarctic Mineral Resource Activities (CRAMRA)," *Polar Record* 26:158 (1990): 195–202.

Blomquist, Robert F. "Ratification Resisted: Understanding America's Response to the Convention on Biological Diversity, 1989–2000," *Golden Gate University Law Review* 32 (2002): 493–586.

Blum, Y. Z. *Historical Titles in International Law.* The Hague, Netherlands: Martinus Nihoff, 1965.

Bondareff, Joan. "Congress Acts to Protect Antarctica," *Territorial Sea Journal* 223 (1990–1991): 223–44.

Borgenson, Scott G. "Arctic Meltdown: The Economic and Security Implications of Global Warming," *Foreign Affairs* 87:2 (2008): 63–77.

Brady, Anne-Marie, ed. *The Emerging Politics of Antarctica*. London: Routledge, 2013.

Braun, Christina, Fritz Hertel, Osama Mostafa, Anja Nordt, Simone Pfeiffer, and Hans-Ulrich Peter. "Environmental Assessment and Management Challenges of the Fildes Peninsula Region." In Tin, *Antarctic Futures*, 169–91.

Bravo, Camila. "Chile en la Antártica: A treinta años del primer chileno nacido en el fin del mundo," *24 Oras*, November 25, 2014, http://www.24horas.cl/tendencias/espectaculosycultura/chile-en-la-antartica-a-treinta-anos-del-primer-chileno-nacido-en-el-fin-del-mundo—1506774.

Brazil. "Bioprospecting Activities of Brazil in Antarctica: A Short Report." Antarctic Treaty Consultative Meeting, Baltimore, USA, April 6–17, 2009.

British Broadcasting Company (BBC). "Japanese Whaling Ships Depart for Antarctic Hunt." December 1, 2015, http://www.bbc.com/news/world-asia-34971637.

Britton, Arthur K. *Extreme Jobs in Extreme Places: Life at a Polar Research Station.* New York: Gareth Stevens Publishing, 2013.

Brooks, Cassandra M. "Competing Values on the Antarctic High Seas: CCAMLR and the Challenge of Marine-Protected Areas," *Polar Journal* 3 (2013): 277–300.

Brooks, William. "Elevated Station Design for the South Pole Redevelopment Project at Amundsen-Scott South Pole Station." Paper presented at the 10th International Cold Regions Conference, August 16–19, 1999.

Broughton, Darcy. "The Potential for Mineral Exploration and Extraction in Antarctica." Postgraduate Certificate in Antarctic Studies 2011/2012, Gateway Antarctica, University of Canterbury, http://www.anta.canterbury.ac.nz/documents/PCAS_14/ProjectReports/Broughton%20-The%20potential%20for%20mineral%20exploration%20and%20extraction%20in%20Antarctica.pdf.

Brown, Alan. "Some Current Issues Facing the Convention on the Conservation of Antarctic Marine Living Resources (CCAMLR)." In Triggs, *Antarctica*, 85–108.

Brownlie, I. *Principles of Public International Law*. 8th ed. Oxford: Oxford University Press, 2012.

Bush, William. "Means and Methods of Implementation of Antarctic Environmental Regimes and National Environmental Instruments: An Exercise in Comparison." In Vidas, *Implementing the Environmental Protection Regime for the Antarctic*, 21–43.

Calder-Steele, Nicole. "Electricity in Isolation: The Progress of Power Generation in Antarctica." PCAS 15 (2012/2013). Critical Literature Review (ANTA602). http://www.anta.canterbury.ac.nz/documents/PCAS_15/Literature%20Review/PCAS_15_Lit_Nicole_Calder-Steele.pdf.

Caramel, Laurence. "Besieged by the Rising Tides of Climate Change, Kiribati Buys Land in Fiji," *The Guardian*, July 1, 2014.

Cattermole, Peter. *Building Planet Earth: Five Billion Years of Earth History*. London: Cambridge University Press, 2000.

*CBC News*. "NASA Map Shows What Antarctica Would Look Like Without Ice." June 7, 2013, http://www.cbc.ca/news/technology/nasa-map-shows-what-antarctica-would-look-like-without-ice-1.1304997.

Center for International Climate and Environmental Research, Norwegian Ministry of the Environment, Norwegian Ministry of Foreign Affairs, and Norwegian Refugee Council. "The Nansen Conference: Climate Change and Displacement in the 21st Century." June 5–7, 2011. http://www.unhcr.org/4ea969729.pdf.

Central Intelligence Agency (CIA). "North America: Greenland." In *The World Factbook*. https://www.cia.gov/library/publications/the-world-factbook/geos/gl.html (last updated June 7, 2016).

Central Intelligence Agency (CIA). *The World Factbook.* Last accessed March 15, 2016, https://www.cia.gov/library/publications/the-world-factbook/.

Chaturvedi, Sanjay. "Biological Prospecting in the South Polar Region: Science-Geopolitics Interface." In Shadian, *Legacies and Challenges in Polar Sciences,* 171–88.

Chaturvedi, Sanjay. "India and Antarctica: Towards Post-Colonial Engagement." In Brady, *The Emerging Politics of Antarctica*, 50–74.

Chen, Maijinn. "Antarctic Architecture," *Architecture Week*, February 16, 2005. http://www.architectureweek.com/2005/0216/building_1-1.html.

China. "Proposal for a New Antarctic Specially Managed Area at Chinese Antarctic Kunlun Station, Dome A." Working Paper 8. Antarctic Treaty Consultative Meeting, Brussels, Belgium, 2013.

Christo, Tiago Malavazi de, Jussara Farias Fardin, Domingos Savio Lyrio Simonetti, Lucas Frizera Encarnacao, and Chistina Engel de Alvarez. "Design and Analysis of Hybrid Energy Systems: The Brazilian Antarctic Station Case," *Renewable Energy* 88 (2016): 236–46.

Clancy, Robert, John Manning, and Henk Brolsma. *Mapping Antarctica: A Five Hundred Year Record of Discovery*. Chichester: Praxis Publishing, 2014.

Clark, Margaret L. "The Antarctic Environmental Protocol: NGOs in the Protection of Antarctica." In Princen, *Environmental NGOs in World Politics*, 160–85.

Climate Action. "Pacific Island Nations Warn of Large-Scale Climate Migration." October 14, 2015, http://www.climateactionprogramme.org/news/pacific_island_nations_warn_of_large_scale_climate_migration.

Cohen, Harlan. "Public Participation in Antarctica: The Role of Non-Governmental and Intergovernmental Organizations." In Berkman, *Science Diplomacy,* 271–76.

Commission for the Conservation of Antarctic Marine Living Resources. "Marine Protected Areas." Last accessed March 15, 2016, https://www.ccamlr.org/en/science/marine-protected-areas-mpas.

Commission for the Conservation of Antarctic Marine Living Resources. "Members." Last accessed March 15, 2016, https://www.ccamlr.org/en/organisation/members.

Conforti, Benedetto. "Territorial Claims in Antarctica: A Modern Way to Deal with an Old Problem," *Cornell International Law Journal* 19 (1986): 249–58.

Conforti, Benedetto, Luigi Ferrari Bravo, Francesco Francioni, Natalino Ronzitti, Giorgio Saerdoti, and Riccardo Pavoni, eds. *Italian Yearbook of International Law*. Vol. XV. Boston: Martinus Nijhoff Publishers, 2005.

Convention on the Regulation of Antarctic Mineral Resource Activities. June 2, 1988. http://sedac.ciesin.org/entri/texts/acrc/cramra.txt.html.

Convey, Peter, Kevin A. Hughes, and Tina Tin. "Continental Governance and Environmental Management Mechanisms Under the Antarctic Treaty System: Sufficient for the Biodiversity Challenges of This Century," *Biodiversity* 13 (2012): 234–48.

Cook, Grahame, ed. *The Future of Antarctica: Exploitation Versus Preservation*. Manchester: Manchester University Press, 1990.

Cool Antarctica. "Antarctica Climate and Weather," Last accessed March 15, 2016, http://www.coolantarctica.com/Antarctica%20fact%20file/antarctica%20environment/climate_weather.php.

Crispini, Lauro, Laura Federico, Giovanni Capponi, and Franco Talarico. "The Dorn Gold Deposit in Northern Victoria Land, Antarctica: Structure, Hydrothermal Alteration, and Implications for the Gondwana Pacific Margin," *Gondwana Research* 19:1 (2011): 128–40.

Council of Managers of National Antarctic Programs. Antarctic Facility List (February 13, 2014), https://www.comnap.aq/Information/SitePages/Home.aspx.

Dadwal, Shebonti Ray. "Arctic: The Next Great Game in Energy Politics." In Sinha, *Arctic,* 46–74.

DailyMail.com. "Maldives Government Highlights the Impact of Climate Change . . . by Meeting Underwater." October 20, 2009, http://www.dailymail.co.uk/news/article-1221021/Maldives-underwater-cabinet-meeting-held-highlight-impact-climate-change.html.

Dastidar, Prabir G., and S. Ramachandran. "Intellectual Structure of Antarctic Science," *Scientometrics* 77:3 (2008): 389–414.

Davis, Ruth. "The Durability of the Antarctic Model and Southern Ocean Governance." In Stephens, *Polar Oceans Governance in an Era of Environmental Change*, 287–307.

Day, David. *Antarctica: A Biography*. Oxford: Oxford University Press, 2013.

*Deep Seabed Mining: Hearing Before the Subcommittee on Oceanography of the Committee on Merchant Marine and Fisheries, House of Representatives, Ninety-Fourth Congress.* London: Forgotten Books, 2015.

Doan, David B. "The Mineral Industry of Antarctica." U.S. Geological Survey—Mineral Information. http://minerals.usgs.gov/minerals/pubs/country/1995/myb3-1995-ay.pdf.

Dodds, Klaus. *The Antarctic: A Very Short Introduction.* Oxford: Oxford University Press, 2012.

Dodds, Klaus. "The Antarctic Peninsula: Territory, Sovereignty Watch and the 'Antarctic Problem.'" In Hemmings, *Antarctic Security in the Twenty-First Century*, 95–115.

Dodds, Klaus. "Post-Colonial Antarctica: An Emerging Engagement," *Polar Record* 42 (2006): 59–70.

Doyle, Alister. "Antarctic Patents Strain Goals of Shared Science." *Reuters*, February 5, 2009. http://www.reuters.com/article/us-antarctica-companies-idUSTRE51503220090206.

Dudeney, John R., and David W. H. Walton, "Leadership in Politics and Science Within the Antarctic Treaty," *Polar Research* 31 (2012): 1–9.

Duhaime, Gérard, and Andrée Caron. "The Economy of the Circumpolar Arctic," *The Economy of the North* 17, http://chaireconditionautochtone.fss.ulaval.ca/documents/pdf/65-Economy-circumpolar-Arctic-2006.pdf.

Dutch Government Information Service. "Treaty Concerning the Archipelago of Spitsbergen, Including Bear Island," Last accessed March 15, 2016, https://verdragenbank.overheid.nl/en/Verdrag/Details/004293#Partijen.

Elzinga, Aant, ed. *Changing Trends in Antarctic Research.* Boston: Kluwer Academic Publishers, 1993.

Elzinga, Aant. "Rallying Around a Flag? On the Persistent Gap in Scientific Internationalism Between Word and Deed." In Brady, *The Emerging Politics of Antarctica*, 193–219.

Environmental Migration Portal. "Press Conference: Human Mobility in the UNFCC: Paris COP21 Draft Text 10th February, 2015, Palais des Nations, Geneva." February 10, 2015. http://environmentalmigration.iom.int/press-conference-human-mobility-unfcc-paris-cop21-draft-text-10th-february-2015-palais-des-nations.

Enzenbacher, Debra. "Antarctic Tourism Policy-Making: Current Challenges and Future Prospects." In Triggs, *Antarctica*, 155–89.

Enzenbacher, Debra. "The Regulation of Antarctic Tourism." In Hall, *Polar Tourism*, 179–215.

Erb, Karl A. "International Collaboration in the Antarctic for Global Science." In Berkman, *Science Diplomacy,* 265–70.

Fayette, Louise Angelique de la. "Responding to Environmental Damage in Antarctica." In Triggs, *Antarctica,* 109–54.

Fitzpatrick, Meagan. "'Science and Sovereignty' Key to New Arctic Research Centre," *CBC News*, August 23, 2012. http://www.cbc.ca/news/politics/science-and-sovereignty-key-to-new-arctic-research-centre-1.1231516.

Fogarty, Elena. "Antarctica: Assessing and Protecting Australia's National Interests." Policy Brief, Lowy Institute for International Policy, August 2011.

Foster, Caroline E. "Antarctic Resources and Human Security." In Hemmings, *Antarctic Security in the Twenty-First Century*, 154–71.

France. "In Search of a Legal Regime for Bioprospecting in Antarctica." Information Paper 13. Antarctic Treaty Consultative Meeting, Edinburgh, UK, 2006.

Franck, Thomas M., Mohamed El Baradei, and George Aron. "The New Poor: Land-Locked, Shelf-Locked, and Other Geographically Disadvantaged States," *New York University Journal of International Law and Politics* 7 (1974): 33–57.

Francioni, Francesco. "Establishment of an Antarctic Treaty Secretariat: Pending Legal Issues." In Vidas, *Implementing the Environmental Protection Regime for the Antarctic*, 125–40.

Francioni, Francesco. "The Madrid Protocol on the Protection of the Antarctic Environment," *Texas International Law Journal* 28 (1993): 47–72.

Francioni, Francesco, and Tulio Scovazzi, eds. *Biotechnology and International Law.* Oxford: Hart Publishing, 2006.

Friend, Scott Harris. "Sustainability and Habitation in Antarctica." Master's thesis, Department of Architecture, Montana State University, 2009.

Gabriel, Dana. "U.S. Arctic Ambitions and the Militarization of the High North," *Global Research Center for Research on Globalization*, July 24, 2013.

Garrison, Christopher. "Beneath the Surface: The Common Heritage of Mankind," *KEStudies* 1 (2007): 7.

Gjelsvik, Tore. "The Mineral Resources of Antarctica: Progress in Their Identification." In Vicuna, *Antarctic Resources Policy*, 61–76.

Global Agenda Council. *Demystifying the Arctic.* Davos: World Economic Forum, 2014.

Golledge, N. R., D. E. Kowalewski, T. R. Naish, R. H. Levy, C. J. Fogwill, E. G. W. Gasson, "The Multi-Millennial Antarctic Commitment to Future Sea-Level Rise," *Nature* 526 (2015): 421–25.

Golitsyn, Vladimir. "Balancing Sovereign Interests Beyond National Jurisdictions." In Berkman, *Science Diplomacy*, 51–58.

Government of Norway. Ministry of Justice and Public Security. "Development of the Local Community in Longyearbyen." Last accessed March 15, 2016, https://www.regjeringen.no/en/find-document/dep/JD/reports-to-the-storting/19992000/report-no-9-to-the-storting-/14/id456932.

Gramling, Carolyn. "Overhaul in the Works for Aging U.S. Antarctic Station," *Science*, December 8, 2015. http://news.sciencemag.org/climate/2015/12/overhaul-works-aging-u-s-antarctic-station.

GRID-Arendal. "Population Distribution in the Circumpolar Arctic, by Country (Including Indigenous Population)." Last accessed March 15, 2016, http://www.grida.no/graphicslib/detail/population-distribution-in-the-circumpolar-arctic-by-country-including-indigenous-population_1282.

Griffin, Andrew. "Norwegian Politicians Propose Putting Refugees on Svalbard—Remote Arctic Islands with More Polar Bears Than People," *Independent*, September 13, 2015.

Groves, Steven. "The US Can Mine the Deep Seabed Without Joining the UN Convention on the Law of the Sea." The Heritage Foundation Backgrounder #2746 on UN, December 4, 2012, http://www.heritage.org/research/reports/2012/12/the-us-can-mine-the-deep-seabed-without-joining-the-un-convention-on-the-law-of-the-sea.

Grydehoj, Adam. "The Globalization of the Arctic: Negotiating Sovereignty and Building Communities in Svalbard, Norway," *Island Studies Journal* 7 (2012): 99–118.

Grydehoj, Adam. "Informal Diplomacy in Norway's Svalbard Policy: The Intersection of Local Community Development and Arctic International Relations," *Global Change, Peace, and Security* 26 (2014): 41–54.

Grydehoj, Adam. "Uninherited Heritage: Tradition and Heritage Production in Shetland, Aland, and Svalbard," *International Journal of Heritage Studies* 16 (2010): 77–89.

Guðmundsdóttir, Anna Mjöll, and Anna Dóra Sæþórsdóttir. "Tourism Management in Wilderness Areas—Svalbard," *Rannsóknir í félagsvísindum* X (2009): 41–53.

Gurney, Alan. *Below the Convergence: Voyages Towards Antarctica, 1699–1839*. New York: Norton, 1997.

Hall, Colin Michael, and Margaret E. Johnston, eds. *Polar Tourism: Tourism in the Arctic and Antarctic Regions*. New York: John Wiley & Sons, 1995.

Hall, C. Michael, and Marisk Wouters. "Issues in Antarctic Tourism." In Hall, *Polar Tourism*, 147–66.

Hanesian, John. "The Antarctic Treaty 1959," *International and Comparative Law Quarterly* 9 (1960): 463–80.

Hanson, James D., and John E. Gordon. *Antarctic Environments and Resources: A Geographical Perspective*. Essex, UK: Addison Wesley Longman Limited, 1998.

Harrison, John. *Forgotten Footprints: Lost Stories in the Discovery of Antarctica*. Cardigan, UK: Parthian Books, 2014.

Harrison, James. "Who Benefits from the Exploitation of Non-Living Resources on the Seabed? Operationalizing the Benefit-Sharing Provisions in the UN Convention on the Law of the Sea." University of Edinburgh, July 1, 2015, http://www.benelexblog.law.ed.ac.uk/2015/07/01/who-benefits-from-the-exploitation-of-non-living-resources-on-the-seabed-operationalizing-the-benefit-sharing-provisions-in-the-un-convention-on-the-law-of-the-sea/.Haward, Marcus. "Marine Resources Management, Security and the Antarctic Treaty System, An Ongoing Agenda?" In Hemmings, *Antarctic Security in the Twenty-First Century*, 215–37.

Haward, Marcus, and Tom Griffiths, eds. *Australia and the Antarctic Treaty System: 50 Years of Influence.* Sydney: University of New South Wales, 2011.

Haward, Marcus, and Julia Jabour. "Environmental Change and Governance Challenges in the Southern Ocean." In Stephens, *Polar Oceans Governance in an Era of Environmental Change*, 21–41.Hayashi, Moritaka. "The Antarctica Question in the UN," *Cornell International Law Journal* 19 (1996): 275–90.

Heim, Barbara Ellen. "Exploring the Last Frontiers for Mineral Resources: A Comparison of International Law Regarding the Deep Seabed, Outer Space, and Antarctica," *Vanderbilt Journal of Transnational Law* 23 (1990–1991): 819–49.

Hemmings, Alan. "After the Party: The Hollowing of the Antarctic Treaty System and the Governance of Antarctica." Symposium on Antarctic Politics, University of Canterbury, Christchurch, New Zealand, July 8–9, 2010.

Hemmings, Alan D. "Beyond Claims: Towards a Non-Territorial Antarctic Security Prism for Australia and New Zealand," *New Zealand Yearbook of International Law* 6 (2008): 77–91.

Hemmings, Alan D. "Biological Prospecting in the Antarctic Treaty Area." Intergovernmental Meeting of Experts on Biological Prospecting in the Antarctic Treaty Area, Baarn, Netherlands, February 3–5, 2009.

Hemmings, Alan D. "Commercial Harvest in Antarctica." In Liggett, *Exploring the Last Continent*, 413–28.

Hemmings, Alan D. "Does Bioprospecting Risk Moral Hazard for Science in the Antarctic Treaty System?" *Ethics in Science and Environmental Politics* 10 (2010): 5–12.

Hemmings, Alan D. "Problems Posed by Attempts to Apply a Claimant's Domestic Legislation Beyond Its Own Nationals in Antarctica," *Asia Pacific Journal of Commonwealth Law* 2 (2008): 207–20.

Hemmings, Alan D. "Re-Justifying the Antarctic Treaty System for the 21st Century: Rights, Expectations and Global Equity." In Powell, *Polar Geopolitics?* 55–73.

Hemmings, Alan D. "Security Beyond Claims." In Hemmings, *Antarctic Security in the Twenty-First Century*, 70–94.

Hemmings, Alan. "Why Did We Get an International Space Station Before an International Antarctic Station?" *Polar Journal* 1:1 (2011): 5–16.

Hemmings, Alan D., and Lorne K. Kriwoken. "High-Level Antarctic EIA Under the Madrid Protocol: State Practice and the Effectiveness of the Comprehensive Environmental Evaluation Process," *International Environmental Agreements: Politics, Law and Economics* 10 (2010): 187–208.

Hemmings, Alan D., Donald R. Rothwell, and Karen N. Scott. "Antarctic Security in a Global Context." In Hemmings, *Antarctic Security in the Twenty-First Century*, 328–36.

Hemmings, Alan D., Donald R. Rothwell, and Karen N. Scott, eds. *Antarctic Security in the Twenty-First Century: Legal and Policy Perspectives.* New York: Routledge, 2012.

Herber, Bernard P. *Protecting the Antarctic Commons: Problems of Economic Efficiency.* Tucson: Udall Center for Studies in Public Policy, University of Arizona, 2007.

Herr, Richard A. "The Changing Role of Non-Governmental Organizations in the Antarctic Treaty System." In Vidas, *Governing the Antarctic*, 91–110.

Hill, David. "A New Master Plan for the McMurdo Research Station in Antarctica: OZ Architecture Attempts to Reinvent a Utilitarian Science Outpost in the Wilds of Antarctica," *Journal of the American Institute of Architects* (October 9, 2013), http://www.architectmagazine.com/design/buildings/a-new-master-plan-for-the-mcmurdo-research-station-in-antarctica_o.

Honneland, Geir. "Compliance in the Fishery Protection Zone Around Svalbard," *Ocean Development and International Law* 29 (1998): 339–60.

Hughes, Kevin. "How Committed Are We to Monitoring Human Impacts in Antarctica?" *Environmental Research Letters* 5 (2010): 1–3.

Hughes, Kevin A., and Paul D. Bridge. "Potential Impacts of Antarctic Bioprospecting and Associated Commercial Activities upon Antarctica Science and Scientists," *Ethics in Science and Environmental Politics* 10 (2010): 13–18.

Hughes, Kevin A., Louise C. Ireland, Peter Convey, and Andrew H. Fleming. "Assessing the Effectiveness of Specially Protected Areas for

Conservation of Antarctica's Botanical Diversity," *Conservation Biology* 30 (2016): 113–20.

*Hurriyet Daily News*. "Turkish Scientists Head for Maiden Antarctica Mission." March 29, 2016. http://www.hurriyetdailynews.com/turkish-scientists-head-for-maiden-antarctica-mission.aspx?pageID=238&nID=97046&NewsCatID=341.

Hyodo, Shinji. "Russia's Strategic Concerns in the Arctic and Its Impact on Japan-Russia Relations." In Sinha, *Arctic*, 94–105.

Intergovernmental Panel on Climate Change. "Climate Change 2014: Synthesis Report Summary for Policymakers," 2014, https://www.ipcc.ch/pdf/assessment-report/ar5/syr/AR5_SYR_FINAL_SPM.pdf.

International Arbitral Awards. *Island of Palmas Case* (Netherlands, USA). Reports of the IAA, vol. II, April 4, 1928, 829–71. http://legal.un.org/riaa/cases/vol_II/829-871.pdf.

International Court of Justice. Communique no. 55/26, May 6, 1955. Last accessed March 15, 2016, http://www.icj-cij.org/docket/files/27/12325.pdf.

International Court of Justice. "Whaling in the Antarctic (Australia vs. Japan: New Zealand Intervening)." Judgment of March 31, 2014. http://www.icj-cij.org/docket/files/148/18136.pdf.

International Organization for Migration. "Migration, Climate Change, and the Environment: A Complex Nexus," Migration Portal, last accessed March 15, 2016, https://www.iom.int/complex-nexus#estimates.

International Organization for Migration. "Migration and the Environment," last accessed March 15, 2016, https://www.iom.int/jahia/webdav/site/myjahiasite/shared/shared/mainsite/microsites/IDM/workshops/evolving_global_economy_2728112007/presentations/presentation_migration_environment.pdf.

International Organization for Migration. "World Migration Report 2010: The Future of Migration: Building Capacities for Change." 2010, http://publications.iom.int/system/files/pdf/wmr_2010_english.pdf.

International Seabed Authority. "Deep Seabed Mineral Contractors," last accessed March 15, 2016, https://www.isa.org.jm/deep-seabed-minerals-contractors?page=1&qt-contractors_tabs_alt=0.

International Seabed Authority. "Legal and Technical Commission: Issues Relating to the Operation of the Enterprise, in Particular the Legal, Technical, and Financial Implications for the Authority and for States Parties," ISBA/20/LTC/12, Twentieth Session, Kingston, Jamaica, July 14–25, 2014.

International Seabed Authority. "Legislative History of the Enterprise," Kingston, Jamaica, 2002, 1–385. https://www.isa.org.jm/sites/default/files/files/documents/enterprise-ae.pdf.

International Seabed Authority. *SPC-EU EDF10 Deep Sea Minerals (DSM) Project.* Information Brochure 15, 2013, http://gsd.spc.int/dsm/images/pdf_files/dsm_brochures/DSM_Brochure15_ISA.

International Tribunal for the Law of the Sea. "Responsibilities and Obligations of States Sponsoring Persons and Entities with Respect to Activities in the Area." Case no. 17, 2010-2011, https://www.itlos.org/cases/list-of-cases/case-no-17.

Jabour, Julia. "Biological Prospecting in the Antarctic: Fair Game?" In Brady, *The Emerging Politics of Antarctica*, 242–57.

Jenkins, Mark. "True Colors: The Changing Face of Greenland," In *National Geographic* June 2010, 34–47.

Jennings, R. Y. *The Acquisition of Territory in International Law.* Manchester: Manchester University Press, 1963.

Jin, Huang. "China's First Polar Fixed-Wing Aircraft Arrives at Research Station in Antarctica." *People's Daily Online Exclusives*, December 3, 2015, http://english.cas.cn/newsroom/china_research/201512/t20151203_156975.shtml.

Johnston, Margaret E., and C. Michael Hall. "Visitor Management and the Future of Tourism in Polar Regions." In Hall, *Polar Tourism,* 297–313.

Josephson, Paul R. *The Conquest of the Russian Arctic.* Cambridge: Harvard University Press, 2014.

Joyner, Christopher. "Bioprospecting as a Challenge to the Antarctic Treaty." In Hemmings, *Antarctica Security in the Twenty-First Century,* 197–214.

Joyner, Christopher. "The Exclusive Economic Zone and Antarctica: The Dilemmas of Non-Sovereign Jurisdiction." *Ocean Development and International Law* 19 (1988): 469–91.

Joyner, Christopher C. *Governing the Frozen Commons: The Antarctic Regime and Environmental Protection.* Columbia: University of South Carolina Press, 1998.

Joyner, Christopher, and Sudhir Chopra, eds. *The Antarctic Legal Regime.* Dordrecht, Netherlands: Martinus Nijhoff, 1988.

Kara Kovalchik, "First Baby Born in Antarctica," *Not-So-Famous Firsts: Infant Edition,* September 3, 2010; http://mentalfloss.com/article/25676/not-so-famous-firsts-infant-edition.

Kaye, Stuart. "IUU Fishing in the Southern Ocean: Challenge and Response." In Triggs, *Antarctica*, 39–59.

Kennicutt II, Mahlon. "New Frontiers and Future Directions of Antarctic Science." In Berkman, *Science Diplomacy*, 153–60.

Kerrest, Armel. "Outer Space as International Space: Lessons from Antarctica." In Berkman, *Science Diplomacy*, 133–42.

Kimball, Lee. "The Role of Non-Governmental Organizations in Antarctic Affairs." In Joyner, *The Antarctic Legal Regime*, 33–64.

Klein, Andrew G., Mahlon C. Kennicutt II, Gary A. Wolff, Steve T. Sweet, Tiffany Bloxom, Diana A. Gielstra, and Marietta Cleckley. "The Historical Development of McMurdo Station, Antarctica: An Environmental Perspective," *Polar Geography* 31:3–4 (2008): 119–44.

Klein, Andrew G., Stephen T. Sweet, Mahlon C. Kennicutt II, Terry L. Wade, Terry A. Palmer, and Paul Montagna. "Long-Term Monitoring of Human Impacts to the Terrestrial Environment at the McMurdo Station." In Tin, *Antarctic Futures*, 213–27.

Kreft, S., D. Ekstein, L. Junghans, C. Kerestan, and U. Hagen. "Global Climate Risk Index 2015: Who Suffers Most from Extreme Weather Events? Weather-Related Loss Events in 2013 and 1994 to 2013." Germanwatch, November 2014, https://germanwatch.org/en/download/10333.pdf.

Kremlin. "Press Statement and Answers to Journalists' Questions Following the APEC Summit," http://en.kremlin.ru/events/president/transcripts/19382.

Kuenning, Kris. "Green Antarctica: Station Greenhouses Produce Fresh Food, Feel-Good Environments," Press release of the National Science Foundation, February 25, 2004, http://www.spaceref.com/news/viewpr.html?pid=13724.

Lasserre, Frédéric, and Sébastien Pelletier. "Polar Super Seaways? Maritime Transport in the Arctic: An Analysis of Shipowners' Intentions," *Journal of Transport Geography* 19 (2011): 1465–73.

Leary, David. "From Hydrocarbons to Psychrophiles: The 'Scramble' for Antarctic and Arctic Resources." In Stephens, *Polar Oceans Governance in an Era of Environmental Change*, 125–45.

Lee, Ricky J. *Law and Regulation of Commercial Mining of Minerals in Outer Space.* New York: Springer, 2012.

Lefeber, Rene. "The Legal Need for an Antarctic Liability Regime." In Vidas, *Implementing the Environmental Protection Regime for the Antarctic*, 181–97.

Liggett, Daniela, Machiel Lamers, Tina Tin, and Patrick Maher. "Conclusions: Multiple Dimensions of Human Engagement with the Antarctic Environment." In Tin, *Antarctic Futures*, 335–51.

Liggett, Daniela, Bryan Storey, Yvonne Cook, and Veronika Meduna, eds. *Exploring the Last Continent: An Introduction to Antarctica*. Cham, Switzerland: Springer International Publishing, 2015.

Lopez, Julyssa. "How to Survive in the Arctic: $3 Million Grant from the National Science Foundation Will Expand Research on Urban Sustainability in Extreme Climate Conditions," *GW Today*, November 30, 2015, http://gwtoday.gwu.edu/how-survive-arctic.

Luck, Michael, Patrick T. Maher, and Emma J. Stewart, eds. *Cruise Tourism in Polar Regions: Promoting Environmental and Social Sustainability*. London: Earthscan, 2010.

Luck, Michael, Patrick T. Maher, and Emma J. Stewart. "Setting the Scene: Polar Cruise Tourism in the 21st Century." In Luck, *Cruise Tourism in Polar Regions*, 1–10.

Lynch, William C. "The Nepal Proposal for a Common Heritage Fund: Panacea or Pipedream," *California Western International Law Journal* 10 (1980): 25–52.

MacLellan, Kylie. "Britain Says to Boost Defense of Disputed Falkland Islands," *Reuters*, March 24, 2015.

Malone, James. "The United States and the Law of the Sea After UNCLOS," *Journal of Law and Contemporary Problems* 46 (Spring 1983): 29–36.

MarEx, *Maritime Executive*. "North Korea Signs the Svalbard Treaty." February 6, 2016, http://www.maritime-executive.com/article/north-korea-signs-svalbard-treaty.

Maxwell, Bob. "The Controversy Around the Proposal and Formation of the South Pole Traverse: Critical Review." Postgraduate Certificate of Antarctic Studies 14, University of Canterbury, 2011/2012.

May, John. *The Greenpeace Book of Antarctica: A New View of the Seventh Continent*. London: Dorling Kindersley, 1988.

Mill, Hugh Robert. *Siege of the South Pole*. New York: Stokes Company, 1905.

Miller, Denzil. "Sustainable Management in the Southern Ocean: CCAMLR Science." In Berkman, *Science Diplomacy*, 103–21.

Milun, Kathryn. *The Political Uncommons: The Cross-Cultural Logic of the Global Commons*. Burlington: Ashgate Publishing Co., 2011.

Mino, Isabel, Carlos Naranjo, and Andres Montero. "Evaluation of Energy Efficiency Measures for the Ecuadorian Antarctic Station Pedro Vicente Maldonado Through a Calibrated Building Energy Model," *Revista Ingenieria de Construccion* 30:2 (2015): 127–34.

Myers, Steven Lee, and Clifford Krauss. "Melting Ice Isn't Opening Arctic to Oil Bonanza," *International New York Times*, September 7, 2015.

National Aeronautics and Space Administration. Landsat Image Mosaic of Antarctica. Last accessed March 15, 2016, http://lima.nasa.gov/antarctica.

National Aeronautics and Space Administration. "What Is Antarctica?" Last accessed March 15, 2016, http://www.nasa.gov/audience/forstudents/5-8/features/nasa-knows/what-is-antarctica-58.html.

National Research Council. *Polar Icebreakers in a Changing World: An Assessment of U.S. Needs*. Washington: National Academies Press, 2007.

National Research Foundation and South African Agency for Science and Technology Advancement. "Biotechnology Research in the Antarctic Region," *Antarctic Biotech Research*, 2009.

National Security Council. "Note by the Executive Secretary to the National Security Council on Antarctica." Doc. 333, June 29, 1957, http://history.state.gov/historicaldocuments/frus1955-57v11/d333.

National Security Council. "Statement of US Policy on Antarctica." Doc. 286, April 7, 1959, https://history.state.gov/historicaldocuments/frus1958-60v02/d286.

National Security Council Planning Board. Report to the National Security Council, doc. 301, June 28, 1954, https://history.state.gov/historicaldocuments/frus1952-54v01p2/d301.

*New World Encyclopedia*. "Greenland." Last accessed March 15, 2016, http://www.newworldencyclopedia.org/entry/Greenland.

Nicol, Stephen, Jacqueline Foster, and So Kawaguchi. "The Fishery for Antarctic Krill—Recent Developments," *Fish and Fisheries* 13 (2012): 30–40.

Nielsen, Hanne. "From Shelter to Showpiece: The Evolution of Scientific Antarctic Stations." University of Canterbury Summer Research Project, November 2013–January 2014.

Norwegian Ministry of Foreign Affairs. "Norwegian Government's High North Strategy." 2005, https://www.regjeringen.no/globalassets/upload/UD/Vedlegg/strategien.pdf.

Notre Dame Global Adaptation Index. Last accessed March 16, 2016, ND-GAIN Index. http://index.gain.org.

Noyes, John E. "The Common Heritage of Mankind: Past, Present, and Future," *Denver Journal of International Law and Policy* 40 (2011–2012): 447–71.

O'Brien, Rosalba. "Argentine Judge Orders Seizure of Falklands Drillers' Assets," *The Telegraph,* June 27, 2015.

O'Connell, D. P. *International Law*. 2nd ed. London: Stevens & Sons Ltd., 1970.

O'Rourke, Ronald. *Coast Guard Polar Icebreaker Modernization: Background and Issues for Congress*. Congressional Research Service Report, September 2, 2015.

*Ondas*. "Argentina y Chile hicieron nacer bebés en la Antartida para reclamar el territorio." December 11, 2008. http://www.radiocable.com/bebes-antartida775.html#sthash.nf5S5Rkz.dpuf.

Orheim, Olav, Anthony Press, and Neil Gilbert. "Managing the Antarctic Environment: The Evolving Role of the Committee for Environmental Protection." In Berkman, *Science Diplomacy*, 209–21.

Owen, Tess. "Frozen Norwegian Archipelago Might Provide a New Home for Syrian Refugees," *Vice News*, September 14, 2015.

Parliament of Australia. House of Representatives (1991). *Antarctica Mining Prohibition Bill: Second Reading,* Last accessed March 15, 2016, http://parlinfo.aph.gov.au/parlInfo/search/display/display.w3p;db=CHAMBER;id=chamber%2Fhansardr%2F1991-03-13%2F0008;query=Id%3A%22chamber%2Fhansardr%2F1991-03-13%2F0059%22.

Pedersen, Torbjorn. "The Constrained Politics of the Svalbard Offshore Area," *Marine Policy* 32 (2008): 913–19.

Pedersen, Torbjorn. "International Law and Politics in U.S. Policymaking: The United States and the Svalbard Dispute," *Ocean Development and International Law* 42 (2011): 120–35.

Pedersen, Torbjorn. "Norway's Rule on Svalbard: Tightening the Grip on Arctic Islands," *Polar Record* 45 (2009): 147–52.

Pedersen, Torbjorn. "The Svalbard Continental Shelf Controversy: Legal Disputes and Political Rivalries," *Ocean Development and International Law* 37 (2006): 339–58.

Pedersen, Torbjorn, and Tore Henriksen. "Svalbard's Maritime Zones: The End of Legal Uncertainty," *International Journal of Marine and Coastal Law* 24 (2009): 141–61.

Perlez, Jane. "China, Pursuing Strategic Interests, Builds Presence in Antarctica," *International New York Times*, May 3, 2015.

Phillips, Ari. "Antarctica Recorded Its Hottest Temperature on Record This Week," *Climate Progress,* March 28, 2015. http://thinkprogress.org/climate/2015/03/28/3640232/antarctica-breaks-thermostat.

Phillips, Janet. "Three Women Manage Antarctic Stations: Go South Young Woman," last accessed March 15, 2016, http://www.southpolestation.com/trivia/90s/janet.html.

Pilcher, Natalie. "The Longest Journey in the World: An Investigation into the Liability Annex and Its Evolution." Projects GCAS 10 (2007–2008), last accessed March 15, 2016, http://www.anta.canterbury.ac.nz/courses/gcas/reports/Projects%20GCAS%2010.shtml.

Pincus, Rebecca, and Saleem H. Ali, eds. *Diplomacy on Ice*. New Haven: Yale University Press, 2015.

Polefka, Shiva. "Icebreakers: Essential Assets for a Changing Arctic." Center for American Progress, March 5, 2015, https://www.americanprogress.org/issues/green/report/2015/03/05/107328/icebreakers-essential-assets-for-a-changing-arctic.

Portland State University. "Larsen Ice Shelf 2002: Warmest Summer on Record Leads to Disintegration." 2002, http://web.pdx.edu/~chulbe/science/Larsen/larsen2002.html.

Powell, Richard C., and Klaus Dodds, eds. *Polar Geopolitics? Knowledges, Resources, and Legal Regimes*. Northampton: Edward Elgar, 2004.

"Press Statement and Answers to Journalists' Questions Following the APEC Summit," October 8, 2013, http://en.kremlin.ru/events/president/transcripts/19382.

Princen, Thomas, and Matthias Finger, eds. *Environmental NGOs in World Politics: Linking the Local and the Global.* London: Routledge, 1994.

Prokosch, Peter. "Svalbard," *Linking Tourism & Conservation*, March 5, 2015. http://www.ltandc.org/svalbard/.

Qiu, Jane. "China Aims High from the Bottom of the World: Country Plans Two World-Class Telescopes for Its Antarctic Observatory," *Nature*, August 29, 2012.

Quinn, Joyce A., and Susan L. Woodward. *Earth's Landscape: An Encyclopedia of the World's Geographic Features.* Santa Barbara: ABC-CLIO, 2015.

Rajan, H. P. "The Legal Regime of the Arctic and India's Role and Options." In Sinha, *Arctic,* 138–46.

Reid, Tyler. "Nuclear Power at McMurdo Station, Antarctica." Submitted as coursework for PH241, Stanford University, Winter 2014. http://large.stanford.edu/courses/2014/ph241/reid2.

Richardson, Mike G. "Regulating Tourism in the Antarctic: Issues of Environment and Jurisdiction." In Vidas, *Implementing the Environmental Protection Regime for the Antarctic*, 71–90.

Rigg, Kelly. "Environmentalists' Perspectives on the Protection of Antarctica." In Cook, *The Future of Antarctica*, 68–80.

Riva, Joseph P., and James E. Mielke. "Polar Energy Resources Potential." Congressional Research Service. U.S. House of Representatives, Ninety-Fourth Congress, Second Session, 1976.

Roldan, Gabriela. "Analyzing the Infrastructure of Antarctic Gateway Cities." University of Canterbury, Gateway Antarctica. Postgraduate Certificate in Antarctic Studies (PCAS) 13 (2010–2011).

Romero, Simon. "Antarctic Life: No Dogs, Few Vegetables, and 'a Little Intense' in the Winter," *International New York Times*, January 6, 2016.
Roseth, Tom. "Russia's China Policy in the Arctic." In Sinha, *Arctic*, 75–93.
Rothwell, Donald R. "The Antarctic Treaty as a Security Construct." In Hemmings, *Antarctic Security in the Twenty-First Century*, 33–50.
Roura, Ricardo M., and Alan D. Hemmings. "Realizing Strategic Environmental Assessment in Antarctica," *Journal of Environmental Assessment Policy and Management* 13 (2011): 483–514.
Ryan, Rowena. "Boeing 757 Commercial Plane Lands on Blue-Ice Runway in Antarctica for the First Time," *news.com.au*, December 2, 2015, http://www.news.com.au/travel/world-travel/antarctica/boeing-757-commercial-plane-lands-on-blueice-runway-in-antarctica-for-the-first-time/news-story/da2b1c0f07bad8dc074d39f800dbdee6.
Sakhuja, Vijay. "The Polar Code and Arctic Navigation." In Sinha, *Arctic*, 37–45.
Salmon, Patrick. "Foreign Policy and National Identity: The Norwegian Integrity Treaty 1907–24," *Forsvarsstudier* 1 (1993): 7–63.
Sample, Ian. "Cold Rush Threatens Pristine Antarctic: Scientists Warn That the World's Least Damaged Environment Could Be Changed Forever by the Hunt for Potentially Lucrative Organisms," *The Guardian*, February 2, 2004.
Sanchez, Rodolfo Andres, and Birgit Njaastad. "Future Challenges in Environmental Management of National Antarctic Programs." In Tin, *Antarctic Futures*, 287–306.
Scientific Committee on Antarctic Research. "Antarctic Climate Change and the Environment—2015 Update," May 4, 2015, http://www.scar.org/scar_media/documents/science/antera/ATCM38_ip092_e.pdf.
Scott, Karen N. "Scientific Rhetoric and Antarctic Security." In Hemmings, *Antarctica Security in the Twenty-First Century*, 284–306.
Scott, Shirley V. "Universalism and Title to Territory in Antarctica," *Nordic Journal of International Law* 66 (1997): 33–53.
Seccombe, Mike. "Keating Warns against Signing Antarctic Treaty," *Sydney Morning Herald*, November 21, 1988.
Secretariat of the Antarctic Treaty. *Compilation of Key Documents of the Antarctic Treaty System*. 2nd ed. Buenos Aires, 2004.
Shadian, Jessica M., and Monica Tennberg, eds. *Legacies and Challenges in Polar Sciences: Historical, Legal and Political Reflections on the International Polar Year*. Burlington: Ashgate Publishing, 2009.
Shaw, Justine D., Aleks Terauds, Martin J. Riddle, Hugh P. Possingham, and Steven L. Chown. "Antarctica's Protected Areas Are Inadequate, Unrepresentative, and at Risk," *PLoS Biology* 12(6) (June 17, 2014).

Singleton, Rebecca L., and Callum M. Roberts. "The Contribution of Very Large Marine Protected Areas to Marine Conservation: Giant Leaps or Smoke and Mirrors," *Marine Pollution Bulletin* 87 (2014): 7–10.

Sinha, Uttam Kumar, and Jo Inge Bekkevold, eds. *Arctic: Commerce, Governance, and Policy*. London: Routledge, 2015.

Staalesen, Atle. "Russia Submits Claim for North Pole," *Barents Observer*, August 4, 2015.

Statistics Norway. "This is Svalbard 2014, What the Figures Say," last accessed March 15, 2016, https://www.ssb.no/en/befolkning/artikler-og-publikasjoner/_attachment/213945?_ts=14aca3dea58.

Statistics Portal. "Distribution of the Global Population 2015, by Continent" and "Forecast of the World Population for 2100, by Continent," last accessed March 15, 2016, http://www.statista.com/statistics/237584/distribution-of-the-world-population-by-continent.

Statistik Sentralbyra. Statistics Norway. "Population of Svalbard, 1 July 2012," last accessed March 15, 2016, https://www.ssb.no/en/befolkning/statistikker/befsvalbard/halvaar/2012-09-20.

Statistik Sentralbyra. Statistics Norway. "Population of Svalbard, 1 July 2014," last accessed March 15, 2016, https://www.ssb.no/en/befolkning/statistikker/befsvalbard/halvaar/2014-09-25.

Statistik Sentralbyra. Statistics Norway. "Population of Svalbard, 1 July 2015," last accessed March 15, 2016, https://www.ssb.no/en/befolkning/statistikker/befsvalbard/halvaar/2015-09-24.

Stefansson Arctic Institute. *Arctic Human Development Report: Regional Processes and Global Linkages.* Copenhagen: Nordic Council of Ministers, 2014, http://library.arcticportal.org/1840/1/AHDRFULLTEXT01.pdf.

Stephens, Tim. "Polar Continental Shelves: Australian and Canadian Challenges and Opportunities." In Stephens, *Polar Oceans Governance in an Era of Environmental Change*, 146–65.

Stephens, Tim, and David L. VanderZwaag, eds. *Polar Oceans Governance in an Era of Environmental Change.* Northampton, UK: Edward Elgar, 2014.

Stix, Gary. "Patents on Ice: Antarctica as a Last Frontier for Bioprospectors—and Their Intellectual Property," *Scientific American*, May 1, 2004. http://www.scientificamerican.com/article/patents-on-ice/.

Stokke, Olav Schram. "Asian Stakes and Arctic Governance." In Sinha, *Arctic*, 4–17.

Stokke, Olav Schram. "The Effectiveness of CCAMLR." In Stokke, *Governing the Antarctic*, 120–51.

Stokke, Olav Schram, and Øystein B. Thommessen, eds. *Yearbook of International Cooperation on Environment and Development 2002/03*. London: Earthscan, 2003.

Stokke, Olav Schram, and Davor Vidas, eds. *Governing the Antarctic: The Effectiveness and Legitimacy of the Antarctic Treaty System*. Cambridge, UK: Cambridge University Press, 1996.

Stonehouse, Bernard, and Kim Crosbie. "Tourist Impacts and Management in the Antarctic Peninsula Area." In Hall, *Polar Tourism*, 217–33.

Svalbard Environmental Protection Act. Last accessed March 15, 2016, https://www.regjeringen.no/en/dokumenter/svalbard-environmental-protection-act/id173945.

Tenenbaum, Ellen S. "A World Park in Antarctica: The Common Heritage of Mankind," *Virginia Environmental Law Journal* 10 (1990–91): 109–36.

Thiede, Jorn. "Modern Research in Polar Regions." In Berkman, *Science Diplomacy,* 161–63.

Third UN Conference on the Law of the Sea. "Letter Dated 5 May 1978 from the Representative of Nepal to the President of the Conference." In *Extract from the Official Records of the Third UN Conference on the Law of the Sea.* Vol. X. Reports of the Committees and Negotiating Groups, Seventh and Resumed Seventh Session, Document A/CONF.62/65, 2009, 176.

Tin, Tina, and Alan D. Hemmings. "Challenges in Protecting the Wilderness of Antarctica," *USDA Forest Service Proceedings*, RMRS-P-64 (2011), 147–52.

Tin, Tina, Daniela Liggett, Patrick T. Maher, and Machiel Lamers, eds. *Antarctic Futures: Human Engagement with the Antarctic Environment*. London: Springer, 2014.

Tin, Tina, Benjamin K. Sovacool, David Blake, Peter Magill, Saad El Naggar, Sven Lidstrom, Kenji Ishizawa et al. "Energy Efficiency and Renewable Energy under Extreme Conditions: Case Studies from Antarctica," *Renewable Energy* 35 (2010): 1715–23.

Tingey, Robert J. *The Geology of Antarctica*. Oxford: Clarendon Press, 1991.

Tonnessen, Johan Nicolay, and Arne Odd Johnsen. *The History of Modern Whaling*. London: C. Hurst & Co., 1982.

Triggs, Gillian D., ed. *The Antarctic Treaty Regime: Law, Environment, and Resources*. Cambridge: Cambridge University Press, 1987.

Triggs, Gillian D. "The Antarctic Treaty System: Some Jurisdictional Problems." In Triggs, *The Antarctic Treaty Regime*, 88–109.

Triggs, Gillian D. "Negotiation of a Minerals Regime." In Triggs, *The Antarctic Treaty Regime*, 182–97.

Triggs, Gillian, and Anna Riddell, eds. *Antarctica: Legal and Environmental Challenges for the Future*. London: British Institute of International and Comparative Law, 2007.

Turner, John, Nicholas E. Barrand, Thomas J. Bracegirdle, Peter Convey, Dominic A. Hodgson, Martin Jarvis, Adrian Jenkins et al. "Antarctic Climate Change and the Environment: An Update." *Polar Record* 50:3 (2014): 237–59.

Turner, John, Robert Bindschadler, Pete Convey, Guido di Prisco, Eberhard Fahrbach, Julian Gutt, Dominic Hodgson et al., eds. *Antarctic Climate Change and the Environment: A Contribution to the International Polar Year 2007–2008*. Scientific Committee on Antarctic Research, November 25, 2009, http://www.scar.org/scar_media/documents/publications/ACCE_25_Nov_2009.pdf.

Turner, John, Steve R. Colwell, Gareth J. Marshall, Tom A. Lacklan-Cope, Andrew M. Carleton, Phil D. Jones, Victor Lagun et al. "Antarctic Climate Change during the Last 50 Years." *International Journal of Climatology* 25 (2005): 279–94.

Tvedt, Morten Walløe. "Patent Law and Bioprospecting in Antarctica," *Polar Record* 46 (2010): 1–10.

Tyree, David, Rear Admiral. "New Era in the Loneliest Continent," *National Geographic* 123:2 (February 1963): 260– 96.

UN News Center. "Antarctica Shows Need for Action on Climate Change, Ban Ki-moon Says," last accessed March 15, 2016, http://www.un.org/apps/news/story.asp?NewsID=24613#.VX7ehuvdKfQ.

Union of Concerned Scientists. "Climate Hot Map: Global Warming Effects Around the World," last accessed March 15, 2016, http://www.climatehotmap.org/global-warming-solutions/polar-regions.html.

United Kingdom. "Improvements to the Antarctic Environmental Impact Assessment Process." Working Paper 24., Brasilia, Brazil 2014.

United Nations. Division for Ocean Affairs and the Law of the Sea. "The UN Law of the Sea: A Historical Perspective," last accessed March 15, 2016, http://www.un.org/depts/los/convention_agreements/convention_historical_perspective.htm#Historical%20Perspective.

United Nations Convention on the Law of the Sea. Last accessed March 15, 2016, http://www.un.org/depts/los/convention_agreements/texts/unclos/unclos_e.pdf.

United Nations Convention on the Law of the Sea. Agreement Relating to the Implementation of Part XI, Last accessed March 15, 2016, http://www.un.org/depts/los/convention_agreements/texts/unclos/closindxAgree.htm.

United Nations Convention on the Law of the Sea. "UNCLOS at 30," last accessed March 15, 2016, http://www.un.org/depts/los/convention_agreements/pamphlet_unclos_at_30.pdf.

United Nations Environment Programme. "Industry Involvement in Antarctic Bioprospecting," Information Paper 106. Committee on Environmental Protection of the Madrid Protocol, Cape Town, 2004

United Nations Environment Programme. *Tourism in the Polar Regions: The Sustainability Challenge*. UNEP's Division of Technology, Industry and Economics. The Sustainable Consumption and Production Branch, 2007. https://www.ecotourism.org/filedepot_private/14/DTIx0938xPA-PolarTourismEN.pdf.

United Nations Environment Programme and Antarctic and Southern Ocean Coalition. "Review of the Implementation of the Madrid Protocol: Inspections by Parties (Article 14)." Antarctic Treaty Consultative Meeting, Hobart, Australia, 2012. http://www.asoc.org/storage/documents/Meetings/ATCM/XXXV/Review_of_the_Implementation_of_the_Madrid_Protocol-_Inspections_by_Parties_Article_14.pdf.

United Nations Framework Convention on Climate Change. Conference of the Parties. "Adoption of the Paris Accord." FCCC/CP/2015/L.9/Rev.1 (December 12, 2015), https://unfccc.int/resource/docs/2015/cop21/eng/l09r01.pdf.

United Nations General Assembly. "Consultations of the Secretary-General on Outstanding Issues Relating to the Deep Seabed Mining Provisions of the UN Convention on the Law of the Sea," Forty-Eighth Session, A/48/950 (June 9, 1994).

United Nations General Assembly. "The Question of Antarctica." General Assembly Resolutions: A/RES/41/88 (December 4, 1986), http://www.un.org/documents/ga/res/41/a41r088.htm; A/RES/46/41 (December 6, 1991), http://www.un.org/documents/ga/res/46/a46r041.htm; A/RES/51/56 (December 10, 1996), http://www.unep.org/dewa/Portals/67/pdf/A_res_51_56.pdf.

United Nations General Assembly. *Question of Antarctica*. Report of the Secretary General, A/39/538 (October 31, 1984). http://www.unep.org/dewa/Portals/67/pdf/A-39-583-Part_I.pdf.

United Nations General Assembly. Twenty-Second Session, 1515th Meeting and 1516th Meeting. "Agenda Item 92: Examination of the Question of the Reservation Exclusively for Peaceful Purposes of the Seabed and Ocean Floor, and the Subsoil Thereof, Underlying the High Seas Beyond the Limits of Present National Jurisdiction, and the Use of Their Resources in the Interest of Mankind, General Debate." November 1,

1967. http://www.un.org/depts/los/convention_agreements/texts/pardo_ga1967.pdf.

United Nations Population Fund. "State of World Population 2014: The Power of 1.8 Billion Adolescents, Youth, and the Transformation of the Future." 2014. https://www.unfpa.org/sites/default/files/pub-pdf/EN-SWOP14-Report_FINAL-web.pdf.

United States Congress. House of Representatives. 101st Congress (1990). *Preserving Antarctica's Ecosystem: Hearing on Title II of H.R. 2984, H. CON Res. 287, and H. CON. Res. 227.* Before the Subcommittee on Human Rights and International Organizations of the Committee on Foreign Affairs, last accessed March 15, 2016, http://babel.hathitrust.org/cgi/pt?id=pur1.32754074682067;view=1up;seq=1.

United States Congress. Office of Technology Assessment. *Polar Prospects: A Minerals Treaty for Antarctica.* Washington, DC: U.S. Government Printing Office, September 1989.

United States Department of the Interior. "Mineral Resources of Antarctica." Edited by N. A. Wright and P. L. Williams. Geological Survey Circular 705 (1974). Last accessed March 15, 2016, http://pubs.usgs.gov/circ/1974/0705/report.pdf.

United States Department of Interior and National Science Foundation. "Petroleum and Mineral Resources of Antarctica." Geological Survey Circular 909 (1983). http://pubs.usgs.gov/circ/1983/0909/report.pdf.

United States Department of the Interior and U.S. Geological Survey. "Explanatory Notes for the Mineral-Resources Map of the Circum-Pacific Region of the Antarctic Ice Sheet." 1998, http://pubs.usgs.gov/cp/42/report.pdf.

United States Department of State. "Polar Regions." Doc. 762 (July 1, 1951), https://history.state.gov/historicaldocuments/frus1951v01/d762.

United States Department of State. Office of the Historian. "Decolonization of Asia and Africa, 1945–1960," last accessed March 15, 2016, https://history.state.gov/milestones/1945-1952/asia-and-africa.

United States Department of State-Joint Chiefs of Staff. Memorandum of discussion, doc. 264 (January 10, 1958), https://history.state.gov/historicaldocuments/frus1958-60v02/d264.

Vesilind, Priit J. "Hunters of the Lost Spirit," *National Geographic* 163:2 (February 1983): 151–97.

Vicuna, Francisco Orrego, ed. *Antarctic Resources Policy: Scientific, Legal, and Political Issues.* Cambridge, UK: Cambridge University Press, 1983.

Vicuna, Francesco Orrego. "Port State Jurisdiction in Antarctica: A New Approach to Inspection, Control, and Enforcement." In Vidas, *Implementing the Environmental Protection Regime for the Antarctic*, 45–69.

Vidas, Davor. "Entry into Force of the Environment Protocol and Implementation Issues: An Overview." In Davor Vidas, *Implementing the Environmental Protection Regime for the Antarctic*, 1–17.

Vidas, Davor, ed. *Implementing the Environmental Protection Regime for the Antarctic*. Boston: Kluwer Academic Publishers, 2000.

Vidas, Davor. "An Overview." In Olav Schram Stokke, *Governing the Antarctic: The Effectiveness and Legitimacy of the Antarctic Treaty System* (Cambridge: Cambridge University Press, 1996).

Vidas, Davor. "The Protocol on Environmental Protection to the Antarctic Treaty: A Ten-Year Review." In Stokke, *Yearbook of International Cooperation on Environment and Development 2002/03*, 51–60.

Vigni, Patrizia. "Antarctic Bioprospecting: Is it Compatible with the Value of Antarctica as a Natural Reserve?" In Francioni, *Biotechnology and International Law,* 111–45.

Vigni, Patrizia. *Concorrenza Fra Norme Internazionali: Il Regime Giuridico dell'Antartide Nel Contesto Globale.* Milan: Giuffre Editore, 2005 [in Italian].

Vigni, Patrizia. "The Governance of the Arctic Environment: The EU and US Contribution." European University Institute Working Papers. Academy of European Law, series/report no. EUI AEL; 2013/07; TRANSWORLD Project.

Vigni, Patrizia. "Liability Regime for Antarctica." In Conforti, *The Italian Yearbook of International Law*, vol. XV, 217–42.

Viken, Arvid. "Tourism, Research, and Governance on Svalbard: A Symbiotic Relationship," *Polar Record* 47 (2011): 335–47.

Wainwright, Oliver. "Antarctic Research Stations: Exposed to the Elements and Now to the Public," *The Guardian*, July 25, 2013.

Wallis, Diana, and Steward Arnold, eds. "The Spitsbergen Treaty: Multilateral Governance in the Arctic," *Arctic Papers* 1 (2011): 5–37.

Walton, David W. H. "The Scientific Committee on Antarctic Research and the Antarctic Treaty." In Berkman, *Science Diplomacy,* 75–88.

Weber, Mel. "Delimitation of the Continental Shelves in the Antarctic Area: Lessons for Regime, Resource, and Environmental Security." In Hemmings, *Antarctic Security in the Twenty-First Century*, 172–96.

Weiss, Kenneth R. "The Making of a Climate Refugee," *Foreign Policy*, 2015, http://foreignpolicy.com/2015/01/28/the-making-of-a-climate-refugee-kiribati-tarawa-teitiota.

Wit, Maarten J. de. *Minerals and Mining in Antarctica: Science and Technology, Economics and Politics*. Oxford: Clarendon Press, 1985.

Whiteman, Marjorie M., ed. *Digest of International Law.* Vol. 1. Washington, DC: U.S. Government Printing Office, 1963.

Winkelmann, R., A. Levermann, A. Ridgwell, and K. Caldeira, "Combustion of Available Fossil Fuel Resources Sufficient to Eliminate the Antarctic Ice Sheet," *Science Advances* 1 (2015): 1–5.

Wolfrum, Rüdiger. *The Convention on the Regulation of Antarctic Mineral Resource Activities: An Attempt to Break New Ground.* Berlin: Springer-Verlag, 1991.

Wood, Michael. "The International Seabed Authority the First Four Years," *Max Planck Yearbook of UN Law* 3 (1999): 173–241.

World Trade Organization (WTO). "Members and Observers of WTO." https://www.wto.org/english/thewto_e/countries_e/org6_map_e.htm.

Wright, Helen S. *The Seventh Continent: A History of the Discovery and Explorations of Antarctica*. Boston: Gordon Press, 1918.

Young, Oran R. "Governing International Spaces: Antarctica and Beyond." In Berkman, *Science Diplomacy*, 287–94.

Zia, Asim, Ilan Kelman, and Michael H. Glantz. "Artic Melting Tests the United Nations Convention on the Law of the Sea." In Pincus, *Diplomacy on Ice*, 128–40.

# Index

Note: *f* refers to figure

200-nautical-mile EEZ, 69–71, 162, 216

AAT. *See* Australian Antarctic Territory (AAT)
ACCE report. *See* Antarctic Climate Change and the Environment (ACCE) report
acronyms/abbreviations, 279–280
Act of 17 July 1925 Relating to Svalbard, 173
Advisory Committee (CRAMRA), 102
AEC. *See* Arctic Economic Council (AEC)
Agreed Measures for the Conservation of Antarctic Fauna and Flora of 1964, 88, 205
Agreement Governing the Activities of States on the Moon and Other Celestial Bodies, 149. *See also* Moon Treaty
Agreement Relating to the Implementation of Part XI of the UN Convention on the Law of the Sea (1994 Agreement), 166
air transport to Antarctica, 41
*Akedemik Federov,* 95
Aldrin, Buzz, 157
Amundsen-Scott South Pole Station, 36, 40, 90, 153
ancient Greeks, 8
Antarctic and Southern Ocean Coalition (ASOC), 206, 216, 228, 229, 238, 239
Antarctic Climate Change and the Environment (ACCE) report, 14
Antarctic Convergence, 85, 203
Antarctic Environmental Protection, Clean-Up, and Liability Act of 1990, 92
Antarctic exceptionalism, 1
Antarctic Gamburstev Province Project, 95–96
Antarctic Geological Drilling Program, 96
Antarctic Heritage Sites, 138
Antarctic krill, 139–140, 215, 233–234
Antarctic Kunlun Station, 221
Antarctic Midwinter Festival, 137
Antarctic Ocean, 84–85
Antarctic ozone hole, 14, 17, 84
Antarctic Peninsula, 16*t,* 18, 19, 21*t,* 92, 93*f,* 120, 154
Antarctic Specially Managed Area (ASMA), 220–222, 224
Antarctic Specially Protected Area (ASPA), 213, 222, 224

Antarctic sponge, 234
Antarctic Treaty, 1; area of application, 203; Cold War mentality, 237; commercial activity, 232; consultative parties, 148; environmental issues, 205; human settlement, 155; maritime areas surrounding Antarctica, 70–71; minerals exploration, 96; official languages, 99; scientific investigation, 83, 84; substantial scientific research activity, 2, 86; successful international treaty, 188–189; territorial sovereignty, 2, 67–68, 155; text, 265–272; United States, 168–169; updating of treaty is required, 155
Antarctic treaty consultative meeting (ATCM), 3, 211, 213, 217, 240
Antarctic Treaty System (ATS), 1, 67, 124, 262, 263
Antarctica: accessibility, 41; architectural design, 39–41; Asian states, 127; atomic reactor, 36; benefits of sovereignty claims, 177–178; births, 154; civil society, 237–240; civilian settlements, 154; climate, 12–20; climate migrants, 188, 264; colder than Arctic, 120–121; coldest continent on Earth, 13; commercial activity, 232–237, 246; continent surrounded by ocean, 120; continental shelf, 121; Convention on Biological Diversity (CBD), 236; decolonization, 155; energy-efficiency measures, 40–41; fishing industry, 139–140; freezing of territorial claims, 67–68, 178; gateway cities, 136–138; geographical regions, 12; geology, 12; global commons, 151; habitability, 41; historical overview, 7–11; human colonies, 261; ice cover, 12, 84, 133; importance to advancement of science, 84; jurisdictional vacuum, 225, 237; militarization, 232; minerals, 20–25; natural reserve, 109; NGOs, 238–240; ozone hole, 14, 17, 84; place names, 154; polar nights and polar days, 13; principal resources, 24*f*; research stations, 35–41; sea ice, 15; special kind of common/disputed common, 152; subglacial lakes, 31; Svalbard model, 240–245; temperature variation, 13*f*, 16*t*; territorial sovereignty claims (*see* Who owns Antarctica?); tourism, 147 (*see also* Tourism); tragedy of the commons, 3; transportation revolution, 41; unclaimed territorial portion, 58, 177; urban plan for most crowded regions, 220; warming trends, 14; World Park, as, 148, 189, 206; world's least environmentally protected continent, 3
"Antarctica: Assessing and Protecting Australia's National Interests," 261
Antichthon, 8
Architectural design, 39–41
Arctic, 119–146; AEC, 132–133; Asian states, 124–129; coastal states, 122; continental shelf, 121; continental shelf claims, 129–130; *Demystifying the Arctic,* 132; disputed areas (islands subject to territorial claims), 170; drivers of warming and environmental change, 120; environmental impact of melting ice, 121; GDP, 123; governance, 124–125; icebreaking fleets, 133–135; imports/exports, 124; indigenous people, 123; international economy, 123; local economy, 123; map, 122*f*; maritime delimitation disputes, 131; militarization, 132; ocean surrounded by land, 120; oil and gas, 129; overlapping claims, 129–132; population, 123; research outposts, 125; Svalbard, 4, 150, 169
Arctic Coal Company, 171, 179
Arctic Council, 1, 124, 125
Arctic drillships, 126
Arctic Economic Council (AEC), 132–133
Arctic economy, 147
Arctic Human Development Report, 119, 120, 123, 125, 131
Arctic littoral, 121
"Arctic Meltdown" (Borgenson), 135

Arctic Ocean, 120
ASMA. *See* Antarctic Specially Managed Area (ASMA)
ASOC. *See* Antarctic and Southern Ocean Coalition (ASOC)
ASPA. *See* Antarctic Specially Protected Area (ASPA)
Astronomical research, 84
ATCM. *See* Antarctic Treaty Consultative Meeting (ATCM)
Atomic reactor, 36
ATS. *See* Antarctic Treaty System (ATS)
Auburn, F. M., 108
Australian Antarctic Territory (AAT), 97
author interviews: anonymous climatologist, 17–19; governor of Svalbard, 243–245; Grydehoj, Adam, 182–185; Hammond, Aleqa, 33–35; Hughes, Kevin, 222–224; Rocard, Michel, 106–107; Store, Jonas, 174, 175–176

*Bacillus* sp., 234
*Bahia Paradiso,* 103, 237
Ban Ki-moon, 75
ban on mining, 110, 233
banking system, 163, 166
Barents, Willem, 171
Barents Sea, 131
Barentsburg, 179, 181
BAT. *See* British Antarctic Territory (BAT)
Beck, Peter, 178
Beeby, Christopher, 98–100
Bellingshausen, Fabian Gottlieb von (Thaddeus), 9, 65
Benefit sharing, 167, 236
Bergin, Anthony, 95, 97
Bering Sea, 131
Bharati Station, 39, 90
bidirectional (double-acting) tankers, 126, 134
bioprospecting, 3, 232–236
Birdlife Norway, 243
births, 154
Biscoe, John, 10
Bleasel, Jim, 88
Boeing 757 aircraft, 41
Borgenson, Scott G., 127, 134, 135
Boutros-Ghali, Boutros, 165
Brazilian Antarctic station, 40–41
British Antarctic Territory (BAT), 57
Broughton, Darcy, 22, 23
Brown, Alan, 226
Bush, William, 226
Bushvel Complex, 23
Byrd, Richard, 8

C-17A Globemaster, 41
Camocho, Juan Pablo, 154
*Candida antarctica,* 234
Cape Town, 136–137
*Caranvon Castle,* 57
Carlini Station, 90
CBD. *See* Convention on Biological Diversity (CBD)
CCAMLR. *See* Convention on the Conservation of Antarctic Marine Living Resources (CCAMLR)
CCAS. *See* Conservation of Antarctic Seals (CCAS); Convention on the Conservation of Antarctic Seals (CCAS)
Chaturvedi, Sanjay, 236
Chilingarov, Artur, 129
China, 125–126, 128
China National Offshore Oil Corporation, 128
China National Petroleum Corporation, 128
Christchurch, 136, 137
civil society, 237–240
civilian settlements, 154
clarins, 234
CLCS. *See* Commission on the Limits of the Continental Shelf (CLCS)
cleanup, 231
climate, 12–20
climate change, 7, 210
climate-induced migration, 26–30, 188, 264
climate refugee visa, 29
climate tourism, 35
Clinton, Hillary, 130
Commandante Ferraz, 40
commercial activity, 232–237, 246

Commission on the Limits of the Continental Shelf (CLCS), 70, 71–72, 129, 130
Committee for Environmental Protection, 89, 211–213, 245
Committee on Merchant Marine and Fisheries, 168
common heritage area, 162
Common Heritage Fund, 162
Common Heritage of Mankind principle: benefit sharing, 236; call for principle's inclusion in any new order for Antarctica, 190; global commons, 152, 190; Moon Treaty, 160; "non-appropriation" of common resources, 159; Pardo proposal, 158; revival of CRAMRA, 168; Svalbard model, 174; what it is, 152
COMNAP. *See* Council of Managers of National Antarctic Programs (COMNAP)
Comprehensive Environmental Evaluation, 217, 219
Concordia Station, 92, 153
condominium proposal, 62–63
Conforti, Benedetto, 154
Conservation of Antarctic Seals (CCAS), 67, 68
consultative parties, 148, 150
Consultative party status, 2, 86, 211
continental shelf extension claims, 71–72, 75
Convention on Biological Diversity (CBD), 213, 236
Convention on the Conservation of Antarctic Marine Living Resources (CCAMLR): area of application, 69, 203; Article IV, 68–69; date of ratification, 215; EEZs, 226; fairly complex fisheries agreement, 214; first agreement to allow mining in Antarctica, 232; FAO, 214; inspection, 227; membership, 227; mining of fish, 235; precursor to CRAMRA, 215; protected areas, 215
Convention on the Conservation of Antarctic Seals (CCAS), 9
Convention on the Regulation of Antarctic Mineral Resource Activity (CRAMRA), 98–109; advisory committee, 102; Australian criticism, 107–108; collapse, 103–109, 149, 156; date of signing, 103; enforcement, 207; environmental criticism, 103–105; environmental protection, 206, 207; external accommodation, 99; general provisions, 100; inspectors (observers), 227; internal accommodation, 99, 101; joint ventures, 103; liability regime, 207, 231; mining activities, 207; negotiating parties, 159–160, 166–167; negotiations, 150, 156; NGO lobbying, 239; nondiscrimination principle, 101; prospecting, 102–103, 202, 207, 235; regulatory committees, 101–102; reviving, 149, 150, 157, 168, 189; territorial sovereignty, 69, 100–101, 103; UN criticism, 105–107; veto power, 101, 207–208; worries about sovereignty issues, 107–109
Cook, James, 8
Cook, Peter, 107, 108
coreless winters, 13
Council of Managers of National Antarctic Programs (COMNAP), 39, 88, 89, 238
Cousteau, Jacques-Yves, 104, 107
CRAMRA. *See* Convention on the Regulation of Antarctic Mineral Resource Activity (CRAMRA)
Crispini, Lauro, 22
cruise tourism, 138
cumulative environmental effects, 241

Dadwal, Shebonti Ray, 131
Daewoo Shipbuilding and Marine Engineering, 126
Daniel Bennet & Sons, 10
Dastidar/Ramachandran study, 94–95
Davis, Bruce, 95
de Wit, Maarten J., 23–25
decolonization of Antarctica, 155
Deep Sea Drilling Project, 20
Deep Seabed Hard Mineral Resources Act, 165

*Demystifying the Arctic,* 132–133
dependent and/or associated ecosystems, 203, 207, 230
disputed common, 152
double-acting tankers, 134
Droning Maud Land, 54
Dufek Massif, 23–25
Dumont d'Urville, Jules, 54
Dumont d'Urville Station, 90, 92
*Durvillea antarctica* algae, 234

EAIS. *See* East Antarctica Ice Sheet (EAIS)
East Antarctica, 12
East Antarctica Ice Sheet (EAIS), 12
EEZ. *See* Exclusive Economic Zone (EEZ)
Eisenhower, Dwight, 66, 106, 169
El Niño, 17
elephant seals, 9
elevated research stations, 39–40
Ellsworth Land, 58
Elzinga, Aant, 86
Emmerson, Charles, 127, 130
Enderby, Charles, 10
Enderby, Samuel, 9
energy-efficiency measures, 40–41
Enterprise, the, 163, 164, 166, 167, 168
environmental emergencies, 229–231
Environmental Impact Assessment, 216–225, 235
environmental migration, 26–30, 188
environmental protection, 201–260; Agreed Measures, 205; Antarctic Treaty, 205; author interview (Kevin Hughes), 222–224; bioprospecting, 232–236; Committee for Environmental Protection, 211–213, 245; CRAMRA, 206 207; different sanctions for same offense, 205; enforcement, 225–226; environmental emergencies, 229–231; Environmental Impact Assessment, 216–225, 235; extremophiles, 233, 234; inspection, 223, 227–228; liability, 229–232; Madrid Protocol, 109–110; NGOs, 203, 238–240; protected areas, 213–216; science, 210, 219; Svalbard model, 240–245; tourism, 237; what Antarctica's geographical borders are, 203–204
Environmental Protocol to the Antarctic Treaty, 201. *See also* Madrid Protocol
environmentally protected areas, 213–216
Escudero Proposal, 66, 75
Esperanza, 154
*Euphausia superba,* 233. *See also* Krill fish
Exclusive Economic Zone (EEZ), 226, 227. *See also* 200-nautical-mile EEZ
*Explorer,* 138
extremophiles, 233, 234
*Exxon Valdez,* 104

Factory ships, 11
"Failure of the National Science Foundation (NSF) to Protect Antarctica, The," 92
Falkland Islands, 132
Falklands War, 4, 132
FAO. *See* Food Agricultural Organization (FAO)
Fayette, Louise Angelique de la, 230–231
Fifth Assessment Report (IPCC), 15, 27
Fildes Peninsula, 220, 228
fishing, 3
fishing industry, 139–140. *See also* krill fish
fixed-wing aircraft, 41
flag of convenience, 226
flag states, 227
Food Agricultural Organization (FAO), 214
Franz Josef Land, 171
Free Trade Agreement (FTA), 128
freedom of the seas doctrine, 158, 162
freezing of territorial sovereignty claims, 67–68, 75, 178
Frenot, Yves, 235
freshwater, 133
Fretwell, Peter, 31
Friend, Scott Harris, 12, 38
Friends of the Earth, 238
Friends of the Earth Norway, 243
Frydenlund, Knut, 187
FTA. *See* Free Trade Agreement (FTA)
fur seals, 9

Garcia, David, 39
Garrison, Christopher, 164
gateway cities, 136–138
GATT. *See* General Agreement on Tariffs and Trade (GATT)
Gemenne, François, 29
General Agreement on Tariffs and Trade (GATT), 166
geological research, 85
geostationary orbit, 167
Gjelsvik, Tory, 23
Global Agenda Council on the Arctic, 119
Global Climate Risk Index, 28
global commons, 151, 152, 157, 264
global migration, 26
global warming, 2
*Glomar Challenger,* 85, 96
Gondwanaland, 8, 12
Greenland, 30–35, 121, 171
"Greenland: Ground Zero for Global Warming," 30
Greenpeace, 238, 239
Greenpeace World Park Base, 94
Grydehoj, Adam, 174, 182–185, 188
*Guvernoren,* 11

Hall, C. Michael, 243
Halley Research Station, 39, 89
Hammond, Aleqa, 33–35, 181
*Handlingsrom,* 186
Hans Island, 131, 171
Harper, Stephen, 125
Hawke, Bob, 106
*Healy,* 135
Heard Island, 227
Hemmings, Alan, 91, 139, 155, 156, 177–178, 210, 214, 217, 218
Heroic Era, 10
high seas, 70, 227
Hill, David, 38
historical overview, 7–11
*HMS Challenger,* 161
Hobart, Australia, 136, 137
"How Committed Are We to Monitoring Human Impacts in Antarctica?" (Hughes), 228
Huber, Max, 151, 174
Hudong-Zhonghun Shipbuilding Co. Ltd., 126
Hughes, Kevin, 221–225, 228
human colonies, 261
Huxley, Julian, 64–65
Hyundai Heavy Industries, 126

IAATO. *See* International Association of Antarctica Tour Operators (IAATO)
ice cover, 12, 84
Ice Lab Exhibition (Glasgow), 39
icebreaking fleets, 133–135
icebreaking transporters, 126
ideological rat, 164
illegal, underreported, and unregulated (IUU) fishing, 3, 140, 226
impact assessment, 216–225, 235
*Improvements to the Antarctic Environmental Impact Assessment Process,* 223
Initial Environmental Evaluation, 217
inland water prospecting, 233
inspection, 223, 227–228
Intergovernmental Panel on Climate Change (IPCC) Fifth Assessment Report, 15, 27
International Antarctic Research Institute, 65
International Association of Antarctica Tour Operators (IAATO), 139, 226, 237–239
International Organization for Migration (IOM), 26–28
International Polar Foundation, 40
International Seabed Authority, 163, 164, 167
International Telecommunications Union, 167
International Tribunal on the Law of the Sea, 167
International Union for the Conservation of Nature (IUCN), 238–239
International Whaling Commission, 11
interviews. *See* author interviews
IOM. *See* International Organization for Migration (IOM)

IPCC report. *See* Intergovernmental Panel on Climate Change (IPCC) Fifth Assessment Report
*Island of Palmas Case,* 151
IUCN. *See* International Union for the Conservation of Nature (IUCN)
IUU. *See* illegal, underreported, and unregulated (IUU) fishing

Jan Mayen, 171
Jang Bogo Station, 90
Japanese shipyards, 126
Johnston, Margaret, 243
Joint Declaration on the Antarctic (Argentina/Chile), 56
Jones, Walter, 92
Joyner, Christopher, 159
Juli Escudero Station, 90
jurisdictional vacuum, 225, 237

Keating, Paul, 107, 108
Kemp, Peter, 10
Kemp Land, 10
King George Island, 92, 220
Kiribati, 29, 167
*Kirkpatrick varialosa,* 234
Klein, Andrew G., 37
Korean Polar Research Institute, 136
krill-derived enzymes and extracts, 234
krill fish, 139–140, 215, 233–234

Lake Vostock, 31, 219
Larsen A, 31
Larsen B, 31
Larsen Ice Shelf, 31
Law-Racovita Station, 92
Lazarev, Mikhail, 65
Lee, Ricky, 167
liability, 229–232
Lindblad, Lars Eric, 138
Little Ice Age, 18
*Lockheed Hercules,* 36
Loftleidir Icelandic Airliner, 41
Longyear, John Munro, 179
Longyearbyen, 179, 180
Longyearbyen Community Council, 181
Lowy Institute for International Policy, 261
Lyng, John, 174

MacTown, 90
Madrid Protocol, 109–110; alternative minerals regime, 149; ambiguity, 203–204; Antarctic Specially Managed Area (ASMA), 220–222, 224; Antarctic Specially Protected Area (ASPA), 213, 222, 224; assumption that main threat to environment comes from within Antarctica, 201; ban on mining, 110, 233; cleanup, 231; climate change, 210; commercial activity, 232; components (articles/annexes), 208; Comprehensive Environmental Evaluation, 217, 219; cut-and-paste operation, 206; dependent and associated ecosystems, 203; designates Antarctica as "natural reserve," 109; Environmental Impact Assessment, 216–225; Initial Environmental Evaluation, 217; inspections, 227–228; irony, 209, 230; liability, 229–232; mistakes of the past left unaddressed, 225; origins, 205; prohibited activities, 232; research stations, 39; response action, 229, 231; supplement to Antarctic Treaty, 211; Svalbard model, 209; veto power, 208; walkout clause, 148; weaknesses, 202, 205–210, 245–246
Mahathir Bin Mohamad, 73
"Making of a Climate Refugee, The," 29
Malaysia, 73, 74
Maldives, 29
Malone, James, 164
manganese deposits, 161
*Mapping Antarctica: A Five Hundred Year Record of Discovery* (Clancy et al.), 8
Marambio Station, 90
Marie Byrd Land, 58
marine bioprospecting, 233–234
marine protected areas, 214–216
*Marinomonas protea* bacteria, 234
maritime areas surrounding Antarctica, 69–72
maritime delimitation disputes, 131, 132
McDonald Island, 227
McMurdo Station, 36–38, 59–60, 89–90
*Mikhail Somov,* 95

militarization, 232
Mineral Resources Act (Greenland), 32
minerals, 20–25; ban on mining, 110, 233; CRAMRA (*see* Convention on the Regulation of Antarctic Mineral Resource Activity [CRAMRA]); exploitation of minerals is hazardous, 121; Greenland, 32; Madrid Protocol, 109–110; manganese deposits, 161; map of principal resources, 24*f*; seabed mining, 161–166, 167, 168; Svalbard, 179; walkout clause, 110
mining. *See* minerals
mining ban, 110, 233
mining of nonliving resources, 235
Mirny Station, 90
Montreal Protocol, 84, 111
Moon Treaty, 149, 157, 161–162, 189
*Mulighetstrom,* 186
Muscovy Company, 172

Nansen Initiative, 27, 28
Nasheed, Mohammed, 29
*Nathaniel Palmer,* 95
*National Geographic* articles: "Greenland: Ground Zero for Global Warming," 30; "New Era in the Loneliest Continent," 36; "People of the Arctic," 31
National Science Foundation (NSF), 37, 135
Nauru, 167
Nepal, 162
"New Era in the Loneliest Continent," 36
New International Economic Order (NIEO), 156–157, 162
NGOs. *See* nongovernmental organizations (NGOs)
NIEO. *See* New International Economic Order (NIEO)
1994 Agreement, 166
nonconsultative parties, 150
nondiscrimination principle, 101
nongovernmental organizations (NGOs), 203, 238–240
Northeast Greenland National Park, 30
Northern Sea Route (NSR), 128
Northwest Passage, 131
Norway, 130, 170, 171, 186. *See also* Svalbard
Notre Dame Global Adaptation Index 2014, 28
Noyes, John, 159
NSF. *See* National Science Foundation (NSF)
NSR. *See* Northern Sea Route (NSR)
nuclear power, 37
Ny-Alesund, 179–180

observers, 227
official Antarctic Treaty languages, 99
Operation Deep Freeze, 36
Outer Space Treaty, 157, 161, 162
ozone hole, 14, 17, 84

Pacific Access Category, 29
Palestinian Liberation Organization (PLO), 163
Palma, Emilio Marcos, 154
*Palmer,* 135
Palmer, Nathaniel, 2
Palmer Station, 60
paper satellite problem, 167, 168
parallel system, 162, 166
Pardo, Arvid, 158
Paris Climate Summit, 28
Pedersen, Torbjorn, 186
Pedro Vicente Maldonado Station, 40
"People of the Arctic" (*National Geographic* article), 31
Perez de Cuellar, Javier, 165
Pioneer Investors, 164
PLO. *See* Palestinian Liberation Organization (PLO)
polar bears, 121
Polar Earth Observing Network, 96
polar nights and polar days, 13
*Polar Sea,* 135
*Polar Star,* 135
Polaris motor toboggan, 36
Pomor hunters, 171
population growth, 26
*Prasiola crispa,* 234
precautionary principle, 241
Press, Anthony, 95
*Primero di Maya,* 57

Princess Elizabeth Station, 40
Principle of Common Heritage. *See* Common Heritage of Mankind principle
Programma Nazionale di Ricerche, 137
Prokosch, Peter, 243
protected areas, 213–216
Protocol on Environmental Protection to the Antarctic Treaty, 109. *See also* Madrid Protocol
Punta Arenas, 136, 137
Putin, Vladimir, 128, 186

*Question of Antarctica,* 53, 76, 98, 114
"Question of Antarctica," 148, 152, 153, 159

Reagan, Ronald, 164
Regulatory Committees, 101–102
renovated or repurposed research stations, 41, 50
*Res communis,* 152
*Res nullius,* 152
*Res publica,* 152
research stations, 35–41, 89–94, 153
reserved areas, 167, 168
Resolution A/Res/41/88, 73, 74
Resolution A/Res/46/41, 90
response action, 229, 231
Rocard, Michel, 105, 106–107
Romero, Simon, 155
Roosevelt, Franklin D., 55
Rosneft, 128
Ross Sea, 215
Rothera Station, 90
Roura, Ricardo M., 218
Rumsfeld, Donald, 164
Russia: Arctic, 126–127, 128, 129, 132, 134; Ross Sea, 215–216; scientific drilling in Lake Vostock, 219; Svalbard, 172
"Russian Discoveries in the Antarctic and Present-Day Interest in the Area" (report), 65

Sahara, 151
Samsung Heavy Industries, 126
Samuel Enderby & Sons, 9
SANAE IV, 40
SCAR. *See* Scientific Committee on Antarctic Research (SCAR)
science, 83–96; Antarctic Treaty, 84, 86; astronomical research, 84; Australia, 95; collaborative efforts, 95–96; COMNAP, 88, 89; Dastidar/Ramachandran study, 94–95; geological research, 85; International Antarctic Research Institute, 65; research stations, 89–94; SCAR, 87–89; United States, 94–95
Scientific Committee on Antarctic Research (SCAR), 14, 15, 87–89, 230, 238
scientific imperialism, 155
scientific research stations, 35–41, 89–94, 153
scientific whaling, 11
Scott, Karen N., 221
sea ice, 15, 133
seabed mining, 161–166, 167, 168
sealing industry, 9–10
Shaw, Justin, 214
Sheffield, Scott D., 129
shipbuilding, 125–126
Shoigu, Sergey, 132
Sichuan earthquake, 26
*Silence of the Sea, The* (film), 107
Solar panels, 41
South Korea, 126, 136
South Orkney Islands, 215
South Orkney Islands Shelf Marine Protected Area, 215
South Pole traverse, 218–219
South Shetland Islands of Antarctica, 154
Southern Ocean, 7, 69–72, 121, 215, 226, 227
sovereignty, 151
special kind of common, 152
Spitsbergen, 179
Spitsbergen Commission, 172
Spitsbergen Treaty, 278. *See also* Svalbard Treaty
SPLOS/72, 71, 72
sponges, 234
station design, 39–40
Store, Jonas, 174, 175–176, 186, 187, 188
Store Norske, 179

subglacial lakes, 31
substantial scientific research activity, 2, 86
Svalbard, 169–188; author interview (Adam Grydehoj), 182–185; author interview (governor of Svalbard), 243–245; author interview (Jonas Store), 174, 175–176; both Arctic and Antarctic nation, 169; company town, 171; condominium approach, 172; discovery, 170; disputes over surrounding waters, 187; economic phases, 170; environmental protection, 240, 241–242; example for Antarctica, as (*see* Svalbard model); governance model, 169–170; ice cover, 170; map, 170*f*; mining, 179; 1920 treaty (*see* Svalbard Treaty); 1925 Act, 173; "normalization" of Longyearbyen, 179, 180; Norway, 170, 171, 186; open to citizens of the world, 185; population, 180; realpolitik, 186–188; Syrian refugees, 188; taxes, 181; temperature, 170; tourism, 180, 242–243, 244; towns (settlements), 179–180; vegetation, 170; what is it?, 169; what life is like, 178–185
Svalbard Environmental Protection Act of 2001, 241–242
Svalbard Environmental Protection Fund, 242, 244–245
Svalbard model, 263–264; advantages, 176–177; environmental protection, 240–245; local democracy, 181; mining, 237; rationalization of scientific activity, 219; realpolitik, 188; reconciling sovereignty with Common Heritage of Mankind principle, 174; sovereignty without exclusivity, 190
Svalbard Treaty: effect of, 173; equal economic rights, 187; main provisions, 173; membership, 174; opposition to, 174; preamble, 172; text, 273–278
Syowa Station, 90
Syrian refugees, 188

Taishan Station, 90
Tasmanian Polar Network, 137
temperature, 13*f*, 16*t*
*Terra Australis Incognita*, 8
*Terra nullius*, 171
Terre Adélie, 54
terrestrial bioprospecting, 233
territorial sovereignty claims. *See* Who owns Antarctica?
Third UN Conference on the Law of the Sea (UNCLOS III), 158, 159, 161. *See also* United Nations Convention on the Law of the Sea (UNCLOS)
Tin, Tina, 214
Todd, Edward, 88
Tonga, 167
tooth fish, 215
tour operators, 226. *See also* International Association of Antarctica Tour Operators (IAATO)
tourism: Antarctica, 3, 35, 138–139, 147, 237, 246; Svalbard, 180, 242–243, 244
tourist cruise ships, 225–226
tragedy of the commons, 3, 153, 263
Transantarctic Mountains, 12, 23
transportation revolution, 41
Treaty of 9 February 1920 Relating to Spitsbergen (Svalbard), 172. *See also* Svalbard Treaty
Treaty of Peace with Japan, 56
Treaty on Principles Governing the Activities of States in the Exploration and Use of Outer Space, Including the Moon and Other Celestial Bodies, 157. *See also* Outer Space Treaty
Tripartite Naval Declaration, 56
Truman, Harry, 158
Trust Arktikugol, 187
trusteeship proposal, 63, 64
Tyree, David, 36

UN Trusteeship, 63, 64
unclaimed territorial portion, 58
UNCLOS. *See* United Nations Convention on the Law of the Sea (UNCLOS)
UNEP. *See* United Nations Environment Program (UNEP)
UNFCCC. *See* United Nations Framework Convention on Climate Change (UNFCCC)

UNHCR. *See* United Nations High Commissioner for Refugees (UNHCR)
Unilever, 234
United Nations Convention on the Law of the Sea (UNCLOS): 200-nautrical-mile EEZ, 162; adopted in 1982/came into force in 1994, 165; Antarctic Treaty, 71; Arctic Ocean, 124–125; Carter administration's complaint about mandate of UNCLOS, 158–159; Common Heritage of Mankind principle, 189; continental shelf, 70; convening of UNCLOS III, 158; Enterprise, 163; 1994 Agreement, 165–166; reserved areas, 167; revival, 165; seabed mining, 161; Southern Ocean, 227; United States, 158, 163, 164, 168
United Nations Environment Program (UNEP), 139
United Nations Framework Convention on Climate Change (UNFCCC), 27
United Nations High Commissioner for Refugees (UNHCR), 28
United Nations University (UNU), 233
United States: Antarctic Treaty, 169–170; anti-UNCLOS lobby, 168; Arctic, 127; Convention on Biological Diversity (CBD), 236; Deep Seabed Hard Mineral Resources Act, 165; icebreaking fleet, 135; liability, 229; main scientific power in Antarctica, 169; place names, 154; science, 94–95; sovereignty claim, 58–62, 153; UNCLOS, 158, 163, 168; U.S. 2013 National Arctic Strategy, 130; walkout clause, 148
United States Antarctic Program (USAP), 137
Unknown Southern Land, 8
UNU. *See* United Nations University (UNU)
Upton, Simon, 210
urban plan for most crowded regions, 220
U.S. Deep Sea Drilling Project, 20
U.S. Geological Survey (USGS), 22, 23, 129
U.S. 2013 National Arctic Strategy, 130
USAP. *See* United States Antarctic Program (USAP)
USGS. *See* U.S. Geological Survey (USGS)
Ushuaia, 136, 137–138

veto power, 101, 207–208
Vicuna, Francesco Orrego, 99, 178, 207, 227
Vidas, Davor, 159, 206, 237
Videla, Gabriel, 57
Viken, Arvid, 242
Villa Las Estrallas, 154, 155
Voronkov, Lev, 127
Vostock Station, 153, 219

WAIS. *See* West Antarctica Ice Sheet (WAIS)
walkout clause, 148
warming trends, 14
Weiss, Kenneth, 29
West Antarctica, 12
West Antarctica Ice Sheet (WAIS), 12
West German Antarctic Program, 22
whaling industry, 10–11
"Where is Antarctica?," 203–204
Who owns Antarctica?, 53–82; 200-nautical-mile EEZ, 69–71; Article IV compromise, 67–68, 75; ATS, 67; CCAMLR, 68–69; claims of territorial sovereignty, 53–56, 149; condominium proposal, 62, 63; continental shelf extension claims, 71–72, 75; Escudero Proposal, 66, 75; explicit rejection of territorial claims, 54–55; freezing of territorial sovereignty claims, 67–68, 75; map of territorial claims, 55*f*; maritime areas surrounding Antarctica, 69–72; skirmishes leading to Antarctic Treaty, 56–58; Soviet position, 65–66; Trusteeship proposal, 63, 64; unclaimed territorial portion, 58; United Nations (UN), 72–75; U.S. negotiating stance, 58–62
"Why Did We Get an International Space Station Before an International Antarctic Station?" (Hemmings), 91
Wilkins Aerodrome, 41

Wilkinson, Peter, 104
Wilkness, Peter, 88
wind, 41
Wolfrum, Rüdiger, 101, 209
World Migration Report 2010, 26
World Park, 148, 189, 206
world population, 26
World Trade Organization (WTO), 177
Wrangel Island, 171
WTO. *See* World Trade Organization (WTO)

*Xuelong* icebreaker, 125
*Xueying,* 41

**About the Author**

**Doaa Abdel-Motaal,** PhD, is the former chief of staff of the United Nations Fund for Agricultural Development, in Rome, Italy, and deputy chief of staff of the World Trade Organization, in Geneva, Switzerland. She has worked on environmental issues and climate change throughout her life, traveling through the Arctic and the Antarctic from 2015 to 2016.

www.ingramcontent.com/pod-product-compliance
Ingram Content Group UK Ltd.
Pitfield, Milton Keynes, MK11 3LW, UK
UKHW021033041026
15470UKWH00014B/1562

9 781440 848032